Charles Seale-Hayne Library
University of Plymouth
(01752) 588 588
LibraryandITenquiries@plymouth.ac.uk

Water Quality Modeling

Volume I
Transport and Surface Exchange in Rivers

Author

Steve C. McCutcheon

Environmental Engineer and Hydrologist
Athens, Georgia

Series Editor

Richard H. French

Associate Executive Director and Research Professor
Water Resources Center
Desert Research Institute
University of Nevada
Las Vegas, Nevada

CRC Press, Inc.
Boca Raton, Florida

Library of Congress Cataloging in Publication Data

Water quality modeling / series editor, Richard H. French.
    p.   cm.
    Includes bibliographical references.
    Contents: v. 1. Transport and surface exchange in rivers / author,
Steve C. McCutcheon.
    ISBN 0-8493-6971-1 (v. 1)
    1. Water quality—Mathematical models. 2. Water chemistry—
Mathematical models. I. French, Richard H. II. McCutcheon,
Steve C.
TD370.W3955   1989
628.1'61—dc20   89-25333
                                                        CIP

    Direct all inquiries to CRC Press, Inc., 2000 Corporate Blvd., N.W., Boca Raton, Florida, 33431.

© 1989 by CRC Press, Inc.

International Standard Book Number 0-8493-6971-1 (v. 1)

Library of Congress Card Number 89-25333
Printed in the United States

SERIES PREFACE

Authoring a technical reference book is a labor of love; for unlike novelists, the technical author has no prospect of becoming rich and famous and his work will be subject to the critical review of his peers. As Series Editor, I salute the authors that have made the series *Water Quality Modeling* a reality. I would also thank the families, colleagues, and staff who supported and encouraged the authors through the creative process.

The ambitious goal of this series is to bring together the collective knowledge and experience of engineers, biologists, chemists, and other scientists to present the state of the art of water quality modeling. The cost of not having available the capability of forecasting the effects of our actions on essential and limited resources such as water is very high in both economic and human terms. Numerical modeling is the only cost-effective method of forecasting the future; and the art of constructing a numerical model causes us to view the environment in which we live as a system of interrelated components.

Water quality modeling is not a static subject, and I would hope that the information and knowledge presented in this series will be a part of the foundation on which new developments are made. Water quality modeling is a complex combination of education, experience, and creativity — it is science and art — and our understanding of the system continues to develop.

Dr. Steve McCutcheon and I were very fortunate in attracting a number of outstanding engineers and scientists to this series. The credit for this series belongs to the authors, not the editor. I am priveleged to be associated with these authors and the publisher.

Richard H. French

PREFACE

This book describes the state of the science and the state of the art for modeling water quality processes. A total of four volumes is planned. Volume I covers river transport and the exchange of heat and dissolved gases that occur mainly at the water surface. Volume II covers biogeochemical processes affecting dissolved oxygen, nutrients, and toxic contaminants.

Coverage of the theoretical basis and the practical approaches makes the book useful to researchers, instructors, graduate students, and practicing engineers. The description of the basic understanding about selected water quality processes for rivers and streams that are important in modeling and the derivation of important mathematical equations from basic principles or important observations makes it possible for researchers and those just entering the field to obtain an understanding of the state of the science. This appreciation of the limitations of current approaches is especially useful in directing future research to improve upon existing empirical or semi-empirical equations for water quality kinetics. Understanding the basis for river modeling will also be very important to practicing engineers. Furthermore, the exploration of model capabilities and limitations makes this a useful reference for graduate courses in water quality modeling. Researchers in the field of stream transport, temperature modeling, gas exchange or reaeration, and allied fields will be able to better define the extent of our understanding. In addition, the critical reviews of existing modeling approaches will assist practicing engineers and others in the selection and development of the appropriate water quality models. Discussion of measurements necessary will assist in the planning of the field studies that are required for all precise studies.

This reference will also be a valuable aid to those faced with difficult modeling problems that do not seem to be amenable to the standard modeling approach or where conditions do not seem to fall within the limitations of standard models. When possible, this book points out the limitations of current modeling practice and should be quite useful for that reason alone. The documentation reports for many current models do not fully address nor explore the impact of limitations of the models in current use and this leads to some misuse.

In this volume, the primary focus will be upon conventional pollutants that influence the dissolved oxygen balance and eutrophication. However, the reviews of the theoretical basis of modeling are useful for understanding other problems as well. For example, the section on reaeration begins with a general review of the two-film theory of volatilization that is also useful for explaining the basis of monitoring volatile organic contaminants. Therefore, selected theoretical derivations should be widely applicable.

The models reviewed in the first section of this volume are those that are of practical use in water quality management studies. Of special interest are studies aimed at the regulation of point source pollution (i.e., treated ot untreated flows of municipal and industrial waste waters into streams). In addition, some methods that can be adapted for studies of non-point source pollution will be addressed when feasible. Non-point source pollution from diffuse sources is becoming more important as point sources are brought under control. Yet, despite the urgency to do so, non-point source problems are not frequently studied using a modeling framework. As a result, only limited information can be presented to assist in the analysis of non-point source pollution problems. Unfortunately, the limited methods available do not constitute a comprehensive and consistent framework.

In organizing this material, it became necessary to determine what contitutes a river or stream to avoid too much overlap with other contributions planned by CRC Press for estuaries, lakes, and reservoirs. In addition, it is important to avoid any gaps in coverage of important topics. Regarding the definition, rivers and streams were simply taken to be any channelized flow moving consistently downstream. The definition is interpreted to exclude water quality modeling in wetlands, lakes, reservoirs, estuaries, and coastal areas. Nevertheless, some overlap is unavoidable since the basic principles are the same in all environmental modeling.

While the definition of what constitutes a river or stream may seem obvious, there are several areas where clarification would be useful. First, boudaries between wetlands and streams are quite indistinct. In this case, the first determining factor of whether stream-type modeling is applicable should be whether the movement of water can be measured or stimulated adequately. Generally, the mass balance approach used in river water quality modeling cannot be successfully applied in wetlands because the flow balance or water balance cannot be measured or predicted. Second, many rivers have been dammed for navigation and flood control; and these reservoirs behave like lakes and rivers at different times. In this case, river models are usually applicable when the run-of-the-river reservoir does not continuously stratify for longer than a couple of days. Third, the boundary between rivers and estuaries cannot always be precisely located. Usually, however, river models can be extended to the point where the estuary stratifies of the flow begins to reverse at times during the tidal cycle. Unfortunately, the conditions that define the demarcation vary dynamically due to changes in river flow rate, meteorological conditions, and tidal conditions. As a practical result, it is usually difficult to determine this transition point; but it is better to slightly overextend river models than to completelt ignore these very important zones of transition between rivers and estuaries. These are important areas related to fisheries productivity, navigation, sedimentation, and other activities that affect man. In the past, these important transition zones have been ignored (D. J. O'Conner, keynote presentation, Meeting of the Association of Environmental Engineering Professors, Washington, D.C., 1978). Therefore, this work will consider tidally affected rivers as an important subset of riverine conditions to be modeled, despite the fact that it is only possible to partially address this need. Tidally affected rivers are difficult to study because dynamic water quality modeling is underdeveloped.

To achieve a practical understanding and to establish the usefulness of the theoretical developments in this book, a limited emphasis will be placed on the measurements required to conduct water quality modeling studies. Existing measurement methods, especially alternative methods, will be cited. Methods that are controversial and in need of additional refinement will be explored in more detail. This is important because current modeling efforts seem to be limited not only by incompleteness of conceptual modeling methods, but also by our inability to always validate new or refined models with information collected in the field. Also, it is simply not possible to collect all the necessary data to make successful model applications in all cases. This cannot be determined beforehand without some knowledge of the measurements involved.

In most water quality studies, data collection is very important because of the semi-empirical nature of water quality kinetics models. All standard models of conventional organic pollution and toxic contaminants presently being used in 1989 require calibration and confirmation testing as well as definition of boundary conditions. This translates into the collection of at least two independent data sets required before model simulations can be treated as predictively valid. The result is that measurement of calibration data and boundary conditions are important enough that any critique of modeling practice and review of the understanding of water quality processes simply cannot overlook matters that are sometimes treated as mundane. Failure to provide at least a minimal background on data collection techniques would make any presentation of the basic principles of water quality modeling less fruitful.

ACKNOWLEDGMENTS

The U.S. Environmental Protection Agency (EPA) has granted this author the permission to complete this work as an outside activity. This permission is gratefully acknowledged. It should be noted that this work is the result of the private efforts of the author and any connection with the U.S. EPA and the federal government should not be inferred.

Sherry McCutcheon drafted and prepared many of the figures for this book and assisted with library searches, typing, correction of drafts, and correspondence. Her role in the preparation of this volume went far beyond the normal encouragement of a spouse that is frequently acknowledged in an effort of this type, and I appreciate this assistance. I regret the domestic disruption that projects like this cause and also thank Sherry and my son, Michael, for their forbearance during the final stages of the writing.

I also appreciate the assistance I have received from CRC Press, Rosi Larrondo, Marsha Baker, and others on the editorial staff were very helpful in the preparation of the manuscript.

During the writing of this book, several individuals were kind enough to take the time to assist me in becoming up to date about some of the diverse topics involved, and I wish to express my appreciation. Edwin Cordes, James Futrell, and Vito Latkovich of the U.S. Geological Survey Hydrologic Instrumentation Facility were kind enough to advise me of the recent developments in methods of measuring water velocity and discharge. Larry Marsh, President of Marsh-McBirney, Inc. was kind enough to appraise me of developments in measuring flows in conduits. Stuart McKenzie of the U.S. Geological Survey office in Portland, OR, followed up our numerous conversations on the conduct of river quality studies and again reviewed the valuable experience gained on the Willamette River study conducted during the 1970s.

I would also note that, like any author, I have benefitted from the exchange and examination of ideas from colleagues. Most recently, I have had the very worthwhile experience to begin an association with Robert Ambrose Jr., Thomas Barnwell Jr., James Martin, and Sandra Bird. Many of our exploration of ideas on how water quality modeling should be pursued have had a notable balance between rigorous scientific skepticism and enthusiastic engineering optimism. Furthermore, my other colleagues at the U.S. EPA Environmental Research Laboratory in Athens, GA, also share in enthusiasm for the persuit of Knowledge, and this creates an atmosphere that readily leads to intensive efforts such as this. I would also note that I have greatly benefitted in my understanding of water quality modeling from my past association with the U.S. Geological Survey. I began my first intensive study of river water quality modeling studies with Marshall Jennings. I have also had interesting and useful advice on modeling from Ronald Rathbun, Verne Schneider, Peter Smith, Phil Curwick, Lewis DeLong, and a number of U.S. Geological Survey district colleagues who have done much in the past to advance our understanding of how water quality surveys should be conducted. Ihave also had the great pleasure to be associated with the editor and originator of this series, Richard French of the Desert Research Institute, University of Nevada. I appreciate the assistance and the encouragement that I have received from Richard throughout our association.

Steve C. McCutcheon
Athens, GA
January 9, 1989

Dedication

———

to
Sherry and Michael

TABLE OF CONTENTS

Chapter 1

INTRODUCTION TO RIVER AND STREAM WATER QUALITY MODELING

I. INTRODUCTION

Models are necessary to both describe and predict water quality conditions. Current *and solve problems on a computer.* modeling practice provides a rational, descriptive framework for the analysis of existing problems and provides limited predictive capability that cannot be achieved by simply monitoring or measuring water quality. Descriptive modeling is most important because it makes it possible to understand the cause-and-effect relationships that govern water quality in a river. Once cause-and-effect relationships are known, management alternatives can be explored and the result of any improvements and changes can be projected. The certainty of any projection depends upon how well the cause-and-effect relationships can be determined. By definition, river models are limited to approximate descriptions of cause and effect and if the utility of these methods is to be understood, these limitations must be understood.

Descriptive modeling can be very useful for extrapolating monitoring data. The use of models to describe water quality conditions in river segments between dispersed sampling locations or for extrapolation over periods between sampling times is superior to any crude linear extrapolation, any statistical analysis, or any qualitative extrapolation. However, descriptive modeling to supplement monitoring has not been fully developed.

In the past, there has been a tendency to think of the collection of monitoring data and modeling as separate approaches to describing water quality. Although it is true that monitoring data are usually collected too infrequently and at locations that are spatially too disperse to support intense model studies, there is a close interdependence of data collection and modeling that could be used to make better use of monitoring data to design intensive water quality studies. Screening-level modeling based on available monitoring data can be used to design refined monitoring studies and should be used to design intensive data collection studies for model calibration. Intensive, synoptic data collection studies are a mandatory step in any precise modeling study, but these studies differ from monitoring only by degree of spatial and temporal sampling intensity. Therefore, monitoring studies and precise modeling studies seem to differ primarily by degree of sampling intensity required.

Calibrated models can be used to easily define cause-and-effect relationships that monitoring studies cannot definitively identify. In addition, models can be used to make limited predictions that are not possible with monitoring programs. Nevertheless, without calibration data, modeling holds little or no advantage over simple monitoring programs.

To aid in the understanding of the importance of modeling, it should be noted that there are a number of different types of studies in which modeling is crucial. Stream water quality modeling is used for screening studies, for planning and design of sewage treatment plants to eliminate pollution, and to guide the management of river basins. Each of these uses involves different types of models or models applied in different ways. Screening-level studies typically use simple models that can only indicate if potential problems may occur. Where problems exist, screening-level models can be useful in formulating a preliminary indication of the major causes. Many important decisions regarding the discharge of organic wastes in municipal and industrial wastewaters are based on a more elaborate, calibrated model of the water quality of the receiving stream. Occasionally, complex models are applied to attempt to refine the operation of sewage treatment plants on a seasonal basis. In addition, modeling has proven to be a useful method of analysis to explore the interactive effects point sources and nonpoint sources on water quality. Point sources are discrete inflows of polluted water that usually enter a stream through a pipe or channel from a wastewater treatment plant. Nonpoint sources are diffuse contributions to the total pollutant load that arise from urban and agricultural areas or natural wetlands. Nonpoint sources enter the stream as overland flow, as groundwater flow, or as flow collected in tributaries. Typically, nonpoint sources cannot be measured. Estimation techniques rely on modeling or simple mass balances.

The usefulness of modeling over the spectrum ranging from simple screening-level modeling to complex operational modeling depends on the reliability of the results and ease of use. The amount and quality of data available for critical locations at important times determine whether the model results are useful. Normally there is little choice for screening data; if adequate data exist to define boundary conditions (i.e., inflows and loads), then screening is normally useful. The results of more detailed modeling for planning and design are usually reliable when adequate data is collected for calibration and validation. However, few post-audit studies have been undertaken to confirm this. (A post-audit study is one that occurs after a sewage treatment plant has been constructed or upgraded to determine if the model projections are valid.) As a result, it should be clear that modeling results are very dependent on field data collection for calibration and definition of the loads and flows entering the stream. Without the proper data collection procedures, modeling studies can only achieve limited results.

From this brief synopsis of why models are important and how they are used, it should be evident that knowledge of various types of models and their data requirements are necessary to understand river water quality modeling. Since it is very difficult to review each of the many models available, this monograph is organized to review the basic modeling concepts that are common to most models in present use. The development and review of the basic modeling concepts are accompanied by selective explanations of the data requirements for model formulations. Just as there are a range of types of models that can be applied, there are a range of data collection techniques for most parameters of interest. As a result, some care in the selection of data collection techniques is very necessary to provide accurate measurements compatible with the approximation of the model chosen. Therefore, this contribution is organized to review important aspects of the state-of-the-science and to develop and derive the state-of-the-art modeling techniques currently used in applications of river-quality models. In addition, the critical measurement techniques are reviewed when those methods are not readily evident because of the state of development.

Specific models are not reviewed to describe the components of each model. Instead, important components of the most practical models are used to illustrate the state-of-the-art in modeling when showing how modeling techniques are derived. Practical models and the state-of-the-art formulations are defined by reviewing the few model evaluations available to define what models are practical and useful and by pointing out which process descriptions are used in the most practical models. As a result, only the important components of some of the best screening-level methods and waste-load allocation models will be covered. Reviews of successful, river-modeling applications are limited to demonstrating the validity of various modeling components and are not intended to provide validation of overall modeling procedures.

In this introductory section, the basic principles of stream water quality modeling are reviewed to determine the theoretical basis and limitation of such models. The basic principles of conservation of mass and momentum are introduced in Part II of this section. Part III reviews important definitions to avoid confusion about terminology. Part IV of this section reviews numerical descriptions of streams and discusses the importance of defining study areas and modeling domains in a careful manner. The basic flow and water-quality equations are derived in Parts V and VI. Part VII discusses models in general use and describes the most practical models available. Part VIII discusses model selection and testing. This arrangement should at least briefly cover the basis for current modeling practice and some of the important model application issues.

This section will be followed by a second section discussing the important physical processes affecting river quality. It is presently planned to follow this volume with a second volume on the other important water quality processes that must be known to define cause-and-effect relationships. The next volume is to cover biochemical and geochemical processes that include biochemical oxygen demand, sediment oxygen demand, and nitrification, plus the phosphorus, nitrogen, and oxygen cycles in streams as well as other topics.

The following section on physical processes covers water balances and stream hydraulics, mass balance techniques for dissolved substances and mixing in streams, heat balances and temperature modeling, and gas exchange. This section should answer basic questions of how water movement and mass transport affect water quality. The parts on heat and gas exchange primarily focus on the surface exchange phenomena and note how they are related.

II. BASIC PRINCIPLES AND LIMITATIONS FOR MODELING

Knowledge of the basic principles of water quality modeling allows modelers to understand how models may be applied and when misapplication may result. The important first principles involved include the conservation of mass and the conservation of momentum (Newton's Law relating mass and force in dynamic system, $F = ma$). Except for the conservation laws, most of the basis of water quality modeling derives from empirical or phenomenological process descriptions. As a result, the limitations of modeling cannot be understood unless the basis of each process model is described.

The most important basic principle underlying water quality modeling is that of conservation of mass. Modeling involves performing a mass balance for defined control volumes over a specified period of time. Essentially, this is an accounting of material of various types in a defined volume of water or a number of water volumes. Typically, material balances involve dissolved and suspended materials such as dissolved oxygen, organic carbon, nitrogen, phosphorus, and suspended sediment, but this principle can be applied to any substance whose transformation kinetics are known.

The mass balance volume may include the surface of the benthic sediments or may involve specification of the flux of material to and from the bottom. On some occasions, separate mass balances are performed for the water column and benthic sediments. The use of a separate mass balance for benthic sediments has primarily been used for the modeling of toxic chemicals that readily attach to sediment and for modeling the nutrient flux controlling eutrophication. Generally, the flux of dissolved oxygen and most nutrients are specified at the benthic interface. It does seem likely, however, that a separate mass balance approach will be necessary in future stream modeling to fully understand the exchange of dissolved oxygen and other critical materials. In general, the mass balance procedure may handle the benthic interaction in a variety of ways, but all precise mass balance modeling includes quantification of the exchange with the atmosphere if volatilization is important and the movement of mass into and out of a defined volume of water because of the flow of water in the river of interest.

The mass balance is performed by accounting for all material that enters and leaves a defined volume of water plus accounting for all changes in mass of a constituent caused by physical, chemical, and biological processes. The accounting for effects of the reactions is one point at which water quality models begin to diverge from first principles (i.e., Newton's Laws of physics, conservation of mass and energy, laws of thermodynamics, etc.). In a number of reactions, it has proven necessary to make use of the accumulated observational evidence to devise empirical or phenomenological equations that describe the process. For instance, many reactions are biologically mediated and the equations describing these processes are, in a number of cases, traceable to the Monod equation that is an empirical analog of the Michaelis-Menten equation for enzyme reactions. In other cases, there are full theoretical descriptions available to describe certain processes, but the full description involves too much computational complexity or requires measurement of physical parameters that are too difficult or expensive to be practically feasible. In such cases, more approximate, limited methods are employed. Unfortunately, the limitations of these semi-empirical and approximate methods are not comprehensively defined for typical stream modeling conditions. The lack of full understanding of the limitations of process formulations is important because it seems to have resulted in at least some inadvertent misuse of models. This misuse does not seem to be widespread, but is serious enough to keep in mind. In addition, the lack of precise understanding of model limitations hampers more extensive application since only experienced modelers seem capable of understanding the vaguely defined limits of applications.

A clear statement of the limitations of model formulations is necessary because modeling techniques are derived from a diverse collection of interdisciplinary methods. It is rare to find a neophyte modeler with the necessary experience in environmental fluid mechanics, aquatic biochemistry, ecology, analytical chemistry, physical chemistry, sanitary engineering, chemical engineering, and the other topical areas from which modeling methods are adopted. In addition, experienced model users do not always take care to determine and report the uncertainty in modeling results in a way that is easily interpreted by resource managers. Therefore, it is important to note the areas where basic principles are used to aid in defining where limited empirical and phenomenological methods are employed.

To guard against the misuse of stream water quality models, there are too few safeguards. Documentation reports for models frequently do not focus on the limitations of the methods proposed. Also it seems that some general purpose models rely on general usage as justification for incorporation of methods into a model. The extent of general usage is largely documented in difficult to obtain engineering reports. Furthermore, in a number of cases, it is not clear if the limitations of the existing empirical methods have been fully defined. In this regard, the desire of engineers to make rational projections has in a few cases outraced the ability of the science to provide the general purpose methods required. As a result, fully successful water quality modeling requires some modeling experience and general knowledge in diverse fields of biology, chemistry, and engineering to note those unique conditions that may be beyond the limitations of current models.

The conservation of mass is not the only first principle employed in water quality modeling. An important aspect of a mass balance involves knowing the effect of water movement. In advanced models, the speed and amount of the water that carries dissolved and suspended materials through a stream are determined from a mass balance of the water and the conservation of momentum principle. The conservation of momentum principle is similar to the mass balance in that accounting is undertaken for fluid momentum in a defined volume of water. In essence, the conservation of momentum equations are a statement about the forces that affect water movement.

Unlike the mass balance where physical, chemical, and biological reactions can only be empirically or phenomenologically represented, the solution of the conservation of momentum equations involves only limited empirical representation of friction losses. Generally friction losses are well understood. However, the approximate representations that are generally employed also involve solving the equations over longer distances or time periods and employing averaging methods to represent the turbulent nature of river and stream flows. These approximate representations range from simple relationships between friction factors and the average velocity to very complex descriptions of forces acting on the flow and the turbulent processes that contribute to resistance to flow in channels. There are, in addition, completely empirical methods of relating flow to depth and velocity in channels. Because the same averaging and approximation techniques are also needed to solve the mass balance equation, the approach taken here will be to first review the derivation of the governing flow equations and then note where the same methods can be used to derive useful forms of the mass balance equation.

The choice of how to solve for the effect of water flow on water quality is dictated by the purpose of the study, i.e., whether prediction or description of the effect of flow is desirable and by how sensitive the water quality parameters of interest are to the effects of flow. If the water quality parameters of interest are not greatly influenced by changes in water flow, the simplest methods of describing the flow may be more than adequate. Furthermore, if critical water quality conditions only occur over a limited range of flow, then description of flow effects may be more than sufficient, especially in the early phases of a study.

More and more, however, the effects of water flow are found to be quite important, and predictively valid methods are required to fully describe the effects of flow. There are also many cases in which it is better to predict the effect of water movement rather than try to measure the effect over the full range of conditions that may be important. In this regard, the extrapolation of a few measurements used to calibrate a flow model can be a very efficient approach to overall data collection and modeling. Flow modeling methods have progressed until it has become feasible in many cases to explore flow effects more fully than in the past.

By contrast with the pollutant mass balance equation that usually involves empirical kinetic descriptions, the flow equations are less divorced from an exact theoretical basis and thus usually require less calibration effort. Precise calculation will in almost all instances require calibration, but the calibrated model may require less confirmation testing and may be more likely to be predictively valid.

As alluded to earlier, practical methods for solving the flow equations involve some element of approximation that must be known and understood to guard against misuse of models when the calculation of water movement is important in making water quality calculations. As a result, it is important to understand the origin of the practical forms of the governing equations representing the movement of water. This understanding can assist in overcoming a lack of well-defined criteria to guide in the selection of approximate methods to treat the full effects of flow. It is also important to understand how to choose empirical factors for the flow equations.

III. DEFINITION AND TERMINOLOGY

To avoid ambiguity, terms like model and computer code are defined in this section. In addition, it is useful to present classifications for models according to use, complexity, and typical data requirements.

The exact definition of what constitutes a river water quality model is not fully agreed upon. Some investigators refer to a model as a computer code calibrated to describe or predict the conditions in a particular stream. Others, including this author, generally refer to a general purpose computer code as a model. The definition of model, however, is broader than this:[1]

Model — An assembly of concepts in the form of one or more mathematical equations that approximate the behavior of a natural system or phenomena.

When the mathematical equations are written in an algorithmic form as part of a computer code, it seems reasonable to also refer to the computer code as a model. To avoid ambiguity, however, it may be best to distinguish between process models or submodels, i.e., equations approximating the behavior of a water quality process from a computerized model that is a collection of the equations in an algorithmic form.

Robert Ambrose, Jr., a reviewer, further points out that it may be useful to employ specific descriptive terms such as "reaeration" model to clearly distinguish a process algorithm. General purpose codified models of a specific type may be best referred to as the "QUAL2e" model, or a general set of codified algorithms common to a number of different codes might be referred to as a "dissolved oxygen" model. When a model has been calibrated to a specific river, then it is recommended that the calibrated code be referred to as the "Chattahoochee River Dissolved Oxygen" Model or "Chattahoochee River" Model. The latter model might refer to a more comprehensive model of the Chattahoochee River. Although this may be a workable approach, it is cumbersome to always add specific descriptors. As a result, some ambiguity will continue until some general consensus arises.

The definition of computer code or program should imply a wider scope than computerized model:[1]

Computer Code or Program — The assembly of algorithms describing water quality processes, the codified numerical solution methods, bookkeeping procedures, and data control and storage procedures that can be executed beginning with the acceptance of data and instructions regarding the processing, interpretation, and analysis of the specified data and any other data that permanently resides within the code, to the reporting and delivery of the results of the computerized analysis.

A collection of two or more separate computer programs that process data to varying degrees and transfer data for further processing or parallel processing of data is sometimes referred to as a modeling system. The WQRRS[2] modeling system illustrated in Figure 1 is a good example of such a system.

Stream water quality models are generally classified according to function, complexity and data requirements, dynamic response, theoretical basis, treatment of inherent variability of processes, and method of solving the governing equations. The primary functional classification distinguishes models that have been devised to primarily simulate conventional pollution (consisting of organic material and nutrients) from models that have been devised to primarily simulate toxic contaminants. Models of conventional pollutants have been in use since at least 1925 when Streeter and Phelps[3] published their classical model for the dissolved oxygen balance. Models of toxic chemicals have only begun to receive widespread use in the last decade or so. Before that time, water quality models referred almost exclusively to models of organic pollution. The shorthand usage of conventional models or models of conventional pollution, therefore, refers to models of organic pollution and nutrients while toxics models refer to models of toxic substances.

Many general purpose models can be applied in a variety of ways and are thus difficult to fully classify. A few models can be only applied to limited conditions. When discussing what should be expected of a model, especially those that can be

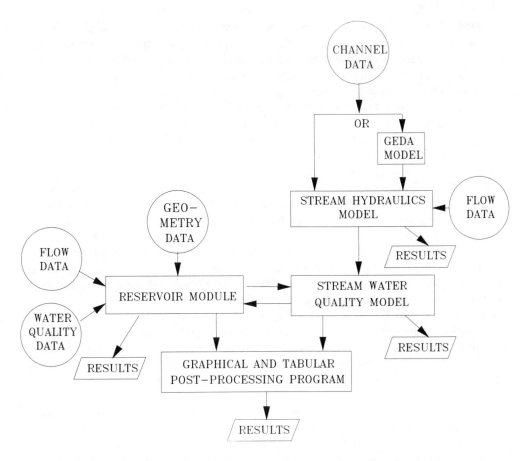

FIGURE 1. Illustration of the components in a modeling system. This particular system is the *Water Quality for River-Reservoir Systems*[2] that involves three primary computer codes that are operated serially or in parallel as physical configuration of the river basin dictates. The primary components are serviced by optional pre- and post-processing programs.

applied in a variety of ways, it is, therefore, useful to develop a classification scheme based on model complexity and data requirements. For this purpose a modified scheme originally developed by Ambrose et al.[4] seems most useful:

Level I — Simple manual or graphical methods based on statistical or deterministic equations that are easily used for crude screening over extensive areas to isolate existing or potential trouble spots for detailed follow-up analysis. Screening involves only readily available data and may be used to identify gaps in monitoring data and identify data needs for more intensive follow-up studies. Important issues and preliminary management options may be suggested by the results, or some preconceptions may be ruled out, but uncertainties are typically large and are not quantified. Mills et al.[5] established a very good example of screening methods for streams and other water bodies. Screening methods are relatively simple but generally require extensive expertise to apply and interpret.

Level II — Simple computerized model used for fine screening or crude planning and assessment for extensive areas over extended periods of time. Model equations are usually deterministic in nature but only approximate the basic processes. As a result, any management projections are somewhat uncertain and wise use requires considerable experience in interpreting the resulting calculations. Formal uncertainty analysis is usually not included. Data collection requirements are usually limited to one preliminary data collection study.

Level III — Computerized model of intermediate complexity used as a fine planning model or a crude engineering design or resource management model. Extensive areas and extended periods of time can be simulated but at significant cost in data collection and preparation, and in computing time. Some approximation of the basic processes limits the applications for design and management. Data collection involves at least two independent data sets to bracket important conditions. Uncertainty analysis is typically included as a part of the computer model or as part of a supplemental analysis. Certain types of studies such as waste load allocations require only modest levels of experience if knowledge of modeling techniques is based on advanced study (i.e., M.S. or Ph.D. degree in environmental engineering).

Level IV — Advanced mechanistic computerized model used for detailed design and management. Data requirements are usually intense and usually involve at least a preliminary data collection study and model screening to design at least two intensive synoptic data collection efforts. Simulations are typically limited to smaller areas and short time periods to avoid extensive data collection and computing costs. Procedures using these models are not defined well enough to specify typical uncertainty analyses required. The incomplete stage of procedural development implies that extensive modeling experience is required to develop models.

Models are also classified according to the way the stream is divided into computational elements and by the way the governing equations are solved. These descriptions are based on the importance of lateral, vertical, and longitudinal variation in water quality constituents and the resulting complexity of the equations required to describe these variations. The following four distinctions have proven useful:

1. **Zero-dimensional model:** A segment of the stream is described by a single computational element, ignoring any lateral, vertical, and longitudinal variation that may occur. The single element is treated like a completely mixed reactor.[6] This approach may be most useful in a screening-level analysis of a mixing zone. It is rarely expected to be used in typical stream studies.
2. **One-dimensional model:** Where lateral and vertical variation is unimportant, the stream is described by a series of computational elements extending downstream and describing the longitudinal gradients that are prevalent in streams. This is the most common approach to describing stream water quality.
3. **Two-dimensional model:** This is a model that describes the variation in two directions. The most useful riverine type describes lateral and longitudinal gradients and assumes that vertical variations are unimportant (i.e., water quality can be depth averaged). These models are occasionally used to define mixing zones in the vicinity of where point sources of pollution enter the stream. Rivers are not frequently stratified; but when this occurs, vertical gradients are usually important as well. Therefore, width-averaged, stratified river models do not seem to be very useful at present.
4. **Three-dimensional model:** This is a model that describes vertical, lateral, and longitudinal gradients of water quality parameters. These models are occasionally used to describe complex mixing zones such as those formed by cooling water discharges from electrical power generation plants where the stratification results in significant vertical gradients.

These four classifications are illustrated in Figures 2, 3, and 4.

One-dimensional models may be further described by the manner in which complex one-dimensional networks can be simulated. The simplest branched networks that can be described using one-dimensional models are dendritic systems shown in Figure 5. The

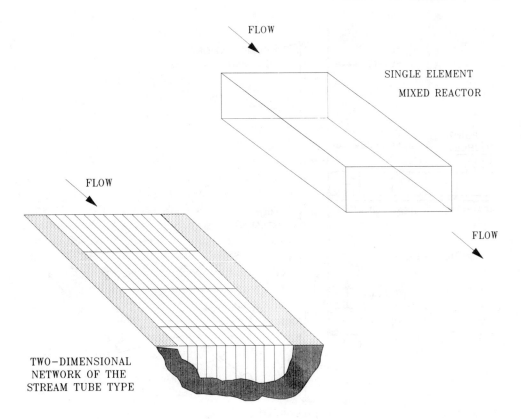

FIGURE 2. Typical zero- and two-dimensional model networks.

flows in two or more branches are combined and flow downstream in a single branch. The total flow in the system leaves in a single channel. It is also possible to apply the linearized one-dimensional mass balance equations to more complex networks that may include parallel channels that loop around islands and merge into a single channel or may diverge into separate channels that do not rejoin the main channel as shown in Figure 5.

Although mixing of contaminants into a stream is a three-dimensional process, it has often proven possible to use one-dimensional models that assume that mixing occurs instantaneously at the point of injection. This is feasible because inflows usually mix quickly and because the effect of many of the important kinetic processes can be averaged over the cross-section without affecting the final concentrations computed at the end of the mixing zone. Therefore, the spatial distinctions of one-, two-, and three-dimensional imply something about how the governing equations are solved but not necessarily anything about the actual process occurring in the stream.

When used to analyze river mixing, models are classified as *near-field* and *far-field*. Near-field models typically include the effects of jet momentum from inflows or describe lateral and vertical concentration gradients. Far-field models are used downstream of near-field models where jet momentum has been dissipated or where lateral and vertical variation in water quality is unimportant. Typically, near-field models are two- or three-dimensional models, and far-field models are one-dimensional models.

Models are also classified as *steady-state* or *dynamic,* according to the manner in which water quality changes with time are described. Steady-state models use averaged loads and flows entering the stream over specified periods of time to compute the

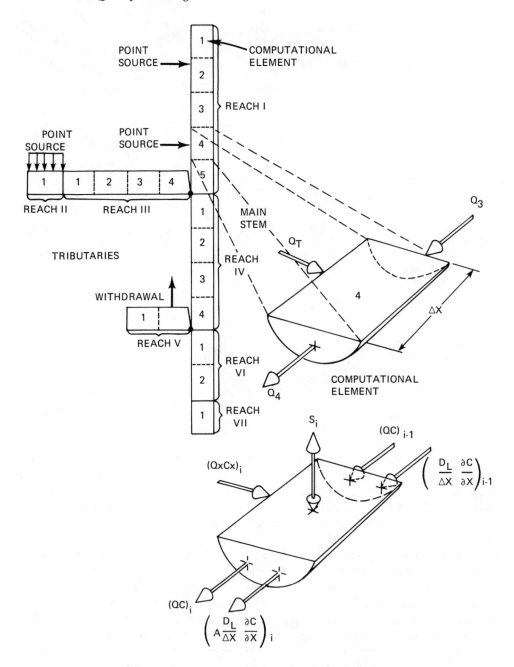

FIGURE 3. Hypothetical illustration of the branched one-dimensional discretization scheme for a finite difference solution scheme.[7] Reaches extend over areas of the stream that have constant hydraulic and biochemical characteristics and does not exactly represent the geometry of tributaries. Q represents flow, C represents concentration, Δx is the computational element length, A is cross-sectional area, D_L is the longitudinal dispersion coefficient, and S_i is the surface flux.

average response in the stream. Typically, flows and loads specified for steady-state models are approximately constant. Occasionally, however, averages of variable flows and loads can be interpreted using steady-state models to determine the average effect. In the simple example given in Figure 6, the loads at the head of the stream and at the point source are specified as constant values, and the results are reported as a constant value at the end of the stream or at any point within the stream. Dynamic models

THREE-DIMENSIONAL MODEL NETWORK

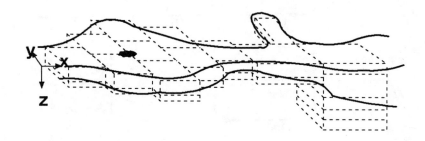

FIGURE 4. Typical three-dimensional network of computational elements for the water column and benthos.[8]

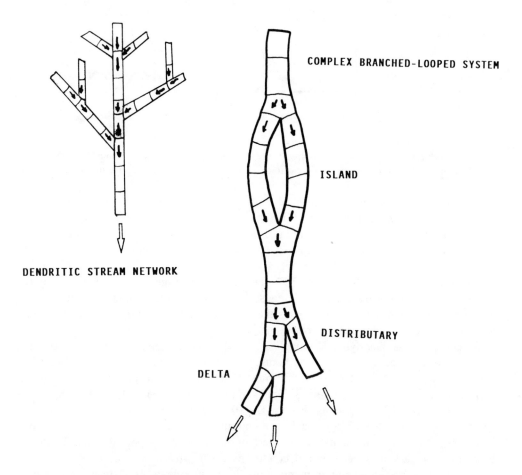

FIGURE 5. Examples of branched networks of one-dimensional elements.

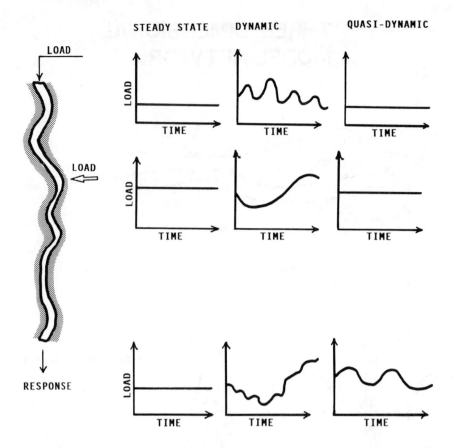

FIGURE 6. Stream response to steady-state, dynamic, and quasi-dynamic boundary conditions (i.e., flows, loads, and meteorological conditions affecting stream water quality).

approximate the response of a stream to time-variable changes in the loads entering the stream. This is illustrated in Figure 6 by showing the composite time variable effect at the end of the reach due to specification of time variable loads entering the stream.

Quasi-dynamic models are those that require that some boundary conditions be constant but allows other conditions to vary. The most useful example of such a model is an option of the QUAL 2e model that bases water-quality calculations on the specification of constant flows and loads, but allows specification of time varying meteorological conditions and simulates the effect of variable sunlight, air temperature, and wind speed on water quality conditions.

As one may imagine, dynamic models are much more complex to apply. The solution of the governing equations is more complex and more data are required. Boundary conditions where loads enter the stream must be described with at least one datum for each time step in the solution period. For example, if the dynamic governing equations are to be solved at the end of each hour for 10 h, then a boundary condition must be specified for each of the 10 h for each load entering the stream. (Short cuts are possible, however, when loads are constant for part of the simulation period.) In addition, it becomes necessary to specify initial conditions in the stream at the beginning of the simulation. Contrast these requirements with the requirements of a steady-state model that only requires that boundary conditions be specified by a single value during the simulation and that initial conditions are not required.

It should be noted that the terms "steady-state" and "dynamic" are not based on strict definitions of the physical processes, but are, to some extent, operational in

FREQUENCY DISTRIBUTION OF OBSERVED AND
CALCULATED VALUES OF A QUALITY VARIABLE

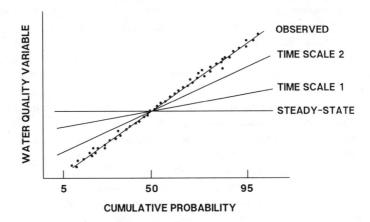

FIGURE 7. Effect of averaging over various time steps or simulation periods on the ability of dynamic models to represent the observed dynamic behavior of a stream water quality variable.[8]

nature. First, steady-state conditions are only approached and never fully achieved in a turbulent environment like a stream. This means that strictly speaking the steady-state approximation should be checked in each application. Fortunately, the steady-state approximation has been found to be widely applicable and can therefore be checked in an indirect way during calibration and confirmation of the model application. If discrepancies between measurements and steady-state model predictions are small, the steady-state approximation is typically assumed to be useful.

Second, a dynamic model, as the term is used in water-quality modeling, does not mean that the model reproduces the exact dynamic response. It is used to describe models that simulate part of the dynamic response. This is illustrated by Ambrose[8] in Figure 7 where the cumulative probability is used to show the effect of averaging over various time periods during a "dynamic" simulation and for averaging over the full simulation period with a steady-state model. When the observed cumulative probability curve (that is only an approximation of true dynamic response in and of itself) is relatively flat, then steady-state approximations are quite adequate in most cases, and dynamic models can approximate most of the observed behavior with calculation over longer time periods.

Other categories of models describe the theoretical basis and the method of solving model algorithms. *Theoretical models* are those based on first principles such as the conservation of mass. These designations usually refer to process models, and except for dissolved solids models and other conservative mass balance models, theoretical models cannot be used in stream water quality modeling. In terms of modeling, theoretical models are superior because they are valid for all conditions encountered.

More frequently, process descriptions in general purpose models are based on *phenomenological* or *semi-empirical models*. These are derived from conceptual approaches that are typically limited to describing a subset of possible stream conditions. For example, there are a number of phenomenological models of the gas transfer process at the water surface. These include a film model, a surface-renewal model, a penetration theory, and a boundary-layer model. Quite often these semi-empirical models are referred to as theories, but this should not imply any direct relationship to first principles. These models are usually limited to specific ranges of stream conditions, but the exact ranges are rarely defined. Overall, these models present the most significant challenge in defining limitations of water-quality models. Typically, phenomenological models have only been validated for very limited ranges of stream conditions and assumed (perhaps incorrectly in most cases) to be valid for most stream conditions encountered.

In addition, process models may also be referred to as *empirical models*. These are models derived by curve fitting or statistical analysis of stream data defining a process of interest. A good example includes equations to describe dissolved oxygen saturation concentrations as a function of temperature and salinity. Also an empirical model is used to relate specific conductance measurements to dissolved solids concentrations. Generally, not many model formulations would be classified as empirical (at least without offending a majority of model developers). However, there are indications that many phenomenological models originally began as empirical equations and were later explained by proposing conceptual models from which the original empirical model could be derived. This is of course a normal part of proposing, testing, and refining hypotheses from the scientific method. Nevertheless, the blurring of the distinction between semi-empirical and empirical models adds some confusion because the limitations of empirical models are clearly defined by the data originally used to derive the algorithm. As indicated above, limitations of semi-empirical models are more difficult to define. Models of biochemical oxygen demand (BOD) seem to be an example of a process model that was originally an empirical model. More recently, the Monod equation of bacterial growth has been used to derive first-order BOD kinetics.

A *mechanistic model* is one formulated from a hypothesis about controlling mechanisms. By contrast a mechanistic model is different from empirical models derived by statistical analysis without regard to controlling mechanisms. As a result, mechanistic models are evidently equivalent to theoretical and phenomenological models. Often little distinction is made between phenomenological and mechanistic models.

A distinction is also made between *stochastic* and *deterministic models*. Stochastic models incorporate in the inherent uncertainty of models by describing the central tendency (typically the mean) and some measure of the variability (typically the standard deviation) of parameters. Deterministic models ignore parameter variability or assume that variability of parameters is understood and taken into account in other ways. As a result, deterministic models only describe relationships between mean values of parameters. Stochastic models tend to use empirical descriptions of parameter variability whereas deterministic models tend to employ mechanistic process descriptions.

The method of solving water-quality equations also leads to two categories of models. These are *analytical* and *numerical models*. Analytical models are those that are based on an analytical solution of the governing equations. The plug flow solution of the dissolved oxygen balance equation, known as the Streeter-Phelps equation, is perhaps the best known analytical model in stream modeling. Like all analytical models, the Streeter-Phelps model is a continuous relationship between dissolved oxygen and location or time of travel in a reach where model parameters are constant.

By contrast, a numerical model is one that requires a finite difference, finite element, or other approximate method (i.e., harmonic analysis or use of error functions) to solve water-quality equations. Because analytical solutions are quite limited, numerical solution techniques are used in most general purpose stream water quality models. Analytical solutions are, however, very useful to verify numerical solution techniques.

Finally, it is useful to describe the types of data used in modeling. These types include data that define boundary conditions and initial conditions, and data used to calibrate and validate models. The four types of data are defined as follows:

Boundary conditions — These are the set of data that describes the mass and energy that enter the model domain (subset of the stream environment being simulated). This usually involves the specification of flow and loads entering the head of the reach from upstream of the model domain and from point sources and nonpoint sources that enter at discrete or diffused locations over the length of the stream segment being simulated. Boundary condition specification also includes specification of other conditions such as meteorological conditions that govern the calculation of the energy flux through the water surface and chemical or biological characteristics that govern the exchange of material (such as dissolved oxygen) between the bed and water column. Flow models also may require specification of flow or depth at the downstream end of the model domain to take into account backwater effects on water movement if a complex model is used. If the downstream end of the stream is affected by highly dispersive reversals of flow caused by changing tidal conditions or lake level variations, then the quality of the water moving upstream of the end of the model domain must also be specified as an additional boundary condition.

Initial conditions — These are data specified for dynamic or quasi-dynamic models to define the water quality condition at the beginning of the simulation period. Flow, depth, and all water quality parameters being simulated are specified for all computational elements in the model domain. These parameters are used to begin iterative solutions of the governing equations and are important for simulations of highly dispersive systems over short periods of time. For streams with limited mixing and residence times (time of travel) that are smaller than the simulation period, the initial conditions are not important at the end of the simulation but may be important near the beginning of the simulation period. These data are not used in steady-state models except when an iterative solution of the governing equations for dynamic conditions are solved for steady state. Accurate specification of initial conditions may be necessary for quick, accurate solutions of the governing equations.

Calibration data — These are flow and water quality data defining boundary conditions and conditions within the model domain during the period of simulation. Conditions within the model domain are measured for the purpose of comparing observed conditions in the stream with predicted conditions. Model coefficients and parameters are adjusted until the predictions match the observations, if the conditions in the stream can be reasonably approximated with the model being calibrated and if the boundary conditions are accurately measured.

Validation or confirmation data — This is an independent set of flow and water quality data collected in the same manner as the calibration data is collected to define internal and boundary conditions when the stream is at a different state of behavior. Model predictions are compared to these data without modification of the calibrated coefficients and parameters to determine if the model has been adequately calibrated for the range of conditions defined by the calibration and confirmation data.

It should be noted that the term "confirmation" or "validation" is the subject of some controversy. The testing of a calibrated model has also been referred to as verification. Verification of a model, however, implies a comparison with the true condition. This would imply that measurements are exact representations of a true condition which is not the case. The use of the term "verification" to describe model testing would further seem to neglect an important part of the calibration and testing procedure that checks for consistency between measurements of the boundary conditions and the conditions measured in the model domain. This is not a concern when benchmark testing against analytical solutions is involved or when initially testing a computer code installed on a new computing system with results obtained on the system on which the code was developed. Therefore, verification testing is usually employed to describe comparisons where exact correspondence or correspondence to within the range of computer round-off error is sought.

As long as the term "validation" is used to indicate that the calibrated model is consistent with the range of conditions defined by the calibration and validation data, then the term "validation" would seem to be consistent with the formal condition.[9] If, however, "validation" is used to suggest that the model being tested may be applicable to all rivers or even all rivers of similar size, then the term would be inconsistent with the present intention of this author and most modelers as well. To avoid any potential misuse of the term "validation", Rechkow and Chapra[10] propose the alternative, confirmation testing, to accurately reflect the intention of testing to determine if the model in use (which is approximate in nature and is not an exact or true representation in itself) can be reasonably applied to the conditions measured in a particular stream without implying that a wider range of application is automatically possible. The intention of the testing, whether referred to as validation or confirmation, should clearly be to test for and guard against the effects of any preconception. Furthermore, validation is only possible for limited ranges and should not imply any comprehensive range of validity.

IV. STUDY DOMAIN, MODEL DOMAIN, AND COMPUTATIONAL NETWORK

There are at least three steps in setting up a model simulation of stream water quality. The steps include:[7]

1. conceptual definition of the study domain
2. definition of exact model domain
3. definition of the computational grid network

The initial definition of the study domain determines on how successful the study of cause and effect relationships may ultimately be. Definition of the model domain is governed by the convenience desired for model implementation and data collection. Definition of the computational network is governed by the method of solving the governing equations.

A. STUDY DOMAIN

Maps or aerial photographs of the stream and any monitoring data available are used to define the limits of the study domain (see Figure 8 for example). The domain encompasses areas in which the water-quality problems of interest are located and also includes the source of the problem distinctly defined by one or more model boundary conditions. Failure to initially include the complete problem area and to adequately include the source of the problem as a well-defined boundary condition or as an internal process usually requires that the domain be expanded later in the study. This may require that some data collection efforts be repeated.

B. MODEL DOMAIN AND BOUNDARY CONDITIONS

Once the study domain is established, the exact boundary of the model is defined. This is illustrated in Figure 8 where the shaded area defines a segment of the stream extending from river mile 235 to 303 (river kilometer 378 to 488). The boundary of the model domain was chosen to correspond with the mouth of each tributary in this particular case study.[11]

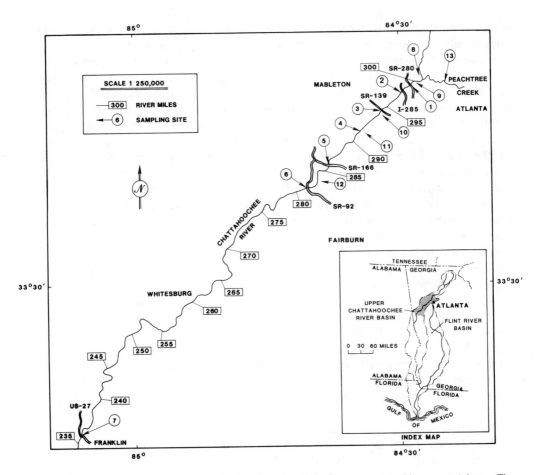

FIGURE 8. Definition of modeling domain for a study of the Chattahoochee River near Atlanta. The domain is the stream channel in the shaded area of the insert blown up in the larger drawing of the channel from River Mile 300 to 235. River miles are measured upstream from the mouth of Chattahoochee River. Note that 1 mi = 1.6 km.

There is little guidance on what stream lengths can be accurately modeled. However, given the present state of modeling, it seems reasonable that the model domain should be limited to stream lengths of approximately 80 to 160 km (50 to 100 mi) for precise modeling. This is an arbitrary criterion based on this author's limited knowledge of modeling studies. Longer reaches have been modeled for screening level basin-wide studies (see for example Hydroscience[12]), but at present, it is not clear that the necessary data can be collected to calibrate models over longer reaches. McKenzie et al.[13] found it necessary to divide a 139-km (86.5-mi) reach into 3 subreaches to make collection of calibration data feasible. In addition, models involve a number of simplifying assumptions that are tailored for short reaches.

Over short reaches, there are many negligible processes that may have significant accumulative effects over longer reaches. For example, conditions such as evaporation and the internal cycling of material are not accounted for very well, if at all, in present models. The accumulation of errors over longer reaches is important because of the effect on calibration parameters and the predictive validity of the calibrated model.

James Martin, a reviewer, also notes that length limitations may be stricter for steady-state models. Steady-state modeling results are unrealistic if the length of the model domain is so long that the time of travel exceeds the period over which inflows typically remain constant. For example, it may be unrealistic to apply a steady-state model to a long reach in which the time of travel is on the order of several months when significant changes in inflow occurs on the order of days or weeks.

Because few if any model documentations discuss limitations on the spatial extent of model applications, it is not clear if many of the assumptions behind existing models have been fully checked. Unfortunately, it seems that many models have been packaged together without giving evidence that the limitations have been fully considered. This should not be taken as an indictment of present modeling practice. Rather, it should be taken as a indication that the limitations of existing models, especially regarding lengths over which they can be applied, are not known well enough to foresee when a modeling study may be unsuccessful. As a result, it may be useful to approach modeling in a more conservative manner until the limitations are better understood.

The most successful modeling has involved evaluating the effects of sewage treatment plant discharges over short reaches. Over short reaches, the treated sewage causes a large perturbation in the material cycling that can usually be easily approximated with a model. Many times, only a few components of the full complex cycle need to be incorporated in a model. Large perturbations in the material cycles are important because they translate into strong gradients of parameters such as biochemical oxygen demand, nitrogen, phosphorus, and dissolved oxygen. These gradients are very useful in modeling because they can be easily measured and can be used with ease to calibrate most standard models. Modeling has been less successful over longer reaches especially far downstream of inflows where loads are largely assimilated and the full, complex cycling of material becomes relatively more important.

In addition, most process models are approximations that cannot be accurately applied over long distances. Rather than solve the water quality equations in a complex manner that incorporates much of the variability that has been observed, the standard practice is to apply semi-empirical approximations of the kinetic reactions (usually mechanistic types) over short discrete reaches. For example, biochemical oxygen demand is modeled over short reaches for which there is some evidence that biomass, pH, and other important controlling factors do not change significantly. Biomass concentrations and other important factors are not normally explicitly related to the important calibration parameter — the deoxygenation rate coefficient. Specification of rate coefficients over short reaches allows changes in specifications as the characteristics of the waste and stream conditions change.

The important element of establishing the model domain is the definition of the exact boundaries at which the model boundary conditions will be specified in the model. Boundary conditions represent information about the energy, water, and dissolved and suspended mass that enter the stream at various times during the study. Generally, the choice of locations at which the boundary conditions should be measured and specified in defining the model domain should coincide. In practice, however, some inexactness in specification of boundary conditions is permissible.

The purposeful introduction of minor discrepancies caused by measuring boundary conditions at some distance from the model boundary becomes necessary because of the difficulty in measuring boundary conditions at locations dictated by the goal of keeping the model framework as simple as possible. It is usually inadvisable to attempt to measure flow and water quality conditions at the confluence of a stream and a tributary because backwater effects and mixing into the tributary preclude accurate measurement. Therefore, the best locations for the measurement of boundary conditions are just upstream of the influence of the receiving stream. At the ideal location, there is frequently only a limited discrepancy with conditions at the actual confluence.

As a practical matter, the boundary conditions can only rarely be measured at the ideal location because of the limited access for field crews. As a result, the boundary condition is often measured at some distance upstream of the confluence. For truly steady-state conditions in the tributary and in the receiving stream, the location of the site to measure a boundary condition is not critical until the travel time from the measurement site is such that parameters begin to change because of decay, settling, gas exchange, or some other process. Travel times between the model boundary point and the measurement site on the order of a few hours seem to be generally acceptable. It would, however, be useful to develop more rigorous time of travel criteria to guide boundary selection.

When dynamic changes are occurring, the location of sites for measuring boundary conditions becomes more important. The added importance depends on the relative magnitude of the load of constituents entering the receiving stream compared to the total load in the stream upstream of the confluence. When the model results are sensitive to the load and the timing of the introduction of a dynamic load into a stream, then the measurement site must correspond as closely as possible with the actual boundary used in the model. The best way to achieve the correspondence required is to extend the model boundary into the tributary to the most convenient location for sampling. The penalty for adding a branch to the model domain is to increase the complexity of the model application. The added complexity, however, is usually trivial compared to the added difficulty and expense in trying to sample at the location that would be ideal according to criteria dictated by modeling convenience alone.

It is also important that the measurement of boundary conditions adequately quantify the distinct sources of pollution problems. Failure to do this limits the effectiveness of the study in addressing all the management issues that may arise. The inability to always distinctly quantify sources is one important reason why screening studies are not definitive.

For stream studies, the definition of individual sources is primarily important when selecting boundaries in the vicinity of tributaries. Definition of nonpoint sources, however, is in an undeveloped state and involves a more complex definition of boundaries than has yet been mastered.

Objectives of the study and the resources available for data collection govern how well individual pollution sources should be resolved as distinct boundary conditions. Tributaries frequently contain loads from more than one important source, and the way that the boundary is selected governs whether the individual sources can be separately identified for investigation of control options. Generally, individual loads can be separated when necessary by extending the model domain into the tributary or headwaters segment of the stream until the important loads are distinctly specified in the model. This will, however, involve greater data collection cost and should not be automatically undertaken without consideration of the effect on the data collection program. Simple preliminary studies may not require that individual sources be quantified.

As an example, consider the hypothetical stream system in Figure 9 where it is suspected that low dissolved oxygen problems in the main stem are caused by two major point sources and nonpoint source pollution in the basin of the major tributary. If the model domain is initially selected to end at the mouth of the major tributary, then the effects of the sources entering the tributary will be combined when measurements of flow and water quality are made at the confluence. This may be best if one of the sources is dominant and the other sources are negligible or if study objectives do not require that individual loads be known. If the other sources cannot be neglected and the relative individual effects must be known for regulatory purposes, then the domain should be extended into the tributary. The extension depends on the importance of all the loads. If the model domain is extended into the tributary to a point between the

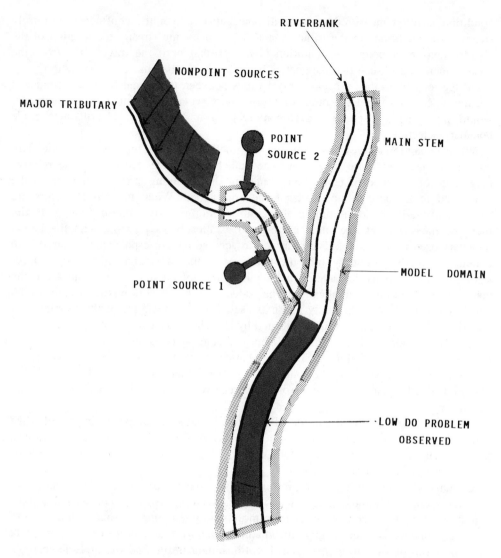

FIGURE 9. Hypothetical definitions of a model domain to distinctly represent the effects of loads on a problem of low dissolved oxygen.

two point sources, then the effect of point source 1 can be distinctly specified. If the domain is extended upstream of point source 2, then it is possible to measure the distinct contributions of each of the three sources for this hypothetical example. Each extension of the domain, however, usually adds one data collection site to the field data collection program.

If the hypothetical problem is complicated by an overlap of the nonpoint source with the point sources as frequently occurs in urban areas, then the same boundaries would be chosen, but more intensive sampling would be necessary. Accurate sampling of the point sources would be more important because instream mass balances could not be used to determine if point sources were adequately sampled. If the nonpoint sources do not enter the stream at distinct locations, such as tributaries, then intensive instream sampling will be necessary to indirectly measure cumulative loads.

C. COMPUTATIONAL NETWORK

The third step in the numerical description of a stream system is the definition of a numerical grid or network. A numerical grid or network is a series or system of

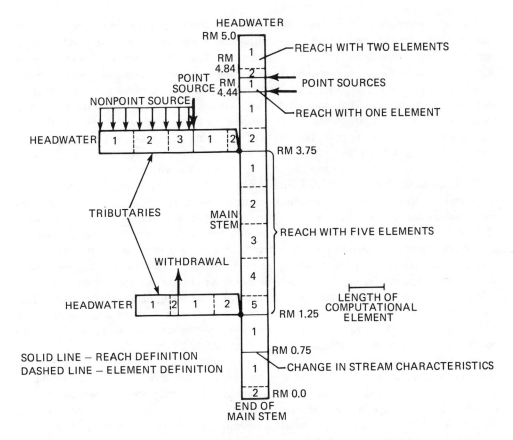

FIGURE 10. Hypothetical illustration of the discretization scheme used for a typical Streeter-Phelps type model.[11] Reaches extend from headwaters to inflows or withdrawals and between point sources. The final reach at the downstream end begins at a point where stream conditions change.

computational elements or nodes at which the governing equations are solved. Volume, velocity, concentration, and other state variables are specified at these control volumes or nodes. Numerical networks are illustrated in Figures 2, 3, 4, 5, and 10.

The definition of the numerical network is often dictated by the procedure used to solve the mass balance and the conservation of momentum equations (when complex flow models are used). Analytical solutions to the mass balance equation may be used for stream reaches that extend from one point where a boundary condition is specified to the point where the next boundary condition is specified. Numerical solution techniques for solving the mass balance equation require that a number of computation elements be established, usually without regard to where calibration data can be measured.

In the section briefly introducing the basic principles, it was noted that it is often necessary to solve the river water quality equations over limited segments. These limits are governed by a number of ill-defined criteria. The criteria are related to requirements of the process descriptions of the physical, chemical, and biological reactions; numerical requirements that arise from the method chosen to solve the mass balance and conservation of momentum equations; and criteria based on the resolution required to address the management issues. In most cases, however, the numerical requirements seem to be the most critical. The limitations of the formulations for kinetic reactions are largely unknown. Criteria to aid in properly resolving the calculations to aid in addressing management issues have not been distilled from the limited experience available.

As the important factors controlling rate constants change, new constants are specified for different conditions. This requires that the model domain be divided finely enough to specify new constants when necessary. The penalty for employing a gross scale division of the model domain is that conditions are represented with average parameters over long components of the domain, introducing error into the nonlinear calculations involved. Also, care must be exercised to not extrapolate the semi-empirical kinetic descriptions beyond the range of the data over which the descriptions have been tested. In this regard, the effects of averaging of nonlinear effects is probably small and negligible, but the limits of the original kinetic descriptions and the limits over which the methods have been applied are not well known.

To assist in defining algorithm limits, Zison et al.,[14] Bowie et al.,[15] and Schnoor et al.[16] have compiled process descriptions along with the values of rates and constants from various applications, but have not provided extensive information about the extent to which these have been tested. For example, it is known that biochemical oxygen demand coefficients change as microbial populations and the nature of the organic carbon change. As a result, the most practical approach tends to involve dividing the model domain into elements as small as practical based on criteria derived for the numerical solution technique without causing the application to be computationally intensive. This usually involves elements that are on the order of about 1 or 2 km in length (order of 1 mi) whereas presently employed algorithms for the kinetics seem to be applicable to reaches of at least on the order of 10 to 20 km (order of 10 mi).

The technique used to solve the mass balance equations and conservation of momentum equations leads to well defined criteria for how the stream should be segmented. Simple models based on an analytical solution of the Streeter-Phelps equation for the mass balance of dissolved oxygen may have long reaches that are numerically useful so long as the boundary conditions do not involve any inflows and the physical, chemical, and biological conditions do not change in the reach. For example, the modified Streeter-Phelps model of Bauer et al.[11] requires that the user define reaches that extend from one inflow or withdrawal to the next or to the point in the stream where the kinetic coefficients change. This is illustrated in Figure 10. The modified Streeter-Phelps model in Figure 10 also allows the user to specify short computation elements to better define nonpoint source inflows, but these shorter elements do not seem to be otherwise necessary.

Other models, such as the QUAL 2e[7] and WQRRS[2] models that do not involve analytical solutions, have stricter criteria dictated by the numerical solution techniques used in solving the water quality equations. Both of these particular models require that reaches (which may be of variable length), be composed of equal-length elements of approximately 0.8 to 3.2 km (0.5 to 2 mi) in length. In the WQRRS model, the computational element length may be varied in separate reaches. Also, a reviewer points out (Thomas Barnwell, Jr.) that versions of the QUAL-II model allows specification of a reach-variable computation element length. Figure 3 indicates that the equal-length criteria for numerical elements can make it more difficult to represent the exact geometry of a stream, especially when representing the exact confluence of tributaries. These discrepancies, however, can be minimized by the use of small computational elements if the length of the domain is relatively short.

The use of smaller computational elements is necessary in representing the water quality conditions more precisely. Figure 11 reproduced from Mills et al.[17] shows that coarser division of the stream can cause significant discrepancies between a numerical solution $(-k_d\Delta t)$ and an exact analytical solution $[1-\exp(-k_d\Delta t)]$. In such a case, the numerical model would require larger calibration values of the deoxygenation coefficient or smaller values of the reaeration coefficient to better match the analytical

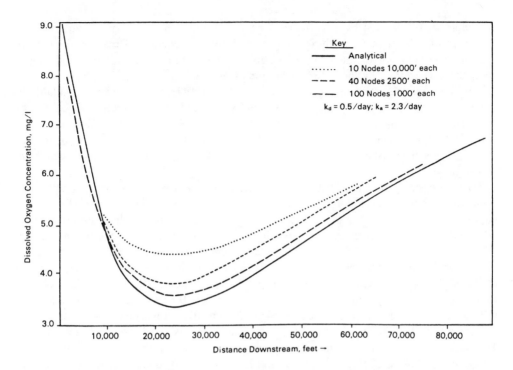

FIGURE 11. The effects of grid spacing between computational nodes on an explicit finite difference solution of the one-dimensional mass balance equation for dissolved oxygen in a stream. k_d is the deoxygenation coefficient and k_a is the reaeration coefficient. Note that the three curves representing the numerical solutions are smoothed between nodes.[17]

solution for a dissolved oxygen balance (dissolved oxygen sag in a stream). This is also an example of the magnitude of the calibration errors that can exist for certain numerical schemes applied to streams where gradients are large.

McCutcheon [18] compared the numerical solution in the QUAL 2e model to an analytical solution of the classical Streeter-Phelps equation (dissolved oxygen balance equation) and found insignificant discrepancies for an extensive range of actual field conditions. Therefore, the effect of the numerical solution technique should be investigated for each model, and this has not been done for all of the models available for use. The penalty for not confirming that the computation network is adequate is that calibration coefficients may not be accurate.

Criteria based on the required resolution of the predictions also becomes important in some instances. Occasionally, gross scale models can force the lumping of the effects of several pollution sources. Finer resolution may be necessary to allow separate specification of each effect. This is usually handled by segmenting the stream into elements that do not exceed the distances between the closest two sources of significance. It may also be necessary to limit element sizes in critical areas to better define minimum or maximum values that are near limiting regulatory standards.

Generally, however, the discretization requirements of numerical solution schemes provide more than enough resolution for providing information to aid management decisions. In fact, the required numerical resolution often provides too much detail from a management perspective, and the details must be properly averaged to provide useful information for the decision-making process.

In the modeling procedure, the choice of the size of the control volume or computational element can be guided to a limited extent by the resolution criteria discussed above. In practice, however, the choice is generally made by a trial-and-error procedure that is guided by experience. Criteria to guide the way control volumes are chosen can be developed, but these are governed by the behavior of mass in the control volume. The behavior of mass in the control volume is unknown until the modeling is completed. Therefore, the need for a trial-and-error procedure arises.

For some models, such as the QUAL 2e model, sufficient experience is available to reduce the procedure to one trial that is usually checked in confirming the validity of the results. In some cases, models have been extensively tested to determine if the numerical approximation errors like that illustrated in Figure 11, can be a problem. For many models there is no clear evidence that numerical discrepancies have been fully investigated.

McCutcheon[18] indicates that the use of element sizes of 0.40 km (0.25 mi) for a small fast moving river and a moderately deep river, and the use of 3.2 km (2 mi) for a large river that varied from shallow and swift moving to deep and sluggish, produce no detectable numerical discrepancies for the implicit finite difference scheme used in QUAL 2e.[7] McCutcheon[18] also found what seemed to be discrepancies between the implicit finite difference scheme used in the WQRRS model and the analytical solution of the Streeter-Phelps equations. Element lengths tested in the WQRRS model were 1.1 to 1.2 km (0.69 to 0.73 mi) for a swift shallow river, 0.84 to 1.22 km (0.52 to 0.76 mi) for a moderately deep river, and 0.84 to 2.8 km (0.52 to 1.77 mi) in a large river varying from shallow and swift to very deep and sluggish. Mills et al.[17] (see Figure 11) indicate that computational spacing of 0.3 km (0.2 mi) for an explicit finite difference solution scheme allows the solution to crudely approximate an analytical solution. This anecdotal evidence indicates that the establishment of the numerical grid can result in significant calibration error (see Figure 11), and from the information available about standard models, it is not clear that the effect of numerical approximation error has been fully addressed for all models. Furthermore, the effect of numerical error on the calibration of a model is rarely investigated in the course of a study. This is unfortunate in light of the lack of clear evidence that numerical error has been fully investigated for all models in general use.

There are also two numerical criteria that can be used for choosing the control volume size so that accurate and stable solutions to the flow and water quality equations can be obtained. One criterion is applicable to solving the flow equation and the other is applicable to the solution of the mass balance equation, when flows are measured or estimated by an empirical method.

For the solution of the flow equations, the Courant condition defines the relationship between the choice of a time step, Δt, over which the equations are solved and the length, Δx, between computational nodes:

$$\Delta t \leq \frac{\Delta x}{|U + \sqrt{gD}|} \tag{1}$$

where U is the average velocity through the control volume over the time step, g is the acceleration of gravity, D is the depth of flow in the control volume and $(gD)^{1/2}$ is the celerity or speed of a gravity wave in the body of water. In addition, French[19] offers a refined stability condition that is usually more restrictive.

The Courant condition requires that the time step be chosen so that a disturbance (i.e., internal wave) cannot move through a control volume of width Δx before the end of the time step. The usual practice is to choose Δx and use Equation 1 to estimate Δt. If Δt is too small for the computing resources available, then Δx is increased. As a result, numerical solution requirements for stability and accuracy influence the definition of the computation network.

A close examination of Equation 1 shows that the application of the Courant condition must involve a trial-and-error procedure as indicated above. U and D must be guessed beforehand to estimate the appropriate values of Δt and Δx. After Δt and Δx are used to solve the flow equations, then the Courant condition should be rechecked to ensure an accurate and stable solution.

A similar criterion can be applied to guide the definition of computational grids for solving the mass balance equation. However, this criterion shows that the solution of the flow equation will always be more restrictive than the solution of the mass balance equation. This criterion is very similar to the Courant condition except that mass transport is not affected by wave celerity. For the transport of conservative substances that are not subject to decay or other change, Fischer et al.[20] gives the appropriate stability criterion as

$$U\Delta t \leq \Delta x \tag{2}$$

where U is again the velocity of water through the control volume. Note that the mass transport condition and the Courant condition are different by the inclusion of $(gD)^{1/2}$, the speed of a wave or disturbance that transfers momentum downstream.

For example, if a flood enters a channel, the flood wave is a disturbance that moves downstream faster than the actual flood waters. The flood wave moves faster because it pushes the water in the channel before it with a difference in pressure caused by differences in water elevation in front of and at the crest of the flood wave. The pressure effect or momentum exchange moves downstream faster than the water can move. When the water in the channel before the flood event is stagnant and of poor quality, and is subject to dilution by cleaner storm waters, this leads to the first flush of contaminated water often observed. The flood wave, therefore, concentrates the existing stagnate water at the front of the flood water to produce the controversial first flush effect because of the difference in the way momentum and mass are transferred downstream.

Early investigators assumed that the speed of the flood wave was the speed of the mass transport. They were off by a factor of about 50% or by $[(gD)^{1/2}]$. Velz[21] resolves this controversy.

When mass is transported through the control volume at the advective velocity, U, and the end of the time step is located at a position between computational nodes (center of a control volume where the mass is assumed to be concentrated as a point mass), the interpolation to assign the transported mass to the nearest 2 nodes leads to numerical errors that are referred to as numerical dispersion. This is discussed by Fischer et al.[20] and is illustrated in Figure 12. In Figure 12, the downstream flow with velocity, U, moves the mass in control volume j-1 a distance U Δt during the time step Δt. The nodes in the center of the control volumes represent the mass in the volume. At the end of the advection step, interpolation is used to divide the mass between nodes that bracket the location to which the mass is moved. In the hypothetical case shown in Figure 12, the mass moved by advection is assigned to nodes j and j+1 by some interpolation scheme. The method of interpolation varies from one solution scheme to another, and this is the operation that gives rise to various degrees of numerical dispersion. The more accurate the interpolation scheme is in assigning mass to adjacent nodes, the smaller numerical dispersion becomes.

In addition, errors in the transport calculation can be minimized by choosing the time step, Δt, and the grid spacing, Δx, such that $u\Delta t/\Delta x = 1$. In this case, numerical dispersion is avoided for the transport calculation. In practice, however, streams are not uniform, and velocity will change for dynamic flows, so that Δt and Δx can not be

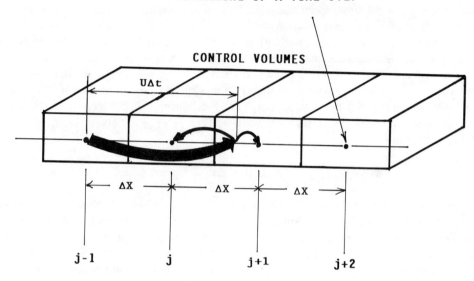

COMPUTATIONAL NODES REPRESENTING
THE MASS IN A CONTROL AT THE
BEGINNING OF A TIME STEP

CONTROL VOLUMES

UΔt

ΔX   ΔX   ΔX

j-1   j   j+1   j+2

FIGURE 12.   Illustration of numerical dispersion in a transport calculation.[20]

chosen to avoid numerical interpolation in every calculation. It should also be noted that choosing very small values of Δt or Δx may be counterproductive when the number of interpolations are increased.

Equation 2 does not involve a trial-and-error procedure because U, the average velocity in the control volume is known beforehand from the solution of flow equation or from measurements, and Δx can be specified to allow the use of reasonable values of Δt. Nevertheless, Equation 2 is overly restrictive because it does not take into account the effect of changes in mass in a control volume. Ambrose[8] notes that a more accurate stability criteria can be expressed as

$$\Delta t \leq \text{minimum} \left(\frac{V_i}{\sum\limits_j Q_{i,j} + \sum\limits_j R_{i,j} + K_i V_i} \right) \tag{3}$$

where V_i is the volume of water in the control volume i, $\Sigma Q_{i,j}$ is the flow entering or leaving the control volume from all the adjoining control volumes j = 1 to n, $\Sigma R_{i,j}$ is the bulk exchange or mixing between all adjacent control volumes j = 1 to n, and K_i is an estimate of the first order decay coefficient describing all mass transformations in the control volume. When the mass reactions are nonlinear, Ambrose recommends that $K_i V_i$ be neglected in Equation 3, and the approximate criteria be used as a guide only. The bulk exchange coefficient R is a measure of the dispersive mixing between adjacent control volumes;

$$R = D_x A/L \tag{4}$$

where D_x is the eddy diffusivity coefficient, A is the area of the control volume between adjacent volumes, and L is the characteristic length over which mixing occurs. L = Δx when all control volumes are of equal size or otherwise, L = $(\Delta x_i + \Delta x_j)/2$.

The appropriate value of Δt to choose is the one for which the term in parenthesis in

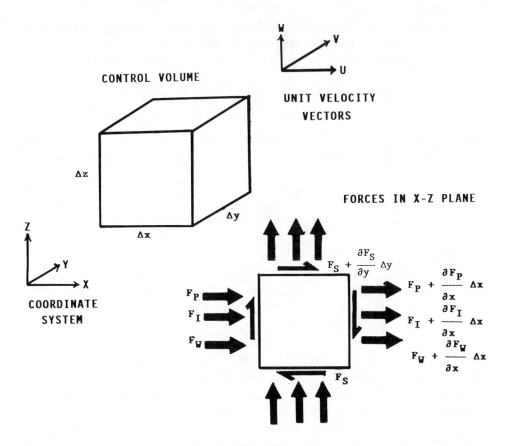

FIGURE 13. Forces due to pressure, inertia, fluid weight, and shear acting on an incremental volume, Δx Δy Δz. The forces in the x-y and y-z planes are similar in nature.

Equation 3 is smallest when all the control volumes in the computational grid are considered. In this regard, Equation 3 is somewhat limited in that it assumes that all control volumes are equal in volume. Note if the effects of mixing and mass transformation are ignored for a computational grid consisting of equal-volume elements (that also have the same area on all sides), then Equation 3 reduces to Equation 2. Given the wider range of applicability, the criteria of Ambrose should be more useful than Equation 2 in checking choices of Δx and Δt.

V. CONSERVATION OF MOMENTUM EQUATION AND FLOW IN CHANNELS

A. DERIVATION OF THE GOVERNING EQUATIONS

Some form of the conservation of momentum equation or an empirical analog is solved to determine how water flows through a channel or other water body. The momentum equation is an expression of the forces acting on a control volume as shown in Figure 13. In Figure 13, only the forces acting in the x coordinate direction are fully described. The forces that affect water movement include forces due to pressure differences, F_p, the inertia of moving water, F_I, the weight of the water, F_w, and shear with adjacent fluid or surfaces, F_s. When the change of these forces over the incremental distances Δx, Δy, and Δz are expressed as shown in Figure 13 and the forces are summed up in all three coordinate directions, Newton's Law that force is equal to mass times acceleration can be used to describe water movement. By also using Newton's phenomenological law of viscosity to relate shear stresses to velocity gradients and the Boussinesq assumption that density differences in the flow have only negligible effects, an approximate, but highly accurate, equation of water motion can be written as[22]

$$\frac{\partial U_i}{\partial t} + U_j \frac{\partial U_i}{\partial x_j} = -\frac{1}{\rho_r}\frac{\partial P}{\partial x_i} + \nu \frac{\partial^2 U_i}{\partial x_j \partial x_j} + g_i \frac{\rho - \rho_r}{\rho_r} \tag{5}$$

where U_i is the instantaneous velocity vector (u, v, w). In terms of the velocity vector (u, v, w), the partial derivatives of U_i with respect to time and spatial changes are as follows:

$$\frac{\partial U_i}{\partial t} = \frac{\partial u}{\partial t} + \frac{\partial v}{\partial t} + \frac{\partial w}{\partial t} \tag{6}$$

$$U_j \frac{\partial U_i}{\partial x_j} = u\left(\frac{\partial u}{\partial x} + \frac{\partial v}{\partial x} + \frac{\partial w}{\partial x}\right) + v\left(\frac{\partial u}{\partial y} + \frac{\partial v}{\partial y} + \frac{\partial w}{\partial y}\right) + w\left(\frac{\partial u}{\partial z} + \frac{\partial v}{\partial z} + \frac{\partial w}{\partial z}\right) \tag{7}$$

$$\frac{\partial^2 U_i}{\partial x_j \partial x_j} = \left(\frac{\partial^2 u}{\partial x^2} + \frac{\partial^2 u}{\partial y^2} + \frac{\partial^2 u}{\partial z^2}\right) + \left(\frac{\partial^2 v}{\partial x^2} + \frac{\partial^2 v}{\partial y^2} + \frac{\partial^2 v}{\partial z^2}\right) + \left(\frac{\partial^2 w}{\partial x^2} + \frac{\partial^2 w}{\partial y^2} + \frac{\partial^2 w}{\partial z^2}\right) \tag{8}$$

In Equation 5, P is the static pressure, ρ is density, ρ_r is a reference density, ν is the kinematic viscosity of water, and g_i is gravitational acceleration.

In addition to the conservation of momentum equation, the equation based on the conservation of water mass is also required to simulate flows in channels. This equation is also known as the continuity equation and is written as

$$\frac{\partial u}{\partial x} + \frac{\partial v}{\partial y} + \frac{\partial w}{\partial z} = 0 \tag{9}$$

These flow equations are written in a general three-dimensional form that is too complicated for practical use. The writing of these equations should illustrate, however, that the basic flow equations are exact. In the set of two equations, there are two unknowns, the velocity vector, U_i, and pressure, P. Density can be taken as a physically measurable quantity like the kinematic viscosity, or density can be expressed using a third equation. This third equation is an equation of state that relates density of water to temperature and salinity or dissolved solids. When density varies because of longitudinal salinity gradients or vertical salinity and temperature gradients, the mass balance equation for salinity or heat balance equation for temperature must be solved simultaneously with the flow equations. Usually this coupling is unnecessary except for cooling water discharges from power plants into rivers. Typically, the flow equations and mass balance equations are solved in a serial fashion, to first determine the velocities in a channel and then incorporate this advective effect with other phenomena that influence the mass balance of a constituent.

Although the flow equations (Equations 5 and 9), known as the Navier-Stokes equations, are an exact form derived from first principles, the general solution is obtained by an approximate, difference technique for discrete time and space increments. The most accurate solutions must take place over very small time and space scales, and as a result, require too much computing time to allow solution of practical environmental fluid mechanics problems. To derive a more useful form of the equations, a statistical technique (called Reynolds averaging after Osborn Reynolds for whom the

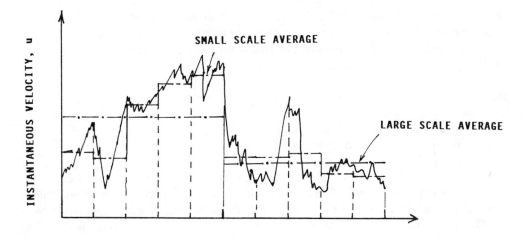

FIGURE 14. The effect of averaging on values of the mean velocity, U, and the average fluctuation of velocity from the mean, u′.

Reynolds number is named), is applied to average the velocity vector over space or over time. This is done by expressing the instantaneous velocity, U_i or (u, v, w) as a function of an average velocity and a fluctuating component such that

$$u = U + u'$$
$$v = V + v'$$
$$w = W + w'$$
$$U_i = \overline{U}_i + u'_i$$
$$P = \overline{P} + P_i \tag{10}$$

This procedure is illustrated in Figure 14 where it can be noted that the length of time over which U and u′ are computed is important. This is also important when looking at the effect of averaging on mixing coefficients.

Substitution of Equation 10 into Equations 5 and 9 yields what are known as the Reynolds equations. Equation 9, the continuity equation, becomes

$$\frac{\partial U}{\partial x} + \frac{\partial V}{\partial y} + \frac{\partial W}{\partial z} = 0 \tag{11}$$

The conservation of momentum equation becomes

$$\frac{\partial \overline{U}_i}{\partial t} + \overline{U}_j \frac{\partial \overline{U}_i}{\partial x_j} = -\frac{1}{\rho_r}\frac{\partial P}{\partial x_j} + \frac{\partial}{\partial x_j}\left(\nu \frac{\partial \overline{U}_i}{\partial x_j} - \overline{u_i u_j}\right) + g_i \frac{\rho - \rho_r}{\rho_r} \tag{12}$$

where U_i is the averaged velocity vector (U, V, W) and u_i is the vector describing the fluctuation of velocity from the mean (u′, v′, w′).

Unfortunately, Equations 11 and 12 now contain three unknown quantities, U_i, $u_i u_j$, and pressure. To solve the resulting equations, the correlation term, $u_i u_j$, must be related to the other unknowns using an empirical or phenomenological relationship. This relationship is necessary to achieve what is known as turbulence closure.

Rodi[22] reviews the more common procedures for turbulence closure of the equations and notes that they are in order of the least complex to the most complex:

1. Zero-equation models — eddy viscosity or mixing length
2. One-equation models
3. Two-equation models — k-∈ model
4. Higher order and subgrid scale models

These closure methods are not of overriding importance except that it should be noted that each method is empirical to some extent. It can be noted that river modeling applications usually involve the eddy-viscosity scheme or even simpler approximations.

There does, however, seem to be two approaches to the closure problem besides the numerous eddy-viscosity applications that have some practical utility. One approach uses the k - ∈ closure model by Rodi.[22] Rodi uses two phenomenological equations that characterize the kinetic energy of the turbulent motion and the dissipation of turbulent energy. This approach seems to be the best available for thermal discharges in rivers but has not been developed to a highly practical state as of yet.

Also useful may be the three-dimensional code of Sheng[23] that uses a higher-order closure scheme. For larger water bodies, Sheng's model requires very little calibration, and is of interest for applications in rivers for this reason. Sheng's model, however, currently ignores vertical accelerations and, as a result, is not highly useful for near field mixing problems.

The eddy-viscosity method relates turbulent shear stress in the fluid (τ_x, τ_y, τ_z) to the gradient of the mean velocity vector based on an assumed analogy between molecular and turbulent motion, such that

$$-\rho \overline{u'v'} = \tau_x = \rho E_x \frac{\partial U}{\partial x} \tag{13}$$

$$-\rho \overline{v'w'} = \tau_y = \rho E_y \frac{\partial V}{\partial y} \tag{14}$$

and

$$-\rho \overline{w'u'} = \tau_z = \rho E_z \frac{\partial W}{\partial z} \tag{15}$$

where E_x, E_y, and E_z are turbulent eddy-viscosity coefficients that are effectively coefficients of proportionality related to the scale of averaging. E_x, E_y, and E_z are the turbulent analogs of molecular viscosity. Note the similarity in the form of Equations 13 to 15 to Newton's law of viscosity. Thus, one should keep in mind that eddy-viscosity values are empirical coefficients that are not only related to the intensity of turbulent mixing, but also are governed by the computational method used to solve the flow equations. The computational method is important because it determines the scale over which the equations are averaged.

The solution of the three-dimensional or two-dimensional form of the flow equations are usually only important in practical design problems when the mixing of a jet flow entering a river or stream is important or when lateral mixing across the stream is important. Typically, jet mixing in rivers is only important for some thermal power plant discharges of cooling water. Lateral mixing is important in very wide rivers in which inflows may not have time to fully mix across the flow.

In most cases, the one-dimensional form of the equations is more than adequate to predict the movement of water in rivers. Vertical stratification and lateral inhomogeneity are normally unimportant. As a result, the governing equation reduces to

$$\frac{\partial U}{\partial t} + U \frac{\partial U}{\partial x} = -\frac{1}{\rho} \frac{\partial}{\partial x} (P + \gamma h) + (\nu + E_x) \frac{\partial^2 U}{\partial x^2} \tag{16}$$

where γ is the specific weight of the water and h is height above a datum.[19]

Equation 16 can be simplified by noting that flows in channels are, almost without exception, turbulent. As a result, the effect of molecular viscosity can be neglected because the eddy viscosity is much larger, $E_x \gg \nu$. Furthermore, if the pressure terms are written in terms of the change in the height of the water surface, and the eddy-viscosity term is written in terms of the channel friction slope, S_f (amount of energy dissipation per unit length of the channel), Equation 16 can be expressed in a simpler form that can also be derived from a consideration of the forces acting on the water in a channel.[19] That simpler form is

$$\frac{\partial U}{\partial t} + U \frac{\partial U}{\partial x} + g \frac{\partial y}{\partial x} - g(S_f - S_o) = 0 \tag{17}$$

where S_o is the slope of the channel bottom. When Equation 17 is solved with a restatement of the continuity equation, given below as Equation 18, then most unsteady flows in rivers can be simulated. The rearranged continuity equation is

$$T \frac{\partial y}{\partial t} + \frac{\partial (UA)}{\partial x} = 0 \tag{18}$$

where T is the top width of the stream, and A is the cross-sectional area.

Tracing the derivation of the Equations 17 and 18 from Newton's law relating force to mass and acceleration is a tedious exercise, but it does demonstrate the theoretical basis for the flow modeling techniques available for water quality modeling and shows when and where semi-empirical methods must be employed to derive a practical method for solving flow in rivers. Next, Equations 17 and 18 can be rearranged to show how most of the practical flow routing methods used in water quality modeling are related.

B. RELATIONSHIP OF PRACTICAL METHODS TO THE GOVERNING EQUATIONS

Brown and Barnwell,[7] French,[19] and Smith[2] give what are probably the most practical methods for stream flow routing for water quality modeling. These methods include:

1. Specification of stage-discharge relationships for steady flow
2. Empirical power law relationships relating discharge, depth, and velocity
3. Trial-and-error solution of the Manning equation
4. Backwater hydraulic solution
5. Muskingum hydrologic routing
6. Modified Puls hydrologic routing

7. Solution of the kinematic wave approximation
8. Solution of the diffusion analogy equations
9. Solution of the full St. Venant equation

Methods 1 through 4 are applicable to steady flow conditions in which discharge is constant or nearly constant. The hydrologic routing methods are semi-empirical techniques for unsteady flow calculations that are only infrequently used for water quality modeling. Methods 7 through 9 are solutions of the governing equations (Equations 17 and 18) or approximations thereof. The Manning equation can be derived from a force balance and is therefore a steady state uniform flow derivative of the basic equations. Furthermore, the power law relationships have been derived in terms of the Manning equation. As a result, the derivation of the basic equations allows one to trace the theoretical basis of the flow equations used in water quality modeling as shown in Figure 15.

French[19] notes that Equation 17 can rewritten in the form of the rating equation that relates discharge, Q, to depth of flow:

$$Q = \Gamma A R^m (S_f)^{1/2} \tag{19}$$

where Γ is an empirical resistance coefficient, R is the hydraulic radius (area divided by the wetted perimeter of the channel), and m is an empirical exponent. In a wide channel that is on the order of 10 times wider than the depth, the hydraulic radius is approximately equal to the depth of flow. French indicates that S_f varies with both the slope of the flood wave moving through the channel and the depth of flow for unsteady flow. If the flow is steady and uniform (the cross section does not vary in shape or area in the longitudinal direction), the normal discharge is

$$Q_n = \Gamma A R^m (S_o)^{1/2} \tag{20}$$

where S_o is the slope of the channel bottom. Normal discharge, Q_n, is the discharge that occurs when the flow is uniform. French uses the relationship between Q and Q_n from Equations 19 and 20 to derive S_f as a function of Q, Q_n, and S_o and substitutes the resulting expression into Equation 17. Solving the result for Q yields an equation that clearly shows the relationship between the complete dynamic equations (St. Venant or conservation momentum equations given above as Equations 17 and 18, and the approximate forms of the equation referred to as the equations based on the diffusion analogy and the kinematic approximation:

$$Q = Q_n \left(1 - \frac{1}{S_o} \frac{\partial y}{\partial x} - \frac{U}{S_o g} \frac{\partial U}{\partial x} - \frac{1}{S_o g} \frac{\partial U}{\partial t} \right)^{1/2} \tag{21}$$

kinematic wave approximation ⌐
diffusion analogy ————————
complete dynamic equations ——————

For example the kinematic wave approximation is $Q = Q_n$ where Q_n is defined in Equation 20.

By writing the governing equations in the form given in Equations 19 to 21, it is possible to see how each practical method is related to the governing basic principles. Chapter 2, section I reviews the practical methods and relates them to the forms given here.

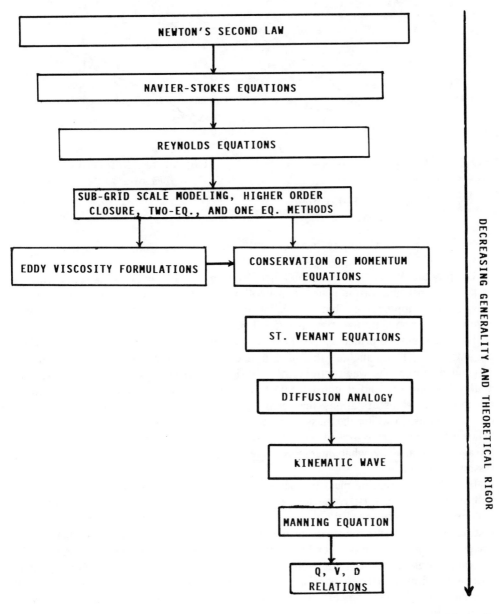

FIGURE 15. Relationship between the first principles for flow models and hydrodynamic and hydraulic modeling techniques. These are methods that can be derived from a balance of the forces or momentum and the conservation of water mass, or can be related to such a method.

VI. MASS BALANCE

Most stream water quality models are primarily based on the conservation of mass principle (or conservation of heat for temperature modeling). In fact, most environmental models for air, groundwaters, and other surface waters have the same basis. As a result, the theoretical understanding and the methods of practical application of mass balance models are well advanced. What is not well characterized in terms of relation to the basic principles are the kinetic reactions that change constituents from one form to another. Water quality kinetics are generally based on phenomenological laws.

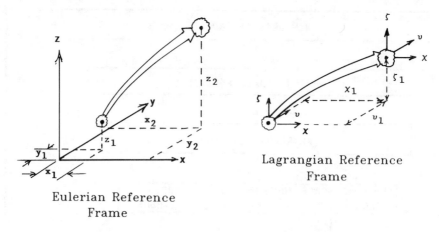

Eulerian Reference
Frame

Lagrangian Reference
Frame

FIGURE 16.   Eulerian and Lagrangian reference systems.

Temperature modeling is based on a heat balance, but this involves the same principle as the mass balance. In the material that follows, reference will be made, primarily, to mass balance applications, but the reader should keep in mind that the heat balance involved in temperature modeling is equivalent.

To understand the principles involved, it is necessary to categorize the processes that are important. These include the advection or movement of mass in a dissolved or suspended form as the water moves, the diffusion and dispersion that mixes and spreads a dissolved or suspended constituent throughout a body of water, and the kinetic reactions and transformations that change constituents to a different form. These different forms, to which the mass is transformed, may be tracked with a different mass balance equation that is linked with the other mass balance equations. The manner of linkage is dictated by what interactions are possible and by what interactions can be reasonably simulated.

To derive the resulting mass balance equations, there are two related approaches that will employed here. Firstly, the long-standing method based on an Eulerian reference system will be followed. This derivation results in the advective-dispersive equation that has been applied extensively in describing and predicting transport of material in streams. Secondly, the more flexible Lagrangian reference system will be employed for the same purpose. An Eulerian reference frame is one that characterizes a dynamic process in terms of a fixed reference point. A Lagrangian reference frame is one that characterizes a dynamic process in terms of a moving reference point. This is illustrated in Figure 16.

The Eulerian reference frame is utilized when an observer stays at a point on the river bank and observes and quantifies water quality as the river flows past. The Lagrangian reference is used when an observer follows the flow in a river with a boat traveling at the average velocity of the flow. In understanding cause-and-effect relationships for river water quality problems and in putting different water quality problems in perspective, many analysts have naturally begun to rely on Lagrangian methods. The use of the hypothetical boat trip by Hines et al.[24] to describe the water quality problems in the Willamette River indicates that the Lagrangian reference frame can also be a very useful conceptual approach for conveying the results of a study. In addition, many studies with limited sampling resources (i.e., Davis et al.[25]) or studies that attempted to optimize sampling resources (i.e., Hines et al.[24] and McCutcheon et al.[26]) have devised Lagrangian sampling programs that were not necessarily aimed at supporting the calibration of Lagrangian models. In fact, it is generally understood that sampling programs designed to collect data to calibrate Eulerian models, such as the QUAL 2e model, can be optimized by staggering sampling periods at locations in the downstream direction by the time of travel between sampling locations.[17,27] Therefore, the use of a Lagrangian perspective has proven very useful in not only devising water quality models, but also in designing sampling programs and interpreting the results of water quality studies even when Eulerian water quality models were used.

It should be noted that if a Lagrangian reference frame is adopted to conceptually and computationally simplify the mass-transport calculation, this does not alleviate the need for an Eulerian reference frame. As of yet, the Lagrangian reference frame has not been mastered to aid in solving the flow equations. Although Heyes[28] seems to report some progress in this regard, and the method of characteristics[19,29] is similar to a Lagrangian approach. Therefore, the calculations of flow and transport will require both reference systems. The types of models that use both reference frames to solve for flow and water quality are referred to as Eulerian-Lagrangian models. Baptista et al.[30] and Cheng et al.[31] give examples of the application of these methods in more complex water bodies. The overall computations using Eulerian-Lagrangian models are still much simpler because of the use of a Lagrangian reference frame for mass transport calculations, but bookkeeping operations become a little more involved when both reference frames must be employed. Bookkeeping operations refer to reading and arranging data for initial conditions, boundary conditions, and geometry, plus the reporting of the results in concentrations at selected locations at various times during and after the simulations.

For steady-state conditions, the overall transport calculations using an Eulerian or Lagrangian reference frame are equally simple. At present, however, use of the Eulerian reference frame seems to be conceptually less demanding. Furthermore, bookkeeping may be less involved or at least easier to understand for Eulerian steady-state computations. Therefore, Lagrangian models may offer little advantage when applied to steady-flow conditions but do offer significant advantages when applied to dynamic conditions.

A. EULERIAN MODELS

The derivation of mass balance equations for an Eulerian reference frame follows a path similar to that followed in deriving the flow equations. Figure 17 shows the conceptual Eulerian framework for quantifying changes in mass in a control volume.

The effect of advection can be described by looking at the change in the mass flux through a fixed incremental volume and noting that it changes with time as a function of the change along the x, y, and z coordinate directions. For example, the change in the x direction goes from uc at the front face to $uc + \partial(uc)/\partial x$ at the back face of the incremental volume shown in Figure 17. If the mass flux through the control volume is summed for all three coordinate directions, then the change in mass concentration or temperature due to advection can be described by the substantial time derivative:

$$\frac{Dc}{Dt} = \frac{\partial c}{\partial t} + \frac{\partial(uc)}{\partial x} + \frac{\partial(vc)}{\partial y} + \frac{\partial(wc)}{\partial z} \qquad (22)$$

Diffusion can be described using Fick's first law based on observations that the diffusive flux along any coordinate axis, x, y, or z (N_x, N_y, N_z), is proportional to the concentration gradient across the plane over which diffusion is occurring:

$$N_x = -D_m \frac{\partial c}{\partial x} \qquad (23)$$

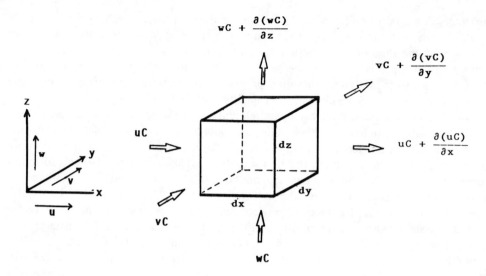

FIGURE 17. Eulerian conceptual framework for defining advection through an incremental volume of dx dy dz.

$$N_y = -D_m \frac{\partial c}{\partial y} \qquad (24)$$

and

$$N_z = -D_m \frac{\partial c}{\partial z} \qquad (25)$$

where the coefficient of proportionality, D_m, is the coefficient of molecular diffusivity with the unit length squared per time. Molecular diffusivity is a physical parameter of the fluid that has been measured for many dilute solutions in water.[32]

The use of Fick's first law is analogous to the use of Newton's law of viscosity in deriving the Navier-Stokes equations for flow. Thus, the similarity in form is not a coincidence. For a heat balance, the same form is necessary to describe the diffusion of heat. In this case, Fourier's law is used where thermal diffusivity would be used in place of molecular diffusivity and c would be taken as $\rho C_p T$ where ρ is the fluid density; C_p is the heat capacity per unit mass in length squared per time squared per degree of temperature, and T is temperature.[33]

If the change of mass in the control volume due to diffusion is to be quantified, then it is necessary to equate the accumulation of mass (i.e., mass in-mass out) due to diffusion to the change in the mass flux through the element. The accumulation of mass, M, is written as

$$\frac{1}{dxdydz} \frac{\partial M}{\partial t} = \frac{\partial c}{\partial t} = \left(D_m \frac{\partial c}{\partial x} \right)_1 - \left(D_m \frac{\partial c}{\partial x} \right)_2 + \left(D_m \frac{\partial c}{\partial y} \right)_1$$

$$- \left(D_m \frac{\partial c}{\partial y} \right)_2 + \left(D_m \frac{\partial c}{\partial z} \right)_1 - \left(D_m \frac{\partial c}{\partial z} \right)_2 \qquad (26)$$

where locations 1 and 2 refer to opposite faces of the control volume. The change in the mass flux across the control volume due to diffusion can be expressed as

$$-\frac{\partial}{\partial x}\left(-D_m\frac{\partial c}{\partial x}\right) - \frac{\partial}{\partial y}\left(-D_m\frac{\partial c}{\partial y}\right) - \frac{\partial}{\partial z}\left(-D_m\frac{\partial c}{\partial z}\right) = \left(D_m\frac{\partial c}{\partial x}\right)_1$$

$$-\left(D_m\frac{\partial c}{\partial x}\right)_2 + \left(D_m\frac{\partial c}{\partial y}\right)_1 - \left(D_m\frac{\partial c}{\partial y}\right)_2 + \left(D_m\frac{\partial c}{\partial z}\right)_1 - \left(D_m\frac{\partial c}{\partial z}\right)_2 \quad (27)$$

When Equations 26 and 27 are combined and it is noted that D_m is a physical constant that is unaffected by the differential operators $\partial/\partial x$, $\partial/\partial y$, and $\partial/\partial z$, the change in concentration due to the diffusive flux can be written as

$$\frac{\partial c}{\partial t} = D_m\left(\frac{\partial^2 c}{\partial x^2} + \frac{\partial^2 c}{\partial y^2} + \frac{\partial^2 c}{\partial z^2}\right) \quad (28)$$

If the effects of advection and diffusion are combined and a term, S, is added to represent sources and sinks of mass or heat caused by physical or biochemical processes, the result is as follows:

$$\frac{\partial c}{\partial t} + \frac{\partial(uc)}{\partial x} + \frac{\partial(vc)}{\partial y} + \frac{\partial(wc)}{\partial z} = D_m\left(\frac{\partial^2 c}{\partial x^2} + \frac{\partial^2 c}{\partial y^2} + \frac{\partial^2 c}{\partial z^2}\right) + S \quad (29)$$

When Equation 29 is combined with Equations 5 and 9 (i.e., the flow equations), the resulting set of equations are exact, but the problem of resolving of turbulent motion remains. That is to say that, Equation 29 exactly represents the transport of mass or heat if S is known exactly or is zero. The practical difficulties of solving the equations at a fine enough scale to resolve the exact transport makes it necessary to apply the Reynolds averaging technique to the mass balance equation as well. As a result, Equation 10 should be rewritten as

$$u = U + u'$$
$$v = V + v'$$
$$w = W + w'$$
$$U_i = \overline{U_i} + u_i'$$
$$P = \overline{P} + p'$$
$$c = C + c' \quad (30)$$

where C is the mean concentration averaged over some time or length and c' is the fluctuating value about the mean. When Equations 29 and 30 are combined the result is as follows:

$$\frac{\partial C}{\partial t} + \frac{\partial(UC)}{\partial x} + \frac{\partial(VC)}{\partial y} + \frac{\partial(WC)}{\partial z} =$$

$$D_m\left(\frac{\partial^2 C}{\partial x^2} + \frac{\partial^2 C}{\partial y^2} + \frac{\partial^2 C}{\partial z^2}\right) - \frac{\partial(\overline{u'c'})}{\partial x} - \frac{\partial(\overline{v'c'})}{\partial y} - \frac{\partial(\overline{w'c'})}{\partial z} + S \quad (31)$$

where $u'c'$, $v'c'$, and $w'c'$ are cross-correlation terms that must be related to the other state variables, U, V, W, and C, to close the governing equations. In the previous section on the flow equations, the methods of turbulence closure were reviewed. It should be noted that the same approach can be used for mass transport assuming that mass and momentum transport are analogous. Therefore, the eddy-diffusivity method, as applied to mass transport, can be written as

$$-(\overline{c'u'}) = D_x \frac{\partial U}{\partial x} \tag{32}$$

$$-(\overline{c'v'}) = D_y \frac{\partial V}{\partial y} \tag{33}$$

and

$$-(\overline{c'w'}) = D_w \frac{\partial W}{\partial w} \tag{34}$$

where D_x, D_y, and D_z are turbulent eddy-diffusivity coefficients that are effectively coefficients of proportionality that are related to the scale of averaging as are the turbulent eddy-viscosity coefficients. D_x, D_y, and D_z are the turbulent analogs of molecular diffusivity but are not physical properties of the fluid. Instead, D_x, D_y, and D_z are properties of the flow governed by turbulent characteristics. In addition, D_x, D_y, and D_z are not equivalent except in very localized areas of isotropic turbulence (i.e., turbulent characteristics are the same in all three coordinate directions: x, y, and z). Furthermore, the values of molecular diffusivity (D_m) are typically much smaller that the values of D_x, D_y, and D_z, and thus molecular diffusion is ignored except in very localized areas within suspended particles and within the channel bed where molecular diffusion is typically more important.[34] Bowie et al.[15] illustrate the observed disparity between turbulent and molecular diffusion in Figure 18. Note the logarithmic scale on the vertical axis.

Although Equation 31 is useful for understanding or conceptualizing how three-dimensional transport may be important in streams, the eddy-viscosity approximation has proven to be of limited usefulness unless a one-dimensional form is used. If it is assumed that $v = w = 0$ and $D_m \ll D_x$, then Equation 31 reduces to a typical form of the advective-dispersive equation used in dynamic stream modeling:

$$\frac{\partial C}{\partial t} = \frac{\partial(UC)}{\partial x} + D_x \frac{\partial^2 C}{\partial x^2} + S \tag{35}$$

In this form, D_x is frequently referred to as the longitudinal dispersion coefficient.

Equation 31 with the eddy diffusion formulation for closure has not proven too useful in stream water quality modeling except for longitudinal mixing problems. The eddy-viscosity and eddy-diffusivity methods are based on an analogy with molecular diffusion and momentum exchange that cannot be fully satisfied except in steady unidirectional fully developed boundary-layer type flows (see for example Rodi[22]).

A steady uni-directional fully developed boundary-layer flow is one that forms when a steady, unidirectional flow is in contact with a solid surface for a sufficient distance to approach a uniform flow, as shown in Figure 19. These are conditions that are approached in most streams that are deep or moderately deep and that are approximately uniform in cross section. Therefore, it is easy to see why Equation 35 has proven so applicable for large streams.

In localized areas, and in shallow streams that are not uniform (i.e. pool-and-riffle

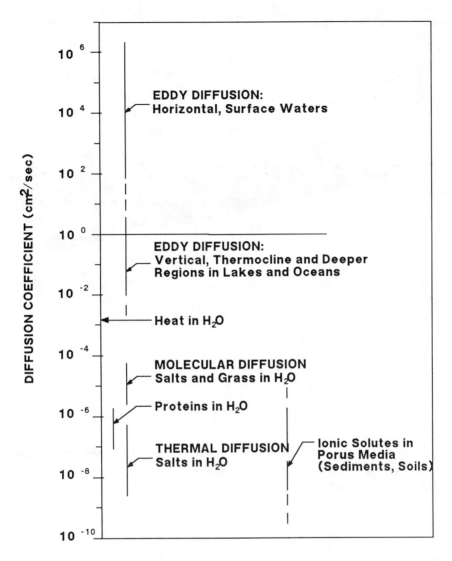

FIGURE 18.   Difference in Magnitude of Molecular and Turbulent Diffusion. (From Bowie et al.;[15] originally from Lerman, A., in *Nonequilibrium Systems in Natural Water Chemistry* [ACS Advances in Chemistry Ser. 106], Hem, J. D., Ed., American Chemical Society, Washington, D.C., 1971. With permission.).

streams), the proper mixing conditions may not be achieved to allow an application of the eddy diffusivity based equations such as Equation 35. In streams, the introduction of jets of cooling water or changes in the channel that cause the flow to accelerate or decelerate are the chief conditions, along with the effects of stratification by dissolved and suspended solids and by heat, that violate the basic assumptions behind the eddy diffusivity approach. In cases of this nature, turbulence is generated and dissipated under different conditions. This is the basic assumption of eddy diffusivity scheme that cannot be achieved in several important types of stream flows. To adequately handle cooling water discharges from power plants, experience shows that higher order closure methods are necessary to simulate the transport and dissipation of turbulence[22] generated in inflow channels and at the highly stratified point of injection. In pool-and-riffle streams, Fickian dispersion models have been modified with "dead-zone" terms to account for nonuniform conditions.

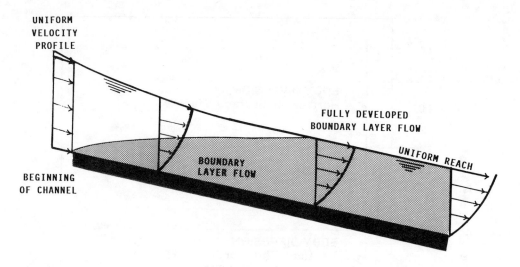

FIGURE 19. Development of unidirectional steady boundary layers in channels. Note that the vertical dimension is exaggerated compared to the longitudinal dimension.

As a specific example of the application of Equation 35, the dynamic mass balance equation for dissolved oxygen can be written as

$$\frac{\partial C_{DO}}{\partial t} + \frac{\partial(UC_{DO})}{\partial x} = D_x \frac{\partial^2 C_{DO}}{\partial x^2} + K_2(C_{SAT} - C_{DO})$$

$$- K_1 L - K_{NH_3}\alpha_1(NH_3\text{-}N) - K_{NO_2}\alpha_2(NO_2\text{-}N) - S_{SOD} + P - R + I \quad (36)$$

where C_{DO} is the concentration of dissolved oxygen, K_2 is the reaeration coefficient, C_{SAT} is the saturation concentration of dissolved oxygen, K_1 is the deoxygenation coefficient, L is biochemical oxygen demand, K_{NH_3} is the ammonia oxidation rate constant, α_1 is the stoichiometric constant for dissolved oxygen uptake per mass of ammonia-nitrogen oxidized, NH_3-N is the ammonia-nitrogen concentration, K_{NO_2} is the nitrite oxidation rate constant, α_2 is the stoichiometric constant for dissolved oxygen uptake per mass of nitrite-nitrogen oxidized, NO_2-N is the nitrite-nitrogen concentration, S_{SOD} is the rate of sediment oxygen demand, P is the rate of dissolved oxygen production by photosynthesis, R is the rate of plant respiration, and I describes amount of dissolved oxygen gained from inflows into the stream. Inflows may include point and nonpoint sources and may also serve as a catch-all term for the effects of minor processes that may effect the dissolved oxygen balance. I is the term that accounts for dilution of wastes entering a stream. In advanced water quality models, mass balances are also written for biochemical oxygen demand (L), ammonia-nitrogen (NH_3-N), nitrate-nitrogen (NO_2-N), plant biomass, and other constituents, and then solved simultaneously with Equation 36.

When flows and loads are steady, i.e., the longitudinal velocity is constant in time at any cross section, Equation 35 is slightly simplified in form but becomes much easier to solve.[7] This results in a quasi-dynamic equation that can simulate a dynamic response of the stream away from the points where boundary conditions must be specified. For example, quite a few streams experience steady flow and loading but respond dynamically to the diel effects of solar radiation on temperature and photosynthesis, and this response seems to occur quickly compared to the time it takes water to move through the stream.

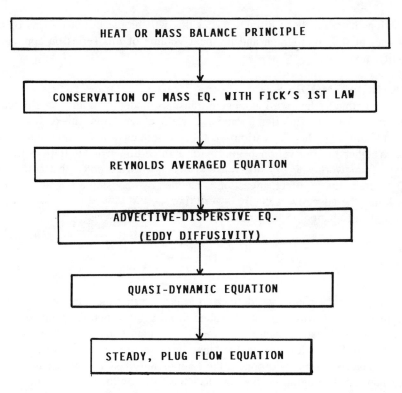

FIGURE 20. Relationship between various formulations of the mass balance equation. The descending order denotes decreasing generality and theoretical rigor, and increasing limitation on application.

If it is also noted that dispersion is generally negligible in streams, such that $D_x = O$, the quasi-dynamic dissolved oxygen equation reduces to a form similar to the classical Streeter-Phelps equation known as a modified Streeter-Phelps equation:

$$\frac{\partial C_{DO}}{\partial t} + \frac{\partial (UC_{DO})}{\partial x} = K_2(C_{SAT} - C_{DO}) - K_1L$$

$$- K_{NH_3}\alpha_1(NH_3\text{-}N) - K_{NO_2}\alpha_2(NO_2\text{-}N) - S_{SOD} + P - R + I \qquad (37)$$

This type of equation is also known as a plug flow type equation to denote that mixing is unimportant.

The original Streeter-Phelps equation only included the effect of reaeration and total biochemical oxygen demand, L_T, that included carbonaceous and nitrogenous demands. That equation is written as

$$\frac{\partial C_{DO}}{\partial t} + \frac{\partial (UC_{DO})}{\partial x} = K_2(C_{SAT} - C_{DO}) - K_1L_T \qquad (38)$$

The fortuitous result of being able to simplify Equation 35 when $D_x = O$ is that the equation can be solved analytically.[35]

The cumulative result, as shown in Figure 20, is that it is possible to firmly relate the practical methods presently being used for mass balance modeling to the basic principle of conservation of mass and identify the empirical approximations involved in the transport equations. To this point, the mechanistic methods used to describe mass transformations have not been discussed, but this will be the primary focus of most of the remainder of this monograph and the next volume of this series on applications to river quality modeling.

B. LAGRANGIAN MODELS

It is also possible to derive a Lagrangian transport model using the same procedure as that used to derive the Eulerian models. Certain observations make it is possible, however, to simplify the derivation of the one-dimensional Lagrangian transport equation. Whether the derivation begins with a three-dimensional framework or one-dimensional framework, the key difference is that it must be noted that advection is to be handled differently. In effect, advection is handled by choosing a Lagrangian reference frame such that the advective velocity is zero (i.e., U = V = W = 0).

If U = 0, then Equation 35 becomes

$$\frac{\partial C}{\partial t} = D_x \frac{\partial^2 C}{\partial \chi^2} + S \tag{39}$$

where χ is the distance coordinate defined in Figure 16 that measures distance from the center of mass of a moving parcel of water. From a mathematical point of view, the difference in the Eulerian (Equation 35) and Lagrangian (Equation 39) mass transport equations is significant despite the fact that the equations are equivalent. Once the troublesome advective term is dropped, the Lagrangian mass transport equation can be solved by numerical integration. By contrast, the form of the Eulerian mass transport equation requires numerical differentiation to solve the equation. This result in itself simplifies the solution procedure and makes possible a more accurate result. In elementary mathematical terms, numerical integration to determine the area under a curve is a simpler, more accurate procedure, in contrast to determining the slope of a curve with numerical integration.

If a coordinate system is defined that moves with a volume of water at the average stream velocity, U, and is related to a Eulerian system from which the flow information is obtained, then

$$\chi = x_{t+\Delta t} - x_t - \int_t^{t+\Delta t} U d\tau \tag{40}$$

where x is distance downstream from a fixed reference point at the beginning of some period of time, t, and at the end of the period of time, t+Δt. At the center of a moving volume of water, $\chi = 0$ as expected.

When the mass balance is written for a typical element as shown in Figure 21, and Reynolds averaging procedure is applied (see Equation 10), the result is

$$\frac{\partial C}{\partial t} + \frac{\partial(\overline{u'c'})}{\partial \chi} = S \tag{41}$$

If an eddy-diffusivity method is used to describe the turbulent mass flux term (see Equation 32), Equation 39 is the result.

If Equation 39 is integrated over some period of time, from t to t+Δt, then the average concentration is

$$C = C_t - \int_t^{t+\Delta t} \frac{\partial}{\partial \chi}\left[D_x \frac{\partial C}{\partial \chi}\right] d\tau + \int_t^{t+\Delta t} S d\tau \tag{42}$$

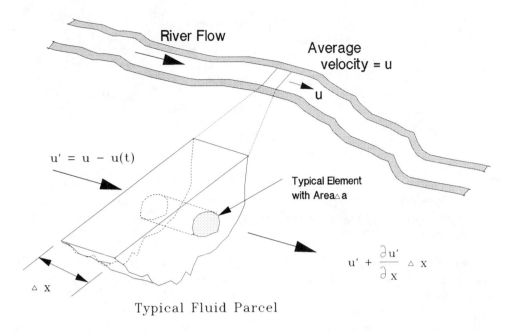

River Flow

Average velocity = u

u

$u' = u - u(t)$

Typical Element with Area $\triangle$ a

$u' + \dfrac{\partial u'}{\partial x} \triangle x$

$\triangle$ x

Typical Fluid Parcel

FIGURE 21. Lagrangian element mass balance for an element with a length, Δx, and moving downstream at a velocity of U.

where C_t is the concentration in the parcel of water at the beginning of the time period.

The dispersive term (first term on the right-hand side) of Equation 42 can be expressed in a simpler form if the actual mixing process is considered in relation to the moving parcels of water. Figure 22 shows that mixing or dispersion in the Lagrangian framework can be treated as a process of exchange between moving parcels of water. This exchange occurs because the vertical velocity profile is nonuniform. Near the water surface and over about 60% of the depth for typical stream conditions, the water is moving faster than the average. At the bottom of the flow, the water is moving slower than average. As two adjoining parcels of water move downstream, the water near the surface from the upstream parcel crosses the boundary between the parcels and becomes part of the downstream parcel. Water in the downstream parcel moves slower and is left behind in the upstream parcel as it overtakes the slower moving bottom water. The amount of water exchanged between particles is exactly equal. Rutherford and McBride[36] show that mass is in fact conserved in Lagrangian schemes. If the exchange flow between parcels is expressed as D_Q which is also taken to be a fraction, f_Q, of the flow in the stream, Q, such that

$$D_Q = f_Q Q \qquad (43)$$

then the change in concentration due to dispersive affects, ΔC, over a period of time, Δt, can be written as

$$\Delta C = \frac{\Delta t}{V_w} [D_Q(K-1)C(K-1) - D_Q(K)C(K)$$

$$+ D_Q(K+1)C(K+1) + D_Q(K+1)C(K)] \qquad (44)$$

where K-1, K, and K+1 refer to the parcels given in Figure 22, and C and D_Q are the

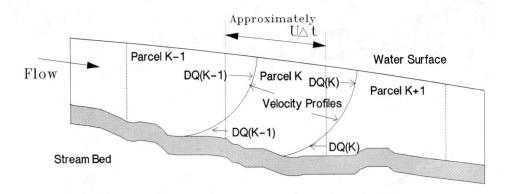

FIGURE 22. Movement of Lagrangian computational elements down a stream channel. DQ is the exchange flow between parcels of water caused by the nonuniform vertical velocity profile.

concentrations and exchange flows of those parcels of water, respectively. V_w is the volume of water exchanged for parcel K.Therefore, the dispersive effect can be determined when the exchange flows are specified and these values of f_Q can be specified for each reach through which the parcels of water move.

To better understand dispersion in a Lagrangian framework, it is possible to relate exchange flows to the Eulerian longitudinal dispersion coefficient defined by Equation 39. For steady flow, $V_w = Q\Delta t$, such that Equation 44 can be expressed as

$$\Delta C = f_Q(K)C(K-1) - [f_Q(K) + f_Q(K+1)]C(K) + f_Q(K+1)C(K+1) \qquad (45)$$

The dispersive term in Equation 39 also expresses the change in concentration due to dispersion and, when written in a finite difference form yields

$$\Delta C = \frac{\Delta t}{\Delta x^2} \{D_x(K)[C(K+1) - C(K)] + D_x(K+1)[C(K+1) - C(K)]\} \qquad (46)$$

where Δx is the length of a parcel, and D_χ is the Lagrangian longitudinal dispersion coefficient at the upstream faces of parcels K and K+1. If it is noted that $Q = AU$ and Equation 46 is set equal to Equation 43, the ratio of exchange flow to stream discharge is written as

$$f_Q = \frac{D_x}{U\Delta x} = \frac{1}{P} \qquad (47)$$

where P is the Peclet number that is the dimensionless ratio of the advective velocity to the dispersive velocity.

Equation 45 is not universally applicable, but in practice it is possible to easily optimize the time step. If the exchange volume is large compared to the smallest parcel volume or the Peclet number is smaller than 2, then Equation 45 is unstable.[37] If the stability condition is satisfied, however, the solution conserves mass and does not involve any numerical dispersion.[36] In practice, stability can be easily achieved. The automatic subdivision of the time step for only those parcels not meeting the stability criteria at only critical locations along the stream can be easily programmed. This allows an optimization of the time step for each computational element at only selected locations. By contrast, the optimization of time steps for the solution of the Eulerian mass balance equation has proven difficult to achieve. Generally, the time step for the

Eulerian solution can only be optimized for the smallest computational element. In addition, it is frequently difficult to adequately represent channel geometry if all Eulerian computational elements must be equal.

The Lagrangian method holds advantage only when dynamic conditions are important. It has already been noted that dispersion and mixing is usually unimportant in streams at steady flow. When mixing is unimportant and advective velocities are uniform, both the Eulerian and Lagrangian mass transport equations reduce to the same form, i.e.,

$$\frac{\partial C}{\partial t} = S \qquad (48)$$

This is the reason that McCutcheon[18] observed that the results of a Lagrangian plug flow model of the Willamette River in Oregon[13] were practically equivalent to the results obtained with an analytical solution of the modified Streeter-Phelps equations for dissolved oxygen[11] in the reaches where the kinetic formulations were equivalent. For a similar reason (i.e., dispersion is unimportant in the river tested) Schoellhamer[37] finds it possible to mimic the results of the Eulerian solution of the QUAL 2e model with a Lagrangian solution.

The Lagrangian mass balance may also be quite useful for two-dimensional and three-dimensional modeling. As of yet, however, it has not proven possible to quantify the vertical and lateral mixing as well as the longitudinal mixing can be described. Nevertheless, Schoellhamer[38,39] has developed a vertically layered Lagrangian model for the two-dimensional case that seems to be quite useful for modeling sediment dispersion in channels.

C. INCORPORATION OF THE SINK AND SOURCE TERMS IN MASS TRANSPORT

In this section, the source and sink term describing kinetic reactions, transformations, and additional and subtraction of material from a stream has been handled in an abstract way by lumping these effects into a single term, S. The exceptions were that examples for dissolved oxygen balance listed the important source and sink terms separately. As one can note in Equation 36, the separate incorporation of each term can be somewhat tedious and difficult to follow. As a result, shorthand methods have been developed to aid in the understanding of the terms and to assist in devising computational schemes.

As implied in Equation 36, most kinetic expressions for conventional pollutants such as dissolved oxygen, biochemical oxygen demand, nitrogen, and phosphorus are approximated with a zero- or first-order expression and most often the mass balance for each constituent is linked directly with the mass balance of other constituents. For example, the mass balance of dissolved oxygen cannot be completely described without writing mass balance equations for biochemical oxygen demand, ammonia, nitrite, and other constituents. In addition, some parameters such as nitrogen are described by a cascading series of mass balances with some cycling between components. The mass balance equation for organic nitrogen must feed mass into the ammonia mass balance equation. The ammonia mass balance equation must feed mass into the nitrite mass balance equation as must the nitrite mass balance be linked to the nitrate mass balance. These interactions are complex but can be described by one general type of equation. That equation is written as

$$\frac{\partial C_i}{\partial t} = \sum_{j=1}^{n} [K_{i,j}(C_j - C_{R,j}) + S_j] \qquad (49)$$

If $C_1 = L$, $C_2 = C_{DO}$, $C_3 = NH_3-N$, and $C_4 = NO_2-N$ in Equation 37, then $K_{2,2} = K_2$, the reaeration rate coefficient, $K_{2,1} = -K_1$, the deoxygenation rate coefficient, $K_{2,3} = -\alpha_1 K_{NH3}$, $K_{2,4} = -\alpha_2 K_{NO2}$, $C_{R,2} = C_{SAT}$, dissolved oxygen saturation concentration, and $S_2 = -S_{SOD} + P - R$, assuming that P and R are constant values. The reference concentrations, $C_{R,1}$, $C_{R,3}$, and $C_{R,4}$ are assumed to be zero because it is expected that biochemical oxygen demand, ammonia, and nitrite decay to zero concentrations. In addition, other interpretations can be made as well to give such an expression great flexibility in representing a system of linear water quality reactions.

VII. MODELS IN GENERAL USE

There are a number of water quality and hydraulics models available for use. In fact, there are so many models that the choice of a model to fit a particular stream system can be overwhelming. Furthermore, untested poorly documented models have been proliferated to the point that it is difficult to determine if an adequate computer code exists for a particular problem or if a new model should be developed. Contributing to the problem is a lack of formal peer review procedures for computerized water quality models.

In this section, four topics will be discussed that are related to models receiving fairly widespread use. In addition, some infrequently used models designed for special conditions will be briefly mentioned. First, the proliferation of models will be discussed to indicate why it is difficult to clearly establish which are the most practical models. Second, the few peer reviews and model evaluation studies available to establish the credibility of the more practical models will be discussed. Third, the more useful water quality models will be briefly reviewed. Finally, this section will cover the important hydraulics models intended to be implemented independently. Other methods of modeling flows, especially the steady-state methods that are usually incorporated directly in water-quality models, will be discussed in the following major section on methods to model water movement (Chapter 2, Section I.F, G, and H).

A. PROLIFERATION OF WATER QUALITY MODELS

Before reviewing the models in general use that seem to be of the most value to the practicing engineer, it is instructive to briefly review a larger set of models from which the more useful models evolved. This may be useful information during the model selection process if unique conditions are encountered. It is also useful in developing a historical perspective on model evolution to see how models have proliferated to unmanageable numbers.

Ambrose et al.[4,40] reviewed the conventional pollutant models in the readily available literature. They found that about 31 such models were in existence in 1982 as shown in Table 1. These models ranged from undocumented models used for a specific river study, to well-documented general purpose models. Most of these models were one-dimensional. Exceptions were the quasi-one-dimensional, link-node EXPLORE-I model and the three-dimensional box models, GENQUAL and WASP, that were designed to also be used in lakes and estuaries.

Ambrose et al.[40] screened the models available using the criteria that the model must be documented and the documentation and code be readily available to users through a public organization. The models chosen also simulated several processes including the major processes affecting nitrogen and phosphorus. The results are listed in Table 1 and a brief resume of each model is given in Appendix I.

The models listed in Table 2 represent a set of useful models that have some credibility. These are not, however, the most useful models. Many have not been thoroughly reviewed for scientific validity and completeness.

TABLE 1
List of Marginally to Fully Useful Stream Models Compiled by Ambrose et al. in 1982[a]

Sophistication level	Category	Models	Abbreviation
I	Manual screening	Simplified mathematical modeling	SMM
		Water quality assessment methodology	WQAM
II	Steady state	DOSAG-3	DOSAG
		G475	G475
		RECEIV-II	REC-II
		RIVSCI	RIVSCI
		SSAMIV	SSAMIV
		WRECEV	WREC
	Quasi-dynamic	AUTO-QUAL,QD	AUTO-Q
		QUAL-II	QUAL-II
		RECEIV-II	REC-II
III	Dynamic, simple hydrodynamics	EXPLORE-I	EXPL
		HSPF	HSPF
		RECEIV-II	REC-II
		RIVSCI	RIVSCI
		WQMM-Chowan	WQMM-CH
		WQRRS	WQRRS
IV	Full hydrodynamics	EXPLORE-1	EXPL
		WQRRS	WQRRS

[a] Adapted from Reference 4.

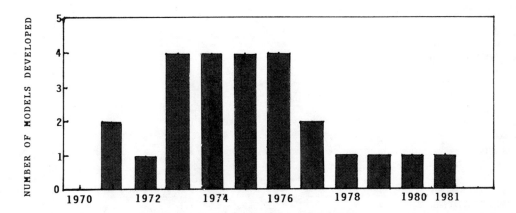

FIGURE 23. Distribution of models developed up to 1981 for conventional pollution problems in rivers.

Since 1981, the development of new models has slowed even more since the end of the proliferation of models that began in the late 1970s. Figure 23 shows that there was a proliferation of new river water quality models that reached a peak in the number of models being developed of 4/year in the mid-1970s. By the late 1970s, model development subsided to a rate of about 1/year that continued into the early 1980s. Since that time, development has slowed even more to about one new model every couple of years.

The distribution in Figure 23 was derived from the models cited by Ambrose et al.[4] in Table 2. The date of development was fixed as the date of the earliest reference cited by Ambrose et al.

TABLE 2
List of Conventional Pollutant Models for Streams and Rivers as of 1982[a]

Model	Spatial domain		Time domain				State variable systems							
	Branching stream	Segmented stream	Steady state	Quasi-dynamic	Dynamic	Hydraulics	Arbitrary pollutant	BOD-DO	Nitrogen	Phosphorus	Carbon	Solids	Biological	Temperature
AUTO-QUAL (41,42)	*	*	*			*	*	*						
AUTO-QD (43)	*	*	*	*		*	*	*						
Bauer and Bennett (44)	*	*		*			*	*						
DOSAG-I (45)	*	*	*			*	*	*						
DOSAG-III (46)	*	*	*			*	*	*	*	*	*			
DOSCI (47)	*	*	*				*	*						
EXPLORE-I (48)	*	*			*	*	*	*	*	*	*	*	*	
GENQUAL (49)	*	*			*		10							
G475 (11)	*	*	*					*	*	*	*			
HSPF (50)	*	*			*	*	*	*	*	*	*	*	*	*
LTM (51)		*			*		10							
MIT-DNM (52,96)	*	*			*	*	*		*					
MTI-DNM (53) (St. Lawrence)	*	*			*	*	*	*	*	*			*	
Overton and Meadows (54)	*	*		*				*	*	*				
PIONEER (55)		*	*					*	*	*		*	*	
QUAL-I (56,57)	*	*	*			*	*	*						*
QUAL-II (58,59,7)	*	*	*	*		*	*	*	*	*			*	*
RECEIV (SWMM) (60—62)	*	*			*	*	*	*						
RECEIV-II (63—65)	*	*	*	*	*	*	*	*	*	*			*	
RIBAM (66—68)	*	*	*	*	*	*	*	*	*	*				
RIVSCI (69,70)	*	*	*		*	*	*	*	*	*			*	
SNSIM (71,72)	*	*	*					*	*					
SMM (73,74)	*	*	*					*	*	*				
SSAM-IV (75)	*	*	*					*	*	*			*	
WASP (76—78)	*	*			*		19							
WASP/SUISAN (79)	*	*			*			*	*	*	*		*	
WIRQAS (Velz) (80—84)	*	*	*					*	*	*				
WRECEV (85)	*	*			*	*	*	*						
WQAM (86)	*	*	*					*	*	*	*			
WQMM/Chowan (87)	*	*			*	*	*	*	*				*	
WQRRS (2)	*	*	*	*	*	*	*	*	*	*	*	*	*	*

Note: Asterisks signify that the model includes the attributes list. The number under "Arbitrary pollutant" denotes the number of user-specified constituents that may include BOD, DO, and other state variables listed.

[a]   Adapted from Reference 40.

Since 1981, there have been few new one-dimensional models of conventional pollutants developed. This writer is only aware of models developed by Bedford, et al.[6,88,89] Martin,[90] Demetracopoloulus and Stefan,[91] and Rutherford and McBride.[36] Therefore, the information available about the development of new models indicates that new development is presently at, or perhaps below, a maintenance level. It may be that new model development is hampered by the excessive proliferation of models in the mid 1970s alluded to by Velz.[92] Many think that the 1970s were a time when the capabilities of models were oversold, and the credibility of modeling studies now suffers as a result.

Figure 23 seems to be clear evidence to support the general perception[92,93] that new model development was indeed excessive in the mid 1970s. From this writer's experience, it is difficult for an engineer, practicing or otherwise, to assimilate and begin to apply, in a serious way, more than 1 or 2 models per year. This writer, learned to use and apply 3 stream models in a one year evaluation study plus gained some working knowledge about three other models, but these studies were conducted using field data collected prior to the study.[18]

Soon after the explosive development of new models in the 1970s, the need for evaluation and critical examination of existing models became apparent. From 1978 to 1980, two significant studies were begun that examined the most likely candidate models to determine if the models were scientifically valid and useful.[18,94] These studies seem to represent a significant part of the limited effort that has been devoted to peer review and critical examination of stream water quality models. As soon as these evaluations began to approach completion, there seemed to be some focusing of resources on the improvement and further development of existing models, especially those models such as QUAL II that received good evaluations. These perceived trends are illustrated in Figure 24 and indicate this writer's perception that resources are now shifting from conventional pollutant model maintenance to development of toxicant fate and transport models. In this regard, also see Beck's[95] work regarding the shift in model development. Most troubling at this time is the indication that toxic chemical models are well into an unnecessary proliferation stage.

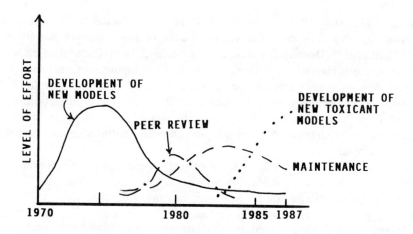

FIGURE 24. Perception of the relative effort being devoted to new model development, model evaluation, and refinement and maintenance of conventional stream water quality models.

B. PEER REVIEWS AND EVALUATION OF MODELS

In determining which models are the most practical and useful, it is helpful to review the model evaluation studies mentioned previously. McCutcheon[18] notes that only about 11 stream models have been evaluated in some detail, and some of these models were found to be impractical, seriously flawed by computer errors, and in some cases, lacking scientific and engineering credibility.

The thorough evaluations of models have focused on level II and III models as defined earlier. Therefore, it has generally been assumed that the manual screening level I methods have received sufficient peer review to justify usage. In addition, it seems to have been recognized that peer review of these method will be less effective unless the methods are codified to avoid interpretation and calculation errors that frequently occur in analyses of this type. The more complex models, especially those in the level IV category, have not been properly peer reviewed beyond the publication of the basic equations in peer reviewed journals. It has proven so difficult to code the complex equations into computer models without coding and conceptual errors that the first step of publishing model algorithms in a peer reviewed journal turns out to be a small part of the overall review necessary.

The models that have been evaluated to some extent are given in Table 3. Table 3 contains several models (i.e., MIT Transient Water Quality Network Model, Lagrangian Model, PIONEER I, QUAL-I, SNSIM, USGS Unsteady-State Water Quality Model, and the Velz method) that are not included in the screening list of Ambrose et al.[40] in Table 2. Evaluations by McCutcheon[18,98] found that the MIT and USGS unsteady models were not fully developed to a practical level. The Lagrangian model was not sufficiently developed in 1982 for the purposes of Ambrose et al. Furthermore, the PIONEER I, QUAL-I, and SNSIM models were, evidently, not readily available. Also, the Velz method has not been codified into a general purpose model that would meet the criteria designed by Ambrose et al.

In addition, the screening list of Ambrose et al. contain models that may not have been evaluated extensively by independent peer reviewers. These include RECEIV-II, RIVSCI, SSAMIV, WRECEV, AUTO-QUAL, EXPLORE-I, HSPF, and WQMM-Chowan. Note that G475 is the USGS Streeter-Phelps type model. Some of these models such as the RECEIV-II model are no longer readily available or useful.

There have been three types of model evaluations. In the early 1970s, model comparisons concentrated on the evaluation of algorithms and equations. Examples

TABLE 3
List of Models that have been Evaluated to Some Extent

Model	Major components	Ref.
DOSAG	Steady-state predictions of DO, CBOD, and NBOD, and includes nonpoint runoff.	45
M.I.T. transient water quality network model	Fully dynamic predictions of discharge, stage, temperature, DO, BOD, NH_3, NO_2, NO_3, phytoplankton-N, zooplankton-N, particulate and dissolved organic N, salinity, and coliforms including the effects of longitudinal dispersion.	96
Lagarangian model	Fully dynamic predictions of any 10 water-quality parameters including the effects of longitudinal dispersion. User specifies any 10 parameters, degree of interaction, and kinetics such as 1st and 2nd order. Hydraulics are separate.	97
PIONEER I	Steady-state predictions of depth; velocity; DO; BOD; NBOD or NH_3, NO_2, NO_3, Algae-N, and organic-N; Phosphorus; coliforms; dissolved solids; total-N; zinc; and one nth order nonconservative including the effects of SOD.	55
QUAL-I	Steady-state predictions of DO, BOD, and NBOD including effects of longitudinal dispersion, nonpoint runoff, and SOD.	56,57
QUAL-II	Quasi-dynamic predictions of temperature, DO, BOD, NH_3, NO_2, NO_3, phytoplankton, PO_4, coliforms, 1 arbitrary nonconservative (1st order) parameter, and conservative parameters including the effects of longitudinal dispersion, nonpoint runoff, and SOD. Discharge, depth, velocity and loads are steady.	58
SNSIM (an EPA model)	Steady-state predictions of DO, BOD, NBOD, and gross photosynthesis-respiration, and including effects of SOD and nonpoint runoff.	72
USGS Streeter-Phelps type model	DO; BOD; NBOD or NH_3, NO_2, NO_3, and organic-N; (2) PO_4; fecal and total coliforms; 3 conservatives; and anoxic conditions including the effects of SOD and nonpoint runoff. Hydraulics are separate.	11
USGS Unsteady-State Water-Quality Model	DO, BOD, and NBOD, including the effects of SOD. Hydraulics are separate.	44
Velz method	Steady-state predictions of DO and BOD with SOD are typical. The user formulates a site specific computer code that can include other parameters such as NBOD, temperature, NH_3, NO_2, and NO_3 for examples. Hydraulics are separate.	21
WQRRS	Dynamic predictions of discharge, stage, temperature, DO, BOD, detritus, NH_3, NO_2, NO_3, several classes of aquatic plants and animals, suspended sediment, organic sediment, pH, alkalinity, carbonate balance, and nonpoint runoff.	2

include the studies of Harper[99] and Lombardo.[100] These studies probably contributed to the standardization of algorithms, but since none of the currently used models have algorithms that have been traced to such studies, it is not clear that these types of studies contributed to establishing the credibility of presently used modeling techniques. Model documentation does not always properly cite proof of algorithm validation, even in the cases where it does exist.

As models proliferated in the 1970s, it became useful to contrast newly developed models with existing models to associate the credibility of the existing model with the new model. Unfortunately, these types of studies did not attempt to limit the redundancy of models by noting where existing models were no longer useful. As examples, Wiley and Huff[101] compared the WQRRS model (see Figure 25) with the DOSAG model (see Figure 26) using limited data available at the time for the Chattahoochee River near Atlanta, and Bauer et al.[102] compared the USGS Steady-State

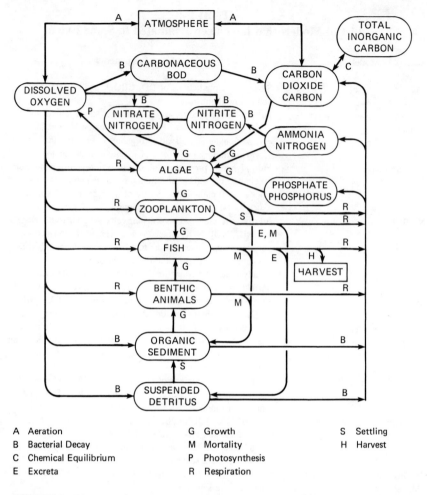

A	Aeration	G	Growth	S	Settling
B	Bacterial Decay	M	Mortality	H	Harvest
C	Chemical Equilibrium	P	Photosynthesis		
E	Excreta	R	Respiration		

FIGURE 25. Water quality and ecological constituents simulated by the WQRRS model.[2]

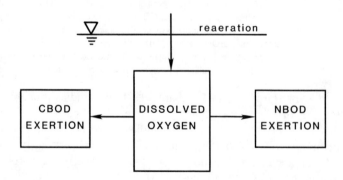

Analytical Solution of the Streeter–Phelps Equations.

Components of the DOSAG Model

FIGURE 26. DOSAG model. Note the lack of interaction with the benthos.

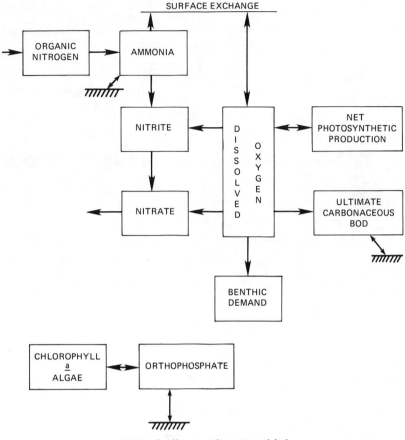

Water-Quality constituents modeled
by the USGS Streeter-Phelps Model

FIGURE 27. U.S. Geological Survey modified Streeter-Phelps model. Note the lack of material cycling. Components for fecal coliform bacteria, three conservative constituents, and nitrogenous biochemical oxygen demand (BOD) are not shown.

model (see Figure 27) to the Pioneer-I model (see Figure 28) in a rigorous manner using calibration and validation data from the Yampa River in Colorado. Wiley and Huff[101] concluded that the WQRRS model sufficiently mimicked the DOSAG model despite the use of different input data. As such, this study did not seem to adequately serve as a verification test against the analytical solution employed in the DOSAG model. The tests of Bauer et al.,[102] however, were notable for the rigorous calibration and validation testing involved.

In the third type of study, several models were evaluated at the same time to establish validity and usefulness. The National Council for Air and Stream Improvement (NCASI) compared the QUAL-I, QUAL-II, DOSAG, and SNSIM models using calibration and validation data from the Ouachita River in Arkansas and Louisiana. Model coefficients were measured independently and the rigorous calibration and validation were judged by statistical criteria rather than relying solely on arbitrary judgments of goodness of fit. The evaluation study was a final component of an overall study that included thorough peer review of the computer codes and assessment of the practical usefulness of the computer codes.

The models evaluated by the NCASI included DOSAG, QUAL-I, SNSIM, and QUAL-II in order of complexity (see Figures 26, 29, 30, and 31). In the first step of

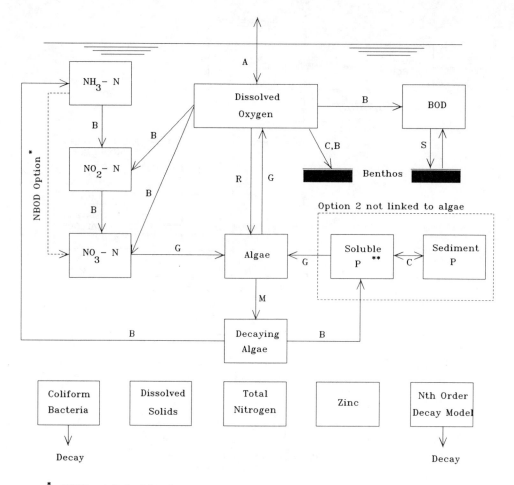

FIGURE 28. Components of the PIONEER-I model. Solution method ignores dispersion and uses a Lagrangian framework. A = aeration, B = bacterial decay, C = chemical equilibrium, G = growth, M = mortality, R = respiration, and S = settling.

the overall study, the DOSAG, QUAL-I, and QUAL-II models received what this writer believes is the only documented independent review of any computer codes for stream water quality modeling.[103,104,105] In each case, the reviewers found significant errors in the codes that had not been detected before. The corrections to the QUAL-II model were so significant and the review was so thorough that the corrected version of the model was soon adopted as the standard version of the QUAL-II model. These reviews were troubling in that they indicated that quality control procedures for computer code development are inadequate or nonexistent. Therefore, an important step in any study must be the confirmation that the computer code is valid. Managers who base decisions on model results should demand that standard, peer-reviewed codes be used or the development and use of of any new code should include arrangements to provide such a peer review. Unfortunately, peer review is a time consuming effort that will inhibit new model development unless care is taken during the calibration procedure to clearly establish when a standard model is inadequate. It is anticipated that

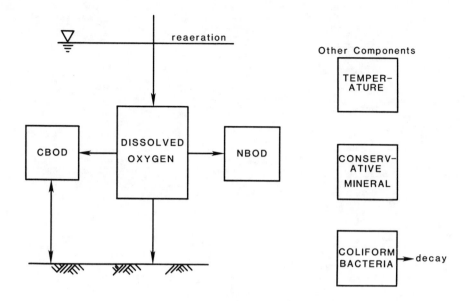

Advective-Dispersive Equation is solved analytically or
by a finite-difference method for steady flow.

Components of the QUAL-IE model

FIGURE 29.   Enhanced QUAL-I model.

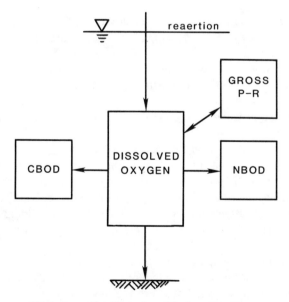

Based on the Streeter-Phelps Equations.

SNSIM model developed by the U. S. EPA

FIGURE 30.   Modified Streeter-Phelps model developed by US Environmental Protection Agency Region
II.

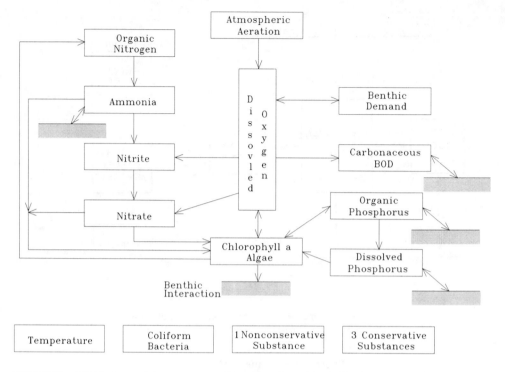

FIGURE 31. QUAL 2e model components. Note that the original QUAL-II model did not include components for organic nitrogen and dissolved phosphorus. Note the material cycling incorporated into this model.

clear documentation will be required to establish the need to develop new models when it is finally accepted that extensive development efforts must be extended even more to provide adequate peer review.

In the second phase of the their study, NCASI[106] compared the applicability of the three models plus they included a fourth model, SNSIM, that provided a level of capability not available in other models. Their evaluation was based on rigorous calibration and validation using very good data for the Ouachita River in Arkansas and Louisiana.

Four surveys were conducted on the Ouachita River in August, September, October, and December 1980 during steady-state conditions to define the dissolved oxygen balance. The data collection locations and frequency were carefully chosen to define the flow and water quality entering the modeled reaches and to define the conditions in the model domain for calibration and validation. A careful quality assurance plan was followed to insure that the data collected were of the best quality. Most important in setting this data collection study above many other careful studies was, however, that the major kinetic parameters, including the deoxygenation rate coefficient, reaeration rate coefficient, sediment oxygen demand rate, and rates of photosynthetic oxygen production and respiration, were measured in the laboratory or in the field independent from the calibration data collection efforts. Sensitivity analysis was used to guide where and when to conduct the parameter measurements.

The calibration and validation tests were also conducted in a careful way to minimize the arbitrary judgment that is inherent in the testing procedure. Rather than rely on the widely used approach of visually inspecting the plots of prediction vs. observation, NCASI developed statistical criteria before the calibration process was undertaken. They estimated that a reasonable calibration would be one that for which a Student's t test showed that there was no significant difference at a 5% confidence level between 95% of the observations and the predictions. They relaxed the validation criterion so that 60% of the observations should not be significantly different from predictions at a 5% confidence level.

In effect, NCASI chose to test the caveat that the simplest applicable model should be used for any study by choosing models that varied in complexity (see Figures 26, 29, 30, and 31 to see the variation in model complexity). They purposefully chose models that did not include some of the processes that were important in the Ouachita River to see if there were procedures in place to avoid the use of a model that was too simplistic. What they found was that the procedures in place are only barely adequate to avoid misuse of simplistic models. In calibrating the models, they found that all four models could be calibrated using different coefficient values. It was only when they attempted to validate the models that they found that the simple models were truly inadequate. In effect, there were no selection criteria that could be used to avoid this misuse of simple models short of trying the model and testing it. This implies a very labor intensive affair if one must iteratively apply more and more complex models until one is found to be applicable. This can be avoided by using more complex models to start with or by using flexible models that can be increased in complexity to model a given set of conditions as needed.

McCutcheon[18] took a different approach to test a number of models using a more extensive data base. All the models tested were reasonably expected to be able to simulate the important processes defined by the data involved, and the models selected were those that were expected to be generally useful and valid. Computer codes were not reviewed, but the tests were set up to detect significant problems. A number of these problems were encountered and corrected.

The data base was selected to include conditions that were generally expected to be steady-state in nature, and the data mainly defined the dissolved oxygen balance. Important components of the dissolved oxygen balance included reaeration, deoxygenation by decay of organic material, nitrification, and sediment oxygen demand. Notably absent were studies that involved a significant effect of photosynthesis.

Studies of three rivers were chosen to provide as much geographic distribution as possible and to cover as much of the range of possible hydraulic and water quality conditions as feasible. The range of conditions covered in the study by McCutcheon[18] are summarized in Table 4. Studies of the Chattahoochee River downstream of Atlanta, the Willamette River in Oregon, and the Arkansas River downstream of Pueblo, Colorado, measured discharge, stream width and depth, deoxygenation rate coefficients, and reaeration rate coefficients that varied by approximately 3 orders of magnitude. Many of the major model parameters such as reaeration coefficients and deoxygenation coefficients were measured independently. In all cases, the data sets were of the best quality available. Two of three of the studies were parts of intensive river quality assessments by the U.S. Geological Survey in which study procedures were refined until sampling procedures were highly accurate. These river quality assessment studies were not biased by any regulatory controversy. The study of the Arkansas River was a high quality study by the U.S. Geological Survey in cooperation with the State of Colorado that made use of the very good sampling techniques developed in the river quality assessment program. In all studies, care was taken in collecting the correct hydrologic data. All information produced in the studies was traceable to standard laboratory or field techniques. At least two data sets were available in each of the three studies for calibration and validation of the models being tested. Only the Arkansas River data had been analyzed prior to this study with one of the models being evaluated. This bias was minimized by reversing the data sets used for calibration and validation, but the bias was not fully eliminated.

TABLE 4

Contrast of Physical, Chemical, and Biological Characteristics from the Chattahoochee, Willamette, and Arkansas Rivers

Study	Discharge (ft³/sec)	Slope	Length (mi)	Width (ft)	Depth (ft)	Travel time (d)	Reaeration coefficient (d⁻¹)	Deoxygenation rate coefficient (d⁻¹)	DO balance components
Chattahoochee River	1100—1800	0.0003	43	234—269	4—6	1.8	0.3—11.0	0.16	Reaeration, deoxygenation, nitrification
Willamette River	6000—8000	0.0005—0.0	83	370—1300	2—60	16.7	0.05—0.4	0.07—0.14	Reaeration, deoxygenation, nitrification, benthic demand
Arkansas River	25—200	0.0015	42	65—190	0.4—1.3	1.7—2.0	6—15	1.5	Reaeration, deoxygenation, nitrification

Note: 1 ft³/sec = 0.028 m³/sec.
    1 mi   = 1.61 km.
    1 ft   = 0.3048 m.

The models tested included a U.S. Geological Survey modified Streeter-Phelps model, the QUAL-II model, the WQRRS model, the MIT Transient Network Model, the USGS Unsteady-State model, and the Velz rational method. More recently, Schoellhamer[37] used the data compiled for the Chattahoochee River to compare results from a Lagrangian model to the QUAL-II model. The Streeter-Phelps type, QUAL-II, WQRRS, and MIT models were chosen because these were, at the time, perceived to be valid and highly useful models. The USGS Unsteady-State model was included in the preliminary evaluation because it was readily available for testing. The Velz method had been used to analyze the data from the Chattahoochee and Willamette River studies so that the results were readily available for comparison.

Preliminary investigations of the models were surprising because the MIT and USGS Unsteady-State models were found to be inoperable. The USGS Unsteady-State model was never completely developed and, as a result of the study by McCutcheon,[18] was abandoned. The MIT model was, evidently, never fully tested before it was delivered to the U.S. EPA and reported to be a useful engineering tool. The verification data supplied with the MIT model are for a simple case that can be run with ease. Attempts to implement the model to analyze more realistic field conditions lead to serious fatal errors in the solution technique. These fatal errors in the MIT model could not be readily determined from the code which was not well documented nor supplemented with commentary statements. These disappointing results obtained with the MIT model were consistent with the unpublished experience of a number of other investigators and served to document a serious problem. Prior to this time the model was thought to be highly useful for unsteady conditions and a number of studies were proposed that were to make use of the model. In a few cases, specialized consultants were in great demand to salvage the studies. In other cases, the studies proposing the use of the MIT model had to rely on other models or failed to be properly concluded.

The primary conclusion from the investigation of the MIT and USGS Unsteady-State models was that there was a distinct lack of peer-review for the computer codes involved. The MIT model was the subject of a number of peer reviewed papers regarding the basic formulation, but the codification of the algorithms was not fully tested by the developers, much less by an independent reviewer until the study by McCutcheon.[18] This was a most unfortunate result because the MIT model was widely regarded as a valid and useful model.

Despite the lack of adequate peer review, the remaining models proved to be valid and useful. This was primarily due to the fact that each model was maintained and corrected as needed by a federal agency interested in river water quality modeling. The U.S. Geological Survey maintained and serviced the modified Streeter-Phelps model. The U.S. EPA Environmental Research Laboratory in Athens, Georgia maintains the QUAL 2e model. The U.S. Army Corps of Engineers Hydrologic Engineering Center in Davis, California, maintains the WQRRS model. Therefore, in the absence of adequate peer review, long-term maintenance seems to be mandatory. In addition, any new development of a model that does not identify and reserve resources for model correction and maintenance should have an adequate independent peer review, or the model development and application should be viewed as a one-time analysis. Any model designed for a one-time, site-specific analysis should not be used for other studies without full justification. To confer the perception that these one-time research models are useful general purpose models is a mistake that contributes to the needless proliferation of models. Proliferation wastes resources on model development, and use of inadequate models wastes resources available for applications.

The importance of providing maintenance was reinforced by the study of McCutcheon[18] and NCASI[106] where it was found that all the evaluated models suffered from significant coding and conceptual errors. For example, the USGS Streeter-Phelps model suffered from errors in computing the dissolved oxygen mass balance at the confluence with tributaries. The QUAL-II model neglected short-wave radiation in computing water temperature and had errors in calculations of the effect of photosynthesis. This explains the low wind speed coefficients required to calibrate the old SEMCOG (*S*outheast *M*ichigan *C*ouncil *o*f *G*overnments) version of QUAL-II. The WQRRS model had errors in the nitrate mass balance that caused the increase in nitrate resulting from ammonia oxidation to be ignored.

These and other errors were corrected by the USGS, NCASI, and the Corps Hydrologic Engineering Center almost as soon as the the study by McCutcheon concluded. In addition, other updates have occurred since that time. The effect on previous studies that used these flawed models was not investigated. However, the USGS, EPA, and Corps of Engineers did make the flaws known to selected users by memoranda, newsletter announcement, or letter to all registered users.

The result shown in Figures 32 to 37 for the calibration and validation of the models for the data from the Chattahoochee River are typical of the comparison by McCutcheon.[18] In comparing the models, the exact same boundary conditions and coefficients were used except in a few cases that were determined to have a negligible impact on any of the comparisons. In Figure 32, the temperature predictions were based on prior estimates of the wind speed coefficients, and in all four cases, the QUAL-II model predictions (based on the corrected model) matched measurements extremely well. The WQRRS model showed some discrepancies for the one set for which the model was calibrated. Similar patterns were observed in fitting the other data for biochemical oxygen demand and ammonia. The predictions of the QUAL-II model matched the analytical predictions of the Streeter-Phelps model very closely, but the WQRRS model predictions showed more discrepancies than expected. The predictions of nitrate showed significant discrepancies because at that time the QUAL-II model did not simulate organic nitrogen, and the WQRRS model had an error in the nitrate calculations at the time of the study.

The validation tests showed that both the Streeter-Phelps and QUAL-II models seemed to fully represent conditions observed except that the large scatter in the data indicated considerable unsteadiness in the instream conditions and loads. The unsteadiness was traceable to the unsteady boundary conditions.[37] The WQRRS model was not validated because it was too cumbersome to apply and contained essentially the same formulations as the Streeter-Phelps and QUAL-II models.

In comparing the results of calibrating and validating the USGS Streeter-Phelps, QUAL-II, and WQRRS models, McCutcheon concluded that all three models were equally valid after some compensation for the coding errors that existed during the comparison. Comparison of the results from the Velz method tended to indicate that this analysis method was also equally valid. The one exception to this conclusion is that the WQRRS model seems to have some numerical problem that is obscured by the flexibility available to calibrate a complex model of this type. Note in Figures 32 to 37 that the WQRRS model does not match the analytical solutions. Nevertheless, the choice of slightly different calibration coefficients would allow the user to almost exactly match the other other model predictions or calibration data. The confirmation of the validity of these modeling approaches was also supported by an evaluation to determine that the documentation was adequate in all cases in explaining what conditions the model was intended to simulate.

The combined result of studies by McCutcheon and NCASI establish the credibility of the QUAL-II model as corrected by NACSI. The studies seem to be a significant reason why the QUAL 2e model is taken as a generally useful and credible model. The

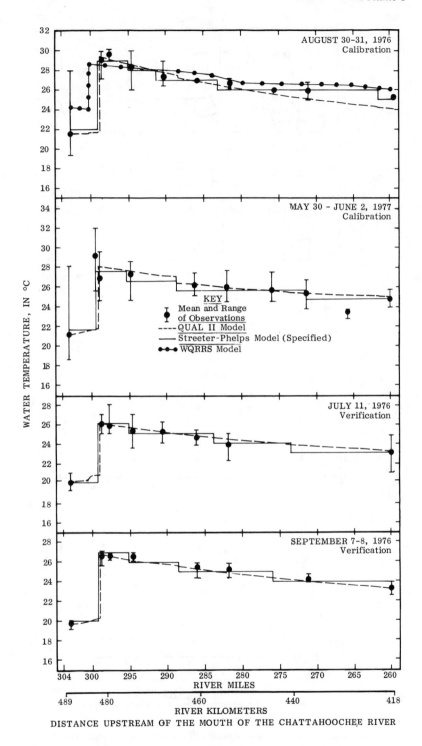

FIGURE 32. Observed, predicted, and specified water temperature in the Chattahoochee River downstream of Atlanta.

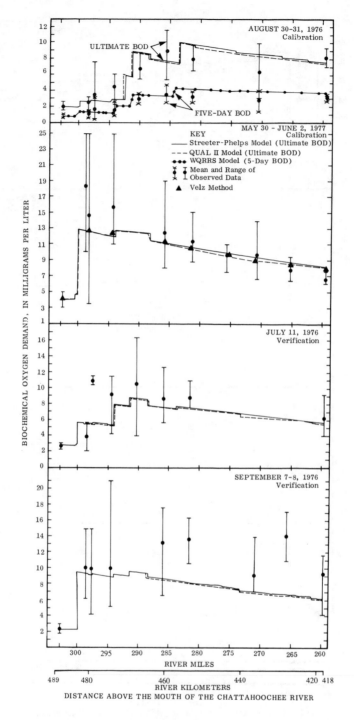

FIGURE 33. Observed and predicted biochemical oxygen demand in the Chattahoochee River downstream of Atlanta.

USGS Streeter-Phelps model was thoroughly checked by independent calculations, but the code was not completely reviewed to establish credibility of the model on par with QUAL-II. Updates to the QUAL-II model[107] that followed the evaluations incorporated most if not all of the advantages of the USGS model. As a result of the redundancy that occured when the QUAL-II model was updated, a comprehensive peer review of the USGS code did not seem useful.

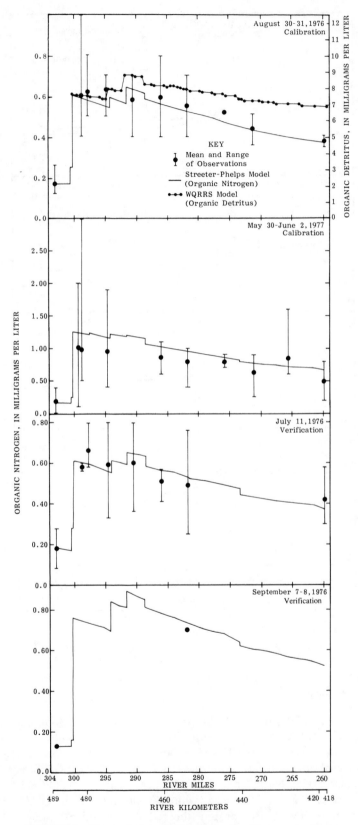

FIGURE 34. Observed and predicted organic nitrogen or organic detritus in the Chattahoochee River downstream of Atlanta.

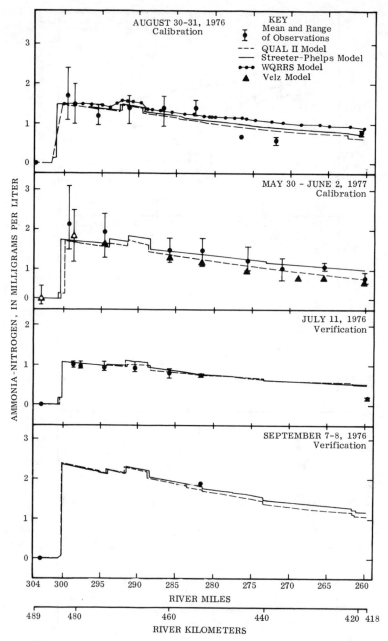

FIGURE 35. Observed and predicted ammonia-nitrogen in the Chattahoochee River downstream of Atlanta.

The past use of the USGS Streeter-Phelps model should not be viewed as lacking less credibility than the QUAL-II model applications. In fact, if the old QUAL-II applications were influenced by the flaws in the old SEMCOG version, they should have little or no credibility. In the future, however, use of the USGS Streeter-Phelps models or any other model that has not been fully peer reviewed should not be applied if the application would be fully redundant with an application of the QUAL 2e model that has been peer reviewed. But despite the redundancy, investigators who have

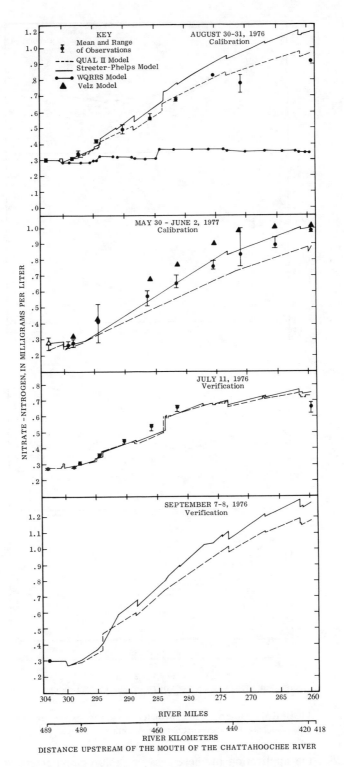

FIGURE 36. Observed and predicted nitrate-nitrogen in the Chattahoochee River downstream of Atlanta.

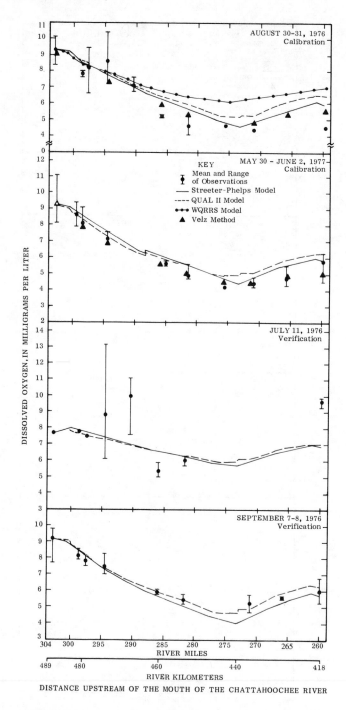

FIGURE 37. Observed and predicted dissolved oxygen in the Chattahoochee River downstream of Atlanta.

experience with the application of these less credible computer codes based on the modified Streeter-Phelps equations will, however, continue to use these redundant codes. From a management perspective, studies involving Streeter-Phelps models or any other inadequately reviewed model should be avoided. As a stopgap measure, individual applications should be checked by independent hand calculations of select, critical conditions. This should be an important component of the process to determine that the computations are valid.

The most significant difference in models occurs when the ease of use is evaluated. McCutcheon[18] found a large difference in the adaptability of models to unique conditions. In addition, data requirements can be quite diverse. In this regard, both the USGS and QUAL-II models were easily calibrated for steady-state conditions. The application of the WQRRS model to steady-state conditions, however, violated the modeling caveat that the simplest applicable model should be used. In these studies, the application of the WQRRS model to steady-state conditions was very tedious. Therefore, unless special conditions arise, very complex dynamic models should not be applied to steady-state conditions.

The Velz method was also very difficult to apply. This method requires writing a new code each time the method of analysis is applied and is therefore not very useful. Both McCutcheon[18] and NCASI[106] found that none of the computer codes evaluated were free of error. Therefore, any method that involves extensive coding should be avoided.

Later comparisons by Schoellhamer[37] indicate that the Lagrangian model of Schoellhamer et al.[51,108] is also equally valid, but as of yet the computer code has not been fully evaluated by independent reviewers. A direct contrast of the Lagrangian model with the WQRRS model for dynamic conditions has not yet been undertaken, but from knowledge of the difference in the effort it requires to apply the two models, the Lagrangian model should be the model of choice. The novice or inexperienced modeler may wish to apply the WQRRS model. The trade-off is flexibility and ease of use for the Lagrangian model vs. use of a well-established kinetic structure that has extensive data requirements. The Lagrangian model must be coupled with a dynamic hydraulics model (one or two are readily available), whereas the WQRRS model is already linked to several types including a solution of the St. Venant equation, the kinematic wave solution, and two hydrologic routing methods. However, even this refined hydrodynamic linkage cannot overcome the tedious process of applying the WQRRS model, and concerns have been expressed in the past regarding the stability of the WQRRS stream hydraulics model.

C. USEFUL WATER QUALITY MODELS

Except for the few model evaluation studies, there is little guidance available on which models are best applied for stream water-quality modeling. In some areas where modeling experience is extensive, it is fairly clear which are the best models despite the redundancy in model development. In other areas, like three-dimension modeling, there is almost no significant experience indicating which models may be the most useful. Therefore, this section will briefly review what seem to be a fairly comprehensive set of stream models, but at best, this guidance is expected to be an incomplete recommendation.

The best guidance on conventional pollutant models available for streams resulting from the evaluations of McCutcheon[18] and NCASI[106] and the accumulated experience with models is compiled in Table 5. The accumulated evidence indicates that the QUAL 2e model is the best available one-dimensional, steady-state model. The evaluation by McCutcheon[18] indicates that the model is numerically accurate to the point that there were no significant differences with the analytical solution of the modified Streeter-Phelps models. The input and output data structures are the standard that all models should follow to facilitate ease of use. Recent modifications of the model[107] included provision for graphical presentation of the data and an updated kinetic structure. In addition, the model was adapted for use on personal computers. Most recently, Brown

TABLE 5
Recommended Models for Stream Conditions

	One dimensional	Two dimensional	Three-dimensional thermal analysis

Steady flow	QUAL 2e[7]	SARAH[109]
		Stream tube models:
		  1. Bauer and Yotsukura[110]
		  2. RIVMIX[111 and 112]
		  3. MIXANDAT, MIXCALBN,
		    MIXCADIF, AND
		    MIXAPPLN[113,114]
			Not applicable

Dynamic flow	Lagrangian model[51,108]
	  + BRANCH[116]
	WASP 4.1 +
	  DYNHYD 4.1[78]
	CE-QUAL-RIV 1[88]
		WASP 4.1[78] + Hydrodynamics?	Rodi k-ϵ model[22,115]
			
			Edinger's model[117,118]

and Barnwell[107] added state-of-the-art uncertainty analysis to assist with sensitivity analysis, first-order error analysis, and Monte Carlo analysis of the calibrated model. Furthermore, Barnwell et al.[119] are designing an expert system to advise users on data collection and preparation of data for model calibration.

In addition to the QUAL 2e model, there are other models such as DOSAG I, QUAL I, and several modified Streeter-Phelps type models that may be be equally valid. However, for reasons already given, other similar models are not as useful. Models may be less useful because of a lack of proven scientific validity and because of difficulty in using the model and interpreting the results. Except for the NCASI corrected versions of DOSAG I and QUAL Ie, no other one-dimensional, steady-state models have been completely peer reviewed and evaluated. In addition, the ease with which the model can be used and the experience obtained with the QUAL 2e model justifies its use over any other simpler model.

Two-dimensional, steady-state models (also see Table 5) are not well developed. As far as this writer is aware, none of the available computer codes have been independently peer reviewed. The models that are available are of two types. The first type involves analytical solutions for various loading scenarios of which the SARAH model[109] seems to be the best available example. The SARAH model was designed for the analysis of mixing of nonpoint and point sources of toxic chemicals from landfills. First-order decay algorithms in the model may be adaptable to simple pollutants like biochemical oxygen demand or conservative substances. The theoretical basis of the SARAH model has been reviewed as part of EPA internal procedures. The code, however, has not been independently reviewed. One important attribute of this model is the support available for applying the model. In addition, the Corps of Engineers manual on *Mixing in Rivers* by E. R. Holley and G. H. Jirka (WES Tech. Rep. E-86-11) may be equally useful.

The second useful type of two-dimensional steady-state model is the stream-tube models that are believed to have been originally developed by Yotsukura and Sayre.[120] The best implementations of the stream-tube approach seem to be available in the RIVMIX model and series of mixing models available from the Ontario Ministry of the Environment. The RIVMIX model[111,112] simulates volatilization and sorption to bottom sediments, as well as lateral mixing. The MIXANDAT, MIXCALBN, MIXCADIF, and MIXAPPLN models are designed to simulate the effects of steady flow and constituents that are conserved or decay exponentially. Different models are used for different pipe and diffuser arrangements for introducing the wastewater into the stream. This set of models has been tested with field data for chlorides, chlorine, and ammonia to determine mixing zones for toxic chemicals being discharged with municipal wastes. These models have also been reported to have been updated and revised by the Ontario Ministry of the Environment. Both the RIVMIX and Ontario models seem to benefit from support and maintenance by Canadian agencies, but it is not clear that the models have received any external peer reviews.

Bauer and Yotsukura[110] have developed a two-dimensional, steady-state stream tube model for excess temperature that can be adapted to the lateral mixing of substances that are subject to exponential decay or are conserved. The Bauer and Yotsukura model, however, has received only limited use and no peer review and evaluation as far as this writer is aware.

The best dynamic models are a Lagrangian box model[51,108] that can be applied to one-dimensional flows and the Eulerian box model, WASP 4.1,[78] that can be applied to one-, two-, and three-dimensional flows. The Lagrangian model has been applied to model the effects of slowly varied flow and loading on the Chattahoochee River dissolved oxygen balance,[37] the effect of flow variation on the oxygen balance in the Ohio River, the effect of diel photosynthesis on the oxygen balance in Spring Creek (Arkansas), and the effect of nonuniform, steady flow on dye movement and diel solar effects on water temperature in the West Fork Trinity River in Fort Worth, Texas. The WASP model has been applied to simulate the highly variable effect of reservoir releases on the dissolved oxygen balance of the Alabama River upstream of Montgomery. In addition, the WASP model has been applied a number of times to simulate dynamic conditions in estuaries and lakes.

For one-dimensional streams, both the WASP 4.1 and Lagrangian models are well adapted for dynamic conditions. Both are very flexible in that both models are designed to allow the user to specify the kinetic structure to be used in a particular simulation. Up to 19 interacting constituents can be specified in the WASP 4.1 model, whereas 10 constituents can be specified in the Lagrangian model. Both models have default kinetic subroutines for the dissolved oxygen balance. Both can mimic the QUAL 2e kinetic structure. The Lagrangian model has a good library subroutine for the heat balance, and the WASP model has very good library subroutines for toxic chemicals and eutrophication. In choosing one of the two models, each unique application must be evaluated to determine which model may be most useful, but overall the WASP 4.1 model seems to be more credible. The WASP 4.1 model is fully maintained by the U.S. EPA Center for Exposure Assessment Modeling at Athens, Georgia (formerly Center for Water Quality Modeling), and has been implemented under a variety of conditions. The Lagrangian model is more of bare-bones, research-type model that has not been fully evaluated and tested. In addition, the Lagrangian model does not seem to be fully supported. Neither computer code seems to have received adequate peer review, however. Furthermore, both models must be linked to flow models for each application.

In addition to the WASP 4.1 and Lagrangian models for dynamic one-dimensional conditions, the U.S. Army Waterways Experiment Station is in the process of completing the development of the CE-QUAL-RIV1 model (the documentation is still in a draft form).[6,88] This writer has not yet had the time available to examine this model, but Bedford et al.[89] indicate that this model may be a well-contained flow and water quality model for dynamic streams. This model may lack flexibility in specifying kinetic descriptions, but seems to involve good unsteady flow modeling algorithms and good general purpose kinetic descriptions even if they are inflexible. Furthermore, the Waterways Experiment Station has developed a good support system to maintain and enhance such a model. As a result, the CE-QUAL-RIV1 model may be most useful for inexperienced users, whereas other users may better appreciate the flexibility in specifying the model kinetic available from the Lagrangian and WASP 4.1 models. The up-to-date hydraulic sub-model contained in the CE-QUAL-RIV1 model also seems to offer a distinct advantage.

Finally, regarding one-dimensional, dynamic models, it should be noted that the WQRRS model can be used for this purpose. The study by McCutcheon,[18] however, indicates that the model is very cumbersome to use and as a result is not very practical. In addition, questions remain about the numerical solution technique.

At present, there has been little activity in the development of dynamic two-dimensional models of which this writer is aware. The WASP 4.1 model will handle the description of the water quality kinetics with ease, but as of yet, has not been linked to the proper hydrodynamic model to simulate unsteady two dimensional flows. When faced with such a problem, the practicing engineer may wish to employ methods used in estuaries or turn to expert assistance to develop a site-specific model.

Three-dimensional models are not frequently used except for thermal discharge modeling. Both the modeling approaches of Rodi[22] and Edinger and Buchak[117,118] have proven useful, but neither model is readily available and these more complex models are not easily used by a practicing engineer. In addition, the hydrodynamics model of Sheng[23] may hold some promise for localized river mixing problems involving stratified flows as long as the vertical acceleration of the flow can be neglected. None of the dynamic models available for stream modeling seem to have been reviewed and independently evaluated.

D. DYNAMIC FLOW MODELS

The flow models that receive some use in solving practical problems tend to be dynamic routing models. Separate models do not seem to exist for steady-state, uniform flows or complex three-dimensional flows. In these cases, the flow and water-quality algorithms are combined in a single code. As a result, this section will review the more practical, readily available hydraulic routing models.

In Table 5, three dynamic one-dimensional hydraulics models were listed as being of some practical usefulness. These include the BRANCH model of the U.S. Geological Survey that has been linked to the Lagrangian model, DYNHYD4.1 of the U.S. EPA that has been linked to the WASP4.1 model, and a solution of the St. Venant equation (Equation 21) contained in the RIV1H sub-model of the CE-QUAL-RIV1 model.

The BRANCH model involves a finite difference solution of a form of the St. Venant equation (Equation 21). This model is well adapted to incorporating channel geometry data in a useful form[51] and seems to have some fail-safe features. The model is, however, more limited in terms of numerical accuracy and flexibility to handle highly unsteady flows. It does not seem to be as fully applicable as other models, but it does seem to be useful and very practical.

The DYNHYD model[78] is very similar to the BRANCH model in that it covers limited conditions in a solid way. This model is a link-node type, finite difference solution of the conservation of momentum equations (Equations 17 and 18) that was originally developed for estuaries and may not be as adaptable to highly nonuniform stream channels and rapidly varied flow (discharge at a location changes drastically with time). Both DYNHYD and BRANCH are supported models that are continually maintained and updated if necessary.

The RIV1H model involves a four-point implicit finite difference solution of the complete St. Venant equation (Equation 21). A Newton-Raphson convergence method is used for nonlinearity.[6] This model is patterned after the National Weather Service model, DAMBRK, by Fread.[6,121]

In addition, the WQRRS model[2] incorporates a series of six options for channel hydraulics that includes a solution of the St. Venant equation and the kinematic wave

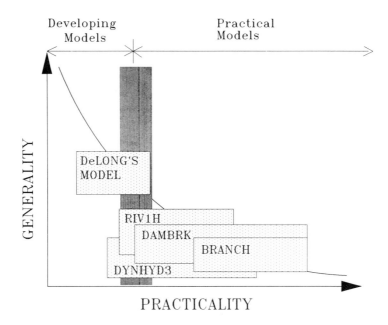

FIGURE 38. Perception of generality vs. practicality for hydraulic modeling.

approximation of the St. Venant equation. Brown[122] reports some difficulty, however, in applying early versions of the St. Venant solution. Since that time, the Hydrologic Engineering Center has continued to apply and improve the model and have developed other dynamic flow routing models as well.

Other one-dimensional hydraulic routing models that may be of use in the future for dynamic water-quality modeling include the dam-break model of Fread[121,123] and the finite-element solution of DeLong.[124] The DAMBRK model is similar to the RIV1H model described above but has been used extensively in dynamic flow routing. DeLong has developed a very accurate solution technique that seems to be applicable to a wider range of flow conditions including very steep flood waves that cause most other models some numerical difficulties. The DeLong model is in a developing state at present, however, and is not very practical at this time.

In the present setting of this monograph, it is somewhat difficult to place the relationship between various hydraulic routing models in context without going into too much detail. As with any set of independently developed models, there are significant redundancies and there are unique attributes that establish at least a minor niche for each model. The overall set of attributes for each model is extensive and only approximately consistent in terms of level of generality and practicality. It does seem possible, however, to develop a subjective relational guidance for the set of more practical hydraulic routing models as shown in Figure 38.

In Figure 38, it is assumed that there is a crude relationship between generality or theoretical rigor and practicality and usefulness for solving engineering problems as shown by the solid line. This assumed relationship is one of convenience because at later stages of model development it can be expected that the most practical model is the most general, in terms of the theoretical basis. In the interim, however, while there are still severe resource tradeoffs in the development of theory vs. development of user friendly application procedures (model designed to make use as easy as possible), models tend to be related in the way shown.

To place the relationship of the models available in context, the subjective relation in Figure 38 is suggested. The reader should note that a different perception of the needs for hydraulic routing models for river water quality modeling will change this proposed relationship. In addition, each model user and model developer will have different perceptions that would affect such a relational ranking. Also, this proposal is based on limited knowledge of the models involved that surely makes the exact boundaries somewhat fuzzy rather than distinct as drawn. Nevertheless, to guide model selection there needs to be some ranking of this type. At the very least, this should serve as an hypothesis that needs to be followed up with a good peer review of the models available so that some of the redundant development can be curtailed if necessary. As the use of hydraulic flow routing models becomes more prevalent in river water quality modeling, the need to review and evaluate routing models becomes more pressing.

VIII. MODEL SELECTION AND TESTING

The selection and use of models for investigating stream water quality problems is governed by the goals and objectives of a study. In many cases the objectives and goals will not remain static but will be refined and refocused as the study progresses. Study objectives govern the degree of approximation that may be useful and dictate what resolution may be necessary. Although scientists and engineers are driven to learn as much about a stream as possible, it usually is necessary to use models that are as simple as possible. From this springs the often quoted caveat that a study should make use of the simplest applicable model. This caveat should not be over zealously applied, however. It is possible to demonstrate that models that are too simplistic can be calibrated but have little validity.[106] When a study starts with an overly simplistic model that cannot be validated for the stream of interest, then the study fails to properly address the management issues if resources are too limited to permit development of a more elaborate model or if the study must enter what is usually an unplanned phase to adopt or develop a more adequate model. It either case, the choice of a model that is too simplistic results in the misuse of resources.

The need to keep modeling as simple as possible arises for several reasons. First, the results from an overly complex model may be difficult to interpret. Second, complex models that require greater amounts of data for calibration may be improperly calibrated unless the proper kinds and amounts of data are collected. Finally, a complex modeling effort may become bogged down in detail and become unable to respond to changing management objectives.

It may be useful in most studies to use a flexible model that can be increased in complexity as objectives are refined and more is learned about the stream of interest. In fact, a phased approach starting with a simple version of a model to guide calibration data collection efforts may be the most efficient and useful approach if the model is indeed flexible and not too cumbersome to use.

The experience of the model user and the perceptions of the resource managers involved are also important to consider. First, the effort it takes a user to learn about and test a new model may lead one to relax the criteria about using the simplest model applicable when the modeler is familiar with a more complex model. However, the model must not be overly complex, require greater data collection efforts, or prevent easy interpretation of the results. Second, it may be better for a user to learn to use a more complex model than conditions may dictate if it is likely that the study objectives may lead to more intensive modeling at a later date. Because existing models do not cover a continuous spectrum, it is frequently necessary anyway to select a model that may be overly complex to at least a slight degree. Third, if simpler models lack the necessary credibility for a given application because of the lack of adequate independent peer review or lack of a management perception that the simple model is valid and useful, then an overly complex model may be necessary. Finally, it is better in many cases to adapt an existing model rather than develop a new computer code regardless of the skill in model development that may be available. It is time consuming to develop a new model and to establish the validity of the code through the proper peer review.

A. MATCHING MODELS WITH EXPECTED CONDITIONS

Once the management objectives and goals of a study have been defined, then a modeling expert must go through the process of determining if these goals and objectives can be formulated into questions that can be answered in a modeling framework. Many questions can be answered without modeling. Some questions cannot be addressed because of the lack of knowledge or because of the lack of a convenient model formulation. The selection of the subset of study questions to be attacked by a modeling approach is based on knowledge of the basic modeling principles. The process of developing modeling objectives should not divert a study from other important objectives and goals that cannot be accommodated with a modeling approach, but it should aid in refining those objectives that can be addressed within a modeling framework. As part of the process of refining objectives, there should be free exchange between those charged with resource management and those charged with modeling. This exchange is very important to avoid the unrealistic expectations that have plagued modeling efforts.[24,125] At this stage, it is very important to avoid underemphasizing other management issues that can only be addressed by nonmodeling approaches such as monitoring or basic investigations. It is also very important to not be overly optimistic about the reliability of the calculations and the uncertainty in the conclusions about the results. Given that no other analysis method can provide significant predictive capability or reduce data collection expense, there is usually no need to oversell a modeling study.

A significant model selection question involves whether one-, two-, or three-dimensional modeling is needed. If conditions in the stream only vary in the downstream direction, then the one-dimensional approximation is very appropriate. If mixing of inflows seems to be important, then the speed at which mixing occurs is a determining factor. If the wastes entering a stream never mix and maximum concentrations are required in resolving management issues, then two- or three-dimensional modeling approaches may be necessary. If the management issues can be resolved using estimates of average concentrations, then the river may not necessarily need to be well mixed in every cross section. This is especially true if the difference in the average time of travel and the actual time of travel of pollutants in an unmixed zone are not significantly different, and additional sampling can be conducted to represent the average condition being used to calibrate the one-dimensional models. The few cases where two- and three-dimensional models have proven necessary generally involve mixing zones, especially for the dilution of toxic substances across shallower streams and the three-dimensional mixing of stratified discharges such as thermal effluents.

The decision of whether dynamic or steady-state models should be used depends on how well parameters must be predicted among other criteria. If maximum or minimum values must be predicted then dynamic models may be necessary. Average values can sometimes best be predicted by a steady-state model. Given the complexity of dynamic models it may also be useful to attempt to see if certain management options can be ruled out with a steady-state model. For example, if dissolved oxygen averages less than the standard and it is not necessary to determine how frequently a standard is violated, then the steady-state approach may work well as a screening model.

It may also be noted that the modeling approach may need to be modified as data are collected. The NCASI[106] found that preliminary data may imply that simple models are useful, but later attempts at calibration and validation may lead to the need for more complex models.

The ability to collect the proper types and number of data to define boundary and initial conditions and calibration and validation conditions may also influence the selection of models and modeling techniques. If a model can not be calibrated because of the lack of data collection resources, then simpler screening may be the only alternative. When data collection is limited by resources available, it is important to note that a modeling approach may fail to provide adequate answers, and this should be understood by the resource managers.

There is only limited guidance available on what form the kinetic structure of a water quality model should have. The developers of models have only the implicit guidance available from historical precedent. Over the years since 1925, water quality models have evolved into more and more useful forms. In this regard, models of the dissolved oxygen balance have evolved from expressions of total biochemical oxygen demand and reaeration into expressions of carbonaceous biochemical oxygen demand, nitrogenous biochemical oxygen demand, sediment oxygen demand, net photosynthesis, and reaeration. Methods of describing the nitrogen and phosphorus mass balance are less well evolved and it is not always clear what form these mass balance equations should take. Toxic chemical models are in an early stage of development and presently these types of models seem to be very simple first-order decay models or complex descriptions of the sum of present knowledge about all possible chemical transformations. Little effort has gone into developing toxic chemical models that are analogous to the highly practical but advanced dissolved oxygen models that have been customized for typical stream conditions. Therefore, model developers have only the vague guidance available from previous studies. These studies, however, rarely evaluate the form and structure of the models applied, making an assessment of the optimum model formulation very difficult to achieve even for well defined classes of problems. Only the reviews of Bowie et al.[15] and Schnoor[16] seem to attack this in a significant way.

In addition, the guidance on the appropriate kinetic structure available to those who apply and use models is also somewhat vague. For conventional pollutants, Krenkel and Novotny[125] describe what processes tend to have an effect on the dissolved oxygen balance in large, swift rivers; moderately large rivers; and shallow rivers. Similar guidance is not available for toxic chemical models. In this regard, model selection is a hypothesis testing exercise where kinetic descriptions or models are initially selected on the basis of any data available for the stream and on the basis of modeling experience in similar streams. The guidance by Bowie et al.[15] and Schnoor et al.[16] are also the most comprehensive compilation of modeling experience related to model application for conventional and toxic pollutants. The result of the limited guidance on model application is that those who apply models must have experience comparable to those who develop models. In fact, the level of experience required should be such that model users can readily become model developers when it becomes clear that models chosen for application are inadequate.

B. MODEL TESTING

Presently, the testing of models seems to be one of the weakest areas in modeling. Most practitioners determine if a calibration is adequate by means of a very subjective judgment that compares model predictions to measurements. On occasion, the calibration measurements may be collected in a very haphazard way. Experts make the judgment that a model is adequately calibrated on the basis of experience and knowledge of the management issues to be resolved. On occasion there are studies that include a sensitivity analysis to demonstrate that the important constituents are reasonably predicted. There are, however, better ways to evaluate the importance of uncertainty and to determine if a model is properly calibrated or not.

To properly evaluate the calibration of a model, it is desirable to determine if discrepancies between measurements and predictions are due to measurement errors, conceptual errors in equations, or approximation errors due to the nature of the model being calibrated. The effect of measurement errors is minimized by optimizing data collection procedures to collect data in the most sensitive locations and by collecting the proper number of replicates. A first-order error analysis or sensitivity analysis[7] can be used to identify critical measurements and sampling locations. To adequately support such an error analysis, one should attempt to determine the variation of measurement by collecting replicate samples at select locations during a study. See McCutcheon et al.[26] in this regard. Once the measurement uncertainty is estimated, it is possible to at least make a qualitative estimate of uncertainty due to approximation. If the model approximations are too uncertain or do not adequately explain the variation in the data (to meet management objectives), then a more refined model can be selected or developed. There is, however, little guidance for making decisions to determine when a more refined model should be employed. Generally, the modeler's experience is used as guidance in these decisions. In some cases, it is possible to use a Monte Carlo analysis[7] to base the decision for further model refinement on a more quantitative basis. Nevertheless, the statistical criteria or method of analysis chosen should not be used blindly to prevent an understanding of the processes and the effect of approximating the processes in a model.

Also important is the testing of a model that should occur before it is released as a general purpose model or proposed as a site specific model. This involves verification testing to determine if the numerical techniques are accurate. This is information that should be reported in the documentation of the model.

Sensitivity testing to determine where more data collection may be necessary involves changing model boundary and initial conditions and calibration parameters by arbitrary amounts to determine the effect on important state variables such as dissolved oxygen. For example, see the results of a sensitivity analysis for a dissolved oxygen model calibrated for the Han River in Seoul, South Korea shown in Figure 39. At this location on the Han River, dissolved oxygen predictions did not seem to be very sensitive to the parameters and loads examined.

The perturbation of parameters and boundary conditions in a sensitivity analysis does not have to involve arbitrary amounts in all cases. It is possible to devise replicate measurements to measure the coefficient of variation[26] and to estimate reasonable expected variation by other means.

There are a number of approaches to design statistical tests for evaluating model calibrations. However, these approaches do not seem to be developed enough to review here. Therefore, the reader may wish to consult the work of Thomann,[127] Reckhow and Chapra,[10] and NCASI.[106]

C. CALIBRATION

The calibration procedure is designed to use field measurements to guide in choosing the empirical coefficients in water-quality models and to determine if the boundary and initial conditions are consistent with instream measurements for the model chosen. A general outline of procedure is given in Table 6.

For simple models in which constituents do not interact, the calibration procedure

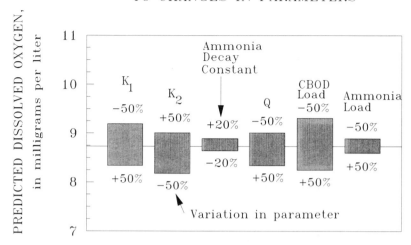

FIGURE 39. Sensitivity analysis of dissolved oxygen in the lower Han River, at station MH-3, near Dugdo, Seoul, Republic of Korea.[126] At this location biochemical oxygen demand (BOD) was predicted to average 1.6 mg/L, dissolved oxygen was predicted to average 8.7 mg/L, ammonia nitrogen was predicted to average 0.13 mg/L, and discharge (Q) was 136 m³/s (4860 ft³/s). In this reach the deoxygenation rate coefficient (K_1) was estimated to be 0.1 to 0.3/d and the reaeration rate coefficient was estimated to be 0.4 to 1.5/d.

TABLE 6
Outline of a General Calibration Procedure for Water Quality Models

Step	Procedure
1.	Calibrate hydraulics or hydrodynamics model by reproducing measurements of discharge, velocity, or stage (depth of flow) at selected sensitive locations. This involves modification of the Manning roughness coefficient, eddy-viscosity coefficients, or empirical flow vs. stage coefficients to predict the proper time of travel through the reach of interest. Dye studies to determine time of travel or average velocity may be used in place of hydraulic measurements for some simpler models.
2.	Select dispersion or mixing coefficients (or eddy diffusivities) to reproduce any dispersive mixing that may be important. Natural tracers or dye clouds may be monitored for this purpose.
3.	Calibrate any process models, such as water temperature, that are not affected by any other water quality constituent.
4.	Calibrate any process model affected by the processes first calibrated. In conventional models this may include biochemical oxygen demand, fecal coliform bacteria, and nitrification.
5.	Finally, calibrate all constituents or material cycles affected by any other process. In conventional models this usually means that the dissolved oxygen balance is calibrated last after biochemical oxygen demand, nitrification, and photosynthesis models are calibrated.

ends with step 3. For cascade-type models (see for example, Thomann[128]) like those used to model nitrification,[11] the first constituent such as organic nitrogen is calibrated, then the next constituent such as ammonia is calibrated, and so on until the final constituent is calibrated. Models of material cycles such as nutrient cycles of nitrogen and phosphorus are more difficult to calibrate. To accurately calibrate these models, all constituents must be modeled and at least one flux measurement must be measured so that in general there are more types of data by one, than there are calibration coefficients. See Wlosinski[129] for a demonstration of the data needed to calibrate material cycles.

D. VALIDATION

Validation testing is designed to confirm that the calibrated model is useful over the limited range of conditions defined by the calibration and validation data sets. As indicated earlier, the procedure is not designed to validate a model as generally useful or even validate the model as useful over an extensive range of conditions for the river of interest. As a result it is important that the calibration and validation data cover the range of conditions over which predictions are desired. As of yet, validation testing procedures have not been advanced enough to give precise definitions of the limits of applications.

Validation testing usually involves an independent data set collected during a second field study. The field study may occur before or after the field study to collect calibration data. For the best results, however, it is useful to collect the validation data after the model has been calibrated. This schedule of calibration and validation ensures that the calibration parameters are fully independent of the validation data. To extend the range of conditions over which the calibrated model is valid, however, it may be useful to save the initial study for validation testing if it is expected that data collected at a later date will provide a less severe test of the calibrated model. For example, low flow conditions or abnormal loading may occur early in the study while later in the study less stressful conditions may be expected. Or it may be possible to schedule a field study of dynamic conditions early in the study but not later. At present, it is very difficult to assemble the necessary resources to conduct as many surveys as needed. Therefore, it is important that surveys be scheduled using as much innovation as possible and that the choice of calibration and validation data sets be as flexible as possible to make the test of the calibrated model as severe as possible.

Regarding validation of dynamic models, the collection of data defining dynamic conditions is quite expensive compared to steady-state data collection. As a result, it is useful in some cases to collect a steady-state data set and calibrate the model, and then attempt validation with a dynamic data set.[6] This may work work well if the dynamic conditions do not cause an insignificant steady-state process such as resuspension to be become important. If this fails and some process does turn out to be important during dynamic events, then a third data set will be necessary for recalibration or, better still, for validation.

Many times, studies are resource-limited to the point that existing data must be used for calibration or it may not be impossible to collect more than one new set of data. In these cases, it is very important that both calibration and validation data be defined. Too many times, resource-limited studies only attempt calibration. This in effect limits the study to describing the conditions during the calibration data collection period and increases the uncertainty associated with the results. To best use existing data, it may be possible to identify a calibration data set and collect validation data under more stressful conditions, or it may be possible to split an existing data set into two sets. A split data set used for calibration and validation is quite limited, but is useful in testing to see if the data are adequate to define conditions required to match model assumptions. For example, a split data set can be useful in testing the adequacy of the steady-state assumption that requires averaging of the effects of loads and flows, plus the averaging of the effects of diel changes in solar radiation. Following the validation testing, the data set can be recombined to determine the best overall calibration if the validation test is successful. In fact, Reckhow and Chapra[10] recommend that this procedure be used in all cases to provide the best predictions with the calibrated model.

There are also innovative opportunities to split a single data set, if only one set can be collected. McCutcheon et al.[26] combined a Lagrangian sampling study that can be used for calibration with an Eulerian sampling program that could be used for validation testing. In addition, it may be possible to conduct a limited Lagrangian study in a critical stream reach during a preliminary study that may serve as calibration data. If the initial full data set validates the preliminary calibration, other data collection may be unnecessary. For this to be successful, however, it is important that clear calibration and validation criteria be defined before the data collection begins.

REFERENCES

1. **American Society of Testing and Materials,** Standard Practice for Evaluating Environmental Fate Models for Chemicals, proposed standard, Subcommittee E-47.06 on Environmental Fate, Committee E-47 on Biological Effects and Environmental Fate, 1983.
2. **Smith, D. J.,** Water Quality For River-Reservoir Systems, U.S. Army Corps of Engineers, Hydrologic Engineering Center, Davis, CA, 1978.
3. **Streeter, H. W. and Phelps, E. B.,** A Study of the Pollution and Natural Purification of the Ohio River, Public Health Bulletin No. 146, U.S. Public Health Service, U.S. Government Printing Office, Washington, D.C., 1925.
4. **Ambrose, R. B., Jr., Najarian, T. O., Bourne, G., and Thatcher, M. L.,** Models for Analyzing Eutrophication in Chesapeake Bay Watersheds: A Selection Methodology, unpublished report, U.S. Environmental Protection Agency Chesapeake Bay Program, Annapolis, MD, 1982.
5. **Mills, W. B., Porcella, D. B., Ungs, M. J., Gherini, S. A., Summers, K. V., Lingfung, M., Rupp, G. L., and Bowie, G. L.,** A Screening Procedure for Toxic and Conventional Pollutants, Parts 1 and 2, Reports EPA/600/6-85/002a and 002b, U.S. Environmental Protection Agency, Athens, GA, 1985.
6. **Dortch, M. and Martin, J. L.,** Water Quality Modeling of Regulated Streams, in *Alternatives in Regulated Flow Management,* Gore, J. A., Ed., CRC Press, Boca Raton, FL, 1988.
7. **Brown, L. and Barnwell, T. O., Jr.,** The Enhanced Stream Water Quality Models QUAL2E and QUAL2E-UNCAS: Documentation and User Manual, Report EPA/600/3-87/007, U.S. Environmental Protection Agency, Athens, GA, 1987.
8. **Ambrose, R. B., Jr.,** Course Notes: Water Quality Analysis Simulation Program—WASP, Chinese Research Academy for the Environmental Sciences, Beijing, China, June 15 to 19, 1987.
9. **Gove, P. B., Ed.,** *Webster's Third New International Dictionary of the English Language Unabridged,* Merriam, Springfield, MA, 1981.
10. **Reckhow, K. H. and Chapra, S. C.,** *Engineering Approaches for Lake Management, Vol. 1: Data Analysis and Empirical Modeling,* Butterworths, Boston, 1983.
11. **Bauer, D. P., Jennings, M. E., and Miller, J. E., Jr.,** One-Dimensional Steady-State Stream Water-Quality Model, U.S. Geological Survey Water Resources Investigations Report 79-45, NSTL, MS, 1979.
12. **Hydroscience, Inc.,** Trinity River Basin Water Quality Management Plan: Water Quality Modeling and Other Analyses in the Trinity River, Texas, report for the Trinity River Authority of Texas, Westwood, NJ, 1974.
13. **McKenzie, S. W., Hines, W. G., Rickert, D. A., and Rinella, F. A.,** Steady-State Dissolved Oxygen Model of the Willamette River, Oregon, U.S. Geological Survey Circular 715-J, Arlington, VA, 1971.
14. **Zison, S. W., Mills, W. B., Deimer, D., and Chen, C. W.,** Rates, Constants, and Kinetics Formulations in Surface Water Quality Modeling, U.S. Environmental Protection Agency Report EPA/600/3-78-105, Athens, GA, 1978.
15. **Bowie, G. L., Pagenhopf, J. R., Rupp, G. L., Johnson, K. M., Chan, W. H., Gherini, S. A., Mills, W. B., Porcella, D. B., and Campbell, C. L.,** Rates, Constants, and Kinetics Formulations in Surface Water Quality Modeling, Second Edition, Report EPA/600/3-85/040, U.S. Environmental Protection Agency, Athens, GA, 1985.
16. **Schnoor, J. L., Sato, C., McKechnie, D., and Sahoo, D.,** Processes, Coefficients, and Models for Simulating Toxic Organics and Heavy Metals in Surface Waters, Report EPA/600/3-87/015, U.S. Environmental Protection Agency, Athens, GA, 1987.

17. **Mills, W. B., Bowie, G. L., Grieb, T. M., and Johnson, K. M.,** Stream Sampling for Waste Load Allocation Applications, Report EPA/625/6-86/013, U.S. Environmental Protection Agency, Washington, D.C., 1986.
18. **McCutcheon, S. C.,** Evaluation of Selected One-Dimensional Stream Water-Quality Models with Field Data, Technical Report E-83-11, U.S. Army Corps. of Engineers Waterways Experiment Station, Vicksburg, MS, 1983.
19. **French, R. H.,** *Open-Channel Hydraulics,* McGraw-Hill, New York, 1985.
20. **Fischer, H. B., List, E. J., Koh, R. C. Y., Imberger, J., and Brooks, N. H.,** *Mixing in Inland and Coastal Waters,* Academic Press, New York, 1979.
21. **Velz, C. J.,** *Applied Stream Sanitation,* 1st ed., John Wiley & Sons, NY, 1970.
22. **Rodi, W.,** *Turbulence Models and Their Application in Hydraulics,* International Assoc. for Hydraulic Research, Delft, The Netherlands, 1980.
23. **Sheng, Y. P.,** Mathematical Modeling of Three-Dimensional Coastal Currents and Sediment Dispersion: Model Development and Application, Technical Report CERC-83-2, U.S. Army Corps. of Engineers Waterways Experimental Station, Vicksburg, MS, 1983.
24. **Hines, G. H., McKenzie, S. W., Rickert, D. A., and Rinella, F. A.,** Dissolved-Oxygen Regimen of the Willamette River, Oregon, Under Conditions of Basinwide Secondary Treatment, U.S. Geological Survey Circular 715-1, Arlington, VA, 1977.
25. **Davis, E. C., Suddath, J. L., Roman-Seda, R. A., McCutcheon, S. C., Sills, J. P., and Thackston, E. L.,** Waste Assimilative Capacity Studies of Streams in the Nashville Area, Technical Report No. 40, Environmental and Water Resources Engineering, Vanderbilt University, Nashville, 1977.
26. **McCutcheon, S. C., et al.,** Water Quality and Streamflow Data for the West Fork Trinity River in Fort Worth, Texas, Water Resources Investigation Report 84-4330, U.S. Geological Survey, NSTL, MS, 1985.
27. **Barnwell, T. O., Jr., Brown, L., Whittemore, R., McCutcheon, S. C., and Walker, W.,** Instruction Materials for Workshop on Stream Quality Routing Model QUAL2E, U.S. Environmental Protection Agency, Athens, GA, 1986.
28. **Heyes, D.,** Simulating Macroscopic Flows, *Nature,* 329, 1, 390, 1987.
29. **Lai, C.,** Flows of Homogeneous Density in Tidal Reaches: Solution by Method of Characteristics, U.S. Geological Survey *Open-File Report,* 1965.
30. **Baptista, A. E., Adams, E. E., and Stolzenbach, K. D.,** Eulerian-Lagrangian Analysis of Pollutant Transport in Shallow Water, Report No. 296, Ralph M. Parsons Laboratory, Aquatic Sciences and Environmental Engineering Dept. of Civil Engineering, MIT, Cambridge, 1984.
31. **Cheng, R. T., Castilli, V., and Milford, S. N.,** Eulerian-Lagrangian solution of the convection-dispersion equation in natural coordinates, *Water Resour. Res.,* 20(7), 944, 1984.
32. **Weast, R. G., Ed.,** *CRC Handbook of Chemistry and Physics,* CRC Press, Boca Raton, FL, 1986.
33. **Bird, R. B., Steward, W. E., and Lightfoot, E. N.,** *Transport Phenomena,* John Wiley & Sons, New York, 1960.
34. **Medine, A. J. and McCutcheon, S. C.,** Fate and Transport of Sediment Associated Contaminants, in *Hazard Assessment of Chemicals — Current Developments,* Vol. 6, Saxena, J., Ed., Hemisphere, Washington, D.C., 1988.
35. **Rich, L. G.,** *Environmental Systems Engineering,* McGraw-Hill, New York, 1973.
36. **Rutherford, C. and McBride, G. B.,** Accurate modeling of river pollutant transport, *J. Environ. Eng.,* 110(4), 808, 1984.
37. **Schoellhamer, D. H.,** Lagrangian transport model with QUAL II kinetics, *J. Environ. Eng.,* 114, 1988.
38. **Schoellhamer, D. H.,** Two-Dimensional Lagrangian Simulation of Suspended Sediment, unpublished manuscript, U.S. Geological Survey, Tampa Bay, FL, 1987.
39. **Schoellhamer, D. H.,** Lagrangian Modeling of a Suspended Sediment Pulse, in *Proc. National Hydraulics Engineering Conf.,* American Society of Civil Engineers, Williamsburg, VA, 1040, 1987.
40. **Ambrose, R. B., Jr., Najarian, T. O., Bourne, G., and Thatcher, M. L.,** Model Sets for Analyzing Eutrophication, unpublished manuscript, U.S. Environmental Protection Agency Athens Environmental Research Laboratory, GA, 1982.
41. **Pheiffer, T. H. and Lovelace, N. L.,** Application of the Auto-Qual Modeling System to the Patuxent River Basin, U.S. Environmental Protection Agency, Annapolis, Md., Report EPA-903/9-74-013, NTIS PB-244 280/4ST, December 1973.
42. **Crim, R. and Lovelace, N. L.,** Auto-Qual Modeling System, Report EPA-440/9-73-003, U.S. Environmental Protection Agency, Washington, D.C., March 1973.

43. **Lovelace, N.,** Auto-Qual Modeling System: Supplement I. Modification for Non-Point Source Loadings, Report EPA-440/9-73-004, U.S. Environmental Protection Agency, Washington, D.C., 1973.
44. **Bauer, D. P. and Bennett, J. P.,** Unsteady State Water Quality Model, Report PB-256, U.S. Geological Survey, NSTL Station, MS, 336, 1976.
45. **Texas Water Development Board,** DOSAG-I — Simulation of Water Quality in Streams and Canals: Program Documentation and Users Manual, Austin, TX, 1970.
46. **Duke, J. H., Jr. and Masch, F. D.,** Computer Program Documentation for the Stream Quality Model, DOSAG-3, Water Resources Engineers, Austin, TX, for U.S. Environmental Protection Agency, Washington, D.C., October 1973.
47. **Shephard, J. L. and Finnemore, E. J.,** Spokane River Basin Model Project: Vol. IV — User's Manual for Steady State Stream Model, U.S. Environmental Protection Agency, Region X, Seattle, 1974.
48. **Baca, R. G., Waddel, W. W., Cole, C. R., Bradstetter, A., and Cearlock, D. B.,** Explore-I: A River Basin Water Quality Model, Pacific Northwest Laboratories of Battelle Memorial Institute, Richland, WA, for U.S. Environmental Protection Agency, Washington, D.C., Contract 68-01-0056, 1973.
49. **Schornick, J. C. and Smith, R. A.,** Preliminary Documentation for GENQUAl Model, U.S. Geological Survey, Quality of Water Branch, unpublished report, Reston, VA, 1984.
50. **Johanson, R. C., Imhoff, J. C., and Davis, H. H., Jr.,** User's Manual for Hydrological Simulation Program-FORTRAN (HSPF), Hydrocomp, Inc., Palo Alto, CA, Report EPA-600/9-80-015, U.S. Environmental Protection Agency, Athens, GA, Apr. 1980.
51. **Schoellhamer, D. H. and Jobson, H. E.,** User's Manual for a One-Dimensional Lagrangian Transport Model, Water Resources Investigation Report 86-4145, U.S. Geological Survey, NSTL, MS, 1986.
52. **Najarian, T. O. and Harleman, D. R. F.,** A Nitrogen Cycle Water Quality Model for Estuaries, Tech. Report No. 204, R.M. Parsons Laboratory for Water Resources and Hydrodynamics, MIT, for U.S. Environmental Protection Agency, Corvallis, OR, 1975.
53. **Thatcher, M. L., Pearson, H. W., and Mayor-Mora, R. E.,** Application of a Dynamic Network Model to Hydraulic and Water Quality Studies of the St. Lawrence River, in *Proc. Second Annual Symposium of the Waterways, Harbors, and Coastal Engineering Division,* American Society of Civil Engineers, San Francisco, 1195, September 1975.
54. **Overton, D. E. and Meadows, M. E.,** Mathematical Modeling for Water Quality Management in Streams Under Unsteady Hydraulic Conditions, Research Report No. 55, Water Resources Research Center, University of Tennessee, Knoxville, 1976.
55. **Waddel, W. W., Cole, C. R., and Baca, R. G.,** A Water Quality Model for the South Platte River Basin — Documentation Report, Battelle Pacific Northwest Laboratories, Richland, WA, for U.S. Environmental Protection Agency, 1974.
56. **Texas Water Development Board,** QUAL-I — Simulation of Water Quality in Streams and Canals: Program Documentation and User's Manual, Austin, TX, 1970.
57. **Texas Water Development Board,** Simulation of Water Quality in Streams and Canals: Theory and Description of QUAL-I Mathematical Modeling System, Report 128, Austin, TX, 1971.
58. **Roesner, L. A., Monser, J. R., and Evenson, D. E.,** Computer Program Documentation for the Stream Quality Model, QUAL-II, U.S. Environmental Protection Agency, Washington, D.C., 1973.
59. **Roesner, L. A., Giguere, P. R., and Evenson, D. E.,** User's Manual for Stream Quality Model (QUAL-II), U.S. Environmental Protection Agency, Athens, GA, 1977.
60. **Huber, W. C., Heaney, J. P., Medina, M. A., Peltz, W. A., Sheikh, H., and Smith, G. F.,** Storm Water Management Model User's Manual, Version II, Report EPA-670/2-75-017, U.S. Environmental Protection Agency, Cincinnati, March 1975.
61. **Huber, W. C., Heaney, J. P., Peltz, W. A., Nix, S. J., and Smolenyak, K. J.,** Interim Documentation, November 1977, Release of EPA SWMM, U.S. Environmental Protection Agency, Cincinnati, Draft Report, Project R-802411, 1977.
62. **University of Florida and Water Resources Engineers, Inc.,** Storm Water Management Model, Volumes 1-4, Metcalf and Eddy, Inc., for U.S. Environmental Protection Agency, Washington, D.C., EPA No. 11024DOC, 1971.
63. **Beckers, C. V., Parker, P. E., Marshall, R. N., and Chamberlain, S. G.,** RECEIV-II, A General Dynamic Planning Model for Water Quality Management, in *Proceedings of EPA Conference on Environmental Modeling and Simulation,* U.S. Environmental Protection Agency, Cincinnati, 1976.

64. **Raytheon Oceanographic and Environmental Services**, New England River Basins Modeling Project — Final Report- Vol. III — Documentation Report — Part I — RECEIV-II Water Quantity and Quality Model, for U.S. Environmental Protection Agency, Washington, D.C., December 1974.

65. **Raytheon Oceanographic and Environmental Services**, RECEIV-II — Documentation Report for Water Quantity and Quality Model, Revised Report, U.S. Environmental Protection Agency, Washington, D.C., 1975.

66. **Marshall, R. N., Chamberlain, S. G., and Beckers, C. V., Jr.**, RIBAM, A Generalized Model for River Basin Water Quality Management Planning, in *Proceedings of EPA Conference on Environmental Modeling and Simulation,* U.S. Environmental Protection Agency, Cincinnati, April 1976.

67. **Raytheon Oceanographic and Environmental Services,** BEBAM — A Mathematical Model of Water Quality for the Beaver River Basin: Vol. III — Documentation Report, for U.S. Environmental Protection Agency, Washington, D.C., 1974.

68. **Raytheon Oceanographic and Environmental Services,** Expanded Development of BEBAM — A Mathematical Model of Water Quality for the Beaver River Basin, for U.S. Environmental Protection Agency, Washington, D.C., 1974.

69. **Raytheon Oceanographic and Environmental Services**, Water Quality Management Plan for the Snohomish River Basin and the Stillagukamish River Basin: Volume III — Methodology, for U.S. Environmental Protection Agency, Washington, D.C., 1974.

70. **Anon.,** Water Quality Management Plan for the Snohomish Basin and the Stillaguamish River Basin: Volume IV — Computer Program Documentation, Part B: Dynamic Estuary Model (SRMSCI), Snohomish County Planning Dept., Everett, WA, 1974.

71. **Braster, R.E., Chapra, S.C., and Nossa, G.A.**, Documentation for SNSIM 1/2: A Computer Program for the Steady-State Water Quality Simulation of A Stream Network, 4th ed., U.S. Environmental Protection Agency, New York, 1975.

72. **Braster, R.E., et al.,** Documentation for SNSIM, U.S. Environmental Protection Agency, Region II, New York, 1978.

73. **Hydroscience, Inc.,** Simplified Mathematical Modeling of Water Quality, Report to Office of Water Programs, U.S. Environmental Protection Agency, Washington, D.C., U.S. Government Printing Office No. 1971-444-367/392, 1971.

74. **Hydroscience, Inc.,** Addendum to Simplified Mathematical Modeling of Water Quality, U.S. Environmental Protection Agency, Washington, D.C., U.S. Government Printing Office No. 1972-484-486/291, 1972.

75. **Grenney, W. T. and Kraszewski, A. K.,** Description and Application of the Stream Simulation and Assessment Model: Version IV (SSAM IV), Instream Flow Information Paper, U.S. Fish and Wildlife Service, Fort Collins, CO, Cooperative Instream Flow Service Group, 1981.

76. **Di Toro, D. M., Fitzpatrick, J. J., and Thomann, R. V.,** Water Quality Analysis Simulation Program (WASP) and Model Verification Program (MVP) — Documentation, Hydroscience, Inc., for U.S. Environmental Protection Agency, Duluth, MN, Contract No. 68-01-3872, 1981.

77. **Thomann, R. V., Di Toro, D. M., Winfield, R. P., and O'Connor, D. J.,** Mathematical Modeling of Phytoplankton in Lake Ontario: Part I. Model Development and Verification, Report EPA-600/3-76-065, U.S. Environmental Protection Agency, Corvallis, OR, 1975.

78. **Ambrose, R. B., Wool, T. A., Connolly, J. P., and Schanz, R. W.,** WASP4, A Hydrodynamic and Water Quality Model — Model Theory, User's Manual, and Programmer's Guide, U.S. Environmental Protection Agency Report, Athens, GA, 1987.

79. **Thomann, R. V., Di Toro, D. M., Winfield, R. P., and O'Connor, D. J.,** Mathematical Modeling of Phytoplankton in Lake Ontario: Part 2. Simulations Using Lake 1 Model, Report EPA-600/3-76-065, U.S. Environmental Protection Agency, Duluth, MN, 1976.

80. **Hines, W. G., Rickert, D. A., McKenzie, S. W., and Bennett, J. P.,** Formulation and Use of Practical Models for River Quality Assessment, U.S. Geological Survey, Circular 715-B, Reston, VA, 1975.

81. **Rickert, D. A. and McKenzie, S. W.,** River-Quality Assessment of the Willamette River Basin, Oregon: Hydrologic Analysis and River-Quality Data Programs, U.S. Geological Survey, Circular 715-D, Reston, VA, 1976.

82. **Jennings, M. E., Shearman, J. O., and Bauer, D. P.,** Selection of Streamflow and Reservoir-Release Models for River-Quality Assessment, U.S. Geological Survey, Reston, VA, Circular 715-E, 1976.

83. **Rickert, D. A., Hines, W. G., and McKenzie, S. W.,** Methods and Data Requirements for River Quality Assessment, *Water Resour. Bull.,* 11(5), 1013, 1975.

84. **Rickert, D. A., Hines, W. G., and McKenzie, S. W.,** Project Development and Data Programs for Assessing the Quality of the Willamette, River, Oregon, Circular 715-C, U.S. Geological Survey, Reston, VA, 1976.

85. **Johnson, A. E. and Duke, J. H., Jr.,** Computer Program Documentation for the Unsteady Flow and Water Quality Model — WRECEV, U.S. Environmental Protection Agency, Washington, D.C., 1976.

86. **Zison, S. W., Haven, K. F., and Mills, W. B.,** Water Quality Assessment, A Screening Method for Nondesignated 208 Area, Report EPA-600/9-77-023, U.S. Environmental Protection Agency, Athens, GA, NTIS No. PB277161/AS, 1977.

87. **Amein, M. and Galler, W. S.,** Water Quality Management Models for the Lower Chowan River, Report UNC-WRRI-79-130, Water Resources Research Institute, Dept. of Civil Engineering, North Carolina State University, Raleigh, 1979.

88. **Bedford, K. W., Sykes, R. M., and Libicki, C. A.,** A Dynamic, One-Dimensional, Riverine Water Quality Model, Draft Documentation, U.S. Army Corps. of Engineers Waterways Experiment Station, Vicksburg, MS, 1982.

89. **Bedford, K. W. Sykes, R. M., and Libicki, C.,** Dynamic advective water quality model for rivers, *J. Environ. Eng.,* 109(3), 535, 1983.

90. **Martin, J. L.,** Simplified, Steady-State Temperature and Dissolved Oxygen Model: User's Guide, Instruction Report E-86-4, U.S. Army Corps. of Engineers Waterways Experiment Station, Vicksburg, MS, 1986.

91. **Demetracopopoulus, A. C. and Stefan, H. G.,** Model of Mississippi River pool: dissolved oxygen, *J. Environ. Eng.,* 109(5), 1020, 1983.

92. **Velz, C. J.,** Dissolved Oxygen/Temperature Modeling, *Workshop on Verification of Water Quality Models,* Thomann, R. V. and Barnwell, T. O., Jr., Eds., U.S. Environmental Protection Agency Report EPA/600/9-80-016, Athens, GA, 1980.

93. **Orlob, G. T.,** Introduction, and Future Directions, in *Mathematical Modeling of Water Quality: Streams, Lakes, and Reservoirs,* Vol. 12, Orlob, G.T., Ed., John Wiley & Sons, New York, 1983, Chapters 1 and 13.

94. **Whittemore, R. C. and Hovis, J.,** A Study of the Selection, Calibration and Verification of Mathematical Water Quality Models, National Council of the Paper Industry for Air and Stream Improvement, Tech. Bull. No. 367, New York, 1982.

95. **Beck, M. B.,** *Water Quality Management: A Review of the Development and Application of Mathematical Models,* No. 11, Lecture Notes in Engineering, Springer-Verlag, New York, 1985.

96. **Harleman, D. R. F., Dailey, J. E., Thatcher, M. L., Najarian, T. O., Brocard, D. N., and Ferrara, R. A.,** User's Manual for the MIT Transient Water Quality Network Model, U.S. Environmental Protection Agency Report EPA-600/3-77-010, Corvallis, OR, 1977.

97. **Jobson, H. E.,** Temperature and Solute-Transport Simulation in Streamflow Using a Lagrangian Reference Frame, U.S. Geological Survey Water Resources Investigations 81-2, NSTL Station, MS, 1980.

98. **McCutcheon, S. C.,** Evaluation of Stream Water Quality Models, in *Proc. 1983 National Conf. on Environmental Engineering,* Am. Soc. of Civil Engineers, New York, 1983, 190.

99. **Harper, M. E.,** Assessment of Mathematical Models Used in Analysis of Water Quality in Streams and Estuaries, Washington State Water Research Center Report Fulfilling Contract OWRR Contract No. 14-01-00011956, Pullman, WA, 1971.

100. **Lombardo, P. S.,** Critical Review of Currently Available Water Quality Models, Hydrocomp, Inc., Report Fulfilling Requirements of OWRR Contract No. 14/31/0001/3571, Mt. View, CA, 1973.

101. **Wiley, R. C. and Huff, D.,** Chattahoochee River Water Quality Analysis, U.S. Army Corps. of Engineers, Hydrologic Engineering Center Report, Davis, CA, 1978.

102. **Bauer, D. P., Steele, T. D., and Anderson, R. D.,** Analysis of Wasteload Assimilative Capacity of the Yampa River, Steamboat Springs to Hayden, Routt County, CO, U.S. Geological Survey Water Resources Investigations 77-119, Reston, VA, 1978.

103. **National Council for Air and Stream Improvement, Inc.,** A Review of the Mathematical Water Quality Model DOSAG and Guidance for Its Use, Stream Improvement Tech. Bulletin No. 327, New York, 1979.

104. **National Council for Air and Stream Improvement, Inc.,** A Review of the Mathematical Water Quality Model QUAL1E and Guidance for Its Use, Stream Improvement Tech. Bulletin No. 331, New York, 1980.

105. **National Council for Air and Stream Improvement, Inc.,** A Review of the Mathematical Model Qual II and Guidance for its Use, National Council for Air and Stream Improvement, Tech. Bulletin No. 331, New York, 1980.

106. **National Council for Air and Stream Improvement, Inc.,** A Study of the Selection, Calibration and Verification of Mathematical Water Quality Models, Tech. Bulletin No. 367, New York, 1982.

107. **Brown, L. C. and Barnwell, T. O., Jr.,** Computer Program Documentation for the Enhanced Stream Water Quality Model QUAL2E, Report EPA/600/3-85/065, U.S. Environmental Protection Agency, Athens, GA, 1985.

108. **Schoellhamer, D. H. and Jobson, H. E.,** Programmer's Manual for a One-Dimensional Lagrangian Transport Model, Water Resources Investigation Report 86-4144, U.S. Geological Survey, NSTL, MS, 1986.

109. **Ambrose, R. B., Jr. and Vandergrift, S.,** SARAH, A Surface Water Assessment Model for Back Calculating Reductions in Abiotic Hazardous Wastes, Report EPA/600/3-86/058, Assessment Branch, Environmental Research Lab., U.S. Environmental Protection Agency, Athens, GA, 1986.

110. **Bauer, D. P. and Yotsukura, N.,** Two-Dimensional Excess Temperature Model for a Thermally Loaded Stream, Report WRD-74-044, U.S. Geological Survey, Water Resources Division, Gulf Coast Hydroscience Center, NSTL, MS, 1974.

111. **Krishnappan, B. G. and Lau, Y. L.,** User's Manual: Prediction of Transverse Mixing in Natural Streams Model-RIVMIX MKl, Environmental Hydraulics Section, Hydraulics Division, National Water Research Institute Canada Centre for Inland Waters, Ontario, Canada, 1982.

112. **Krishnappan, B. G. and Lau, Y. L.,** User's Manual: Prediction of Transverse Mixing in Natural Streams Model-RIVMIX MK2, Environmental Hydraulics Section, Hydraulics Division, National Water Research Institute Canada Centre for Inland Waters, Ontario, Canada, 1985.

113. **Gowda, T. P.,** Water quality prediction in mixing zones of rivers, *J. Environ. Eng.,* 110(4), 751, 1984.

114. **Gowda, T. P.,** Stream Tube Model for Water Quality Prediction in Mixing Zones of Shallow Rivers, Water Resources Paper 14, Water Resources Branch, Ontario Ministry of the Environment, Toronto, Canada, 1980.

115. **Demuren, A. O. and Rodi, W.,** Side discharge into open channels: mathematical model, *J. Hydraul. Eng.,* 109(12), 707, 1983.

116. **Schaffranek, R., Baltzer, R. A., and Goldberg, D. E.,** A Model for Simulation of Flow in Singular and Interconnected channels: Techniques of Water-Resources Investigations of the U.S. Geological Survey, Book 7, Chapter C3, U.S. Government Printing Office, Washington, D.C., 1981.

117. **Edinger, J. E. and Buchak, E. M.,** 4.0. Waterbody Dynamics, Document No. 85-36-R, J.E. Edinger Assoc., Wayne, PA, 1985.

118. **Edinger, J. E. and Buchak, E. M.,** Introduction to Numerical Waterbody Dynamics, Document No. 84-20-R, J.E. Edinger Assoc., Wayne, PA, 1984.

119. **Barnwell, T. O., Jr., Brown, L. C., and Marek, W.,** Development of a Prototype Expert Advisor for the Enhanced Stream Water Quality Model QUAL2E, Internal Report, Environmental Research Lab., U.S. Environmental Protection Agency, Athens, GA, 1987.

120. **Yotsukura, N. and Sayre, W. W.,** Transverse mixing in natural channels, *Water Resour. Res.,* 12(4), 1976.

121. **Fread, D. L.,** DAMBRK: The NWS Dam Break Flood Forecasting Model, Office of Hydrology, National Weather Service, Silver Spring, MD, 1978.

122. **Brown, J.,** personal communication, class presentation for Rivers and Harbors Engineering, EWRE 345, Vanderbilt University, Nashville, 1978.

123. **Fread, D.L.,** Numerical Properties of Four-Point Finite Difference Equations of Unsteady Flow, U.S. National Weather Service Rep. HYDRO-18, 1974.

124. **DeLong, L.L.,** Extension of the Unsteady One Dimensional Open Channel Flow Equations for Flow Simulations in Meandering Channels With Flood Plains, U.S. Geological Survey Water Supply Paper 2290, U.S. Government Printing Office, Washington, D.C., 1986.

125. **Krenkel, P.A. and Novotny, V.,** *Water Quality Management,* Academic Press, New York, 1980.

126. **Engineering Science, Inc., Hyundai Engineering Co., and Hysong Engineering Co.,** Han River Basin Environmental Master Plan: Vol. I to IV (Water Quality Sector), Environmental Administration, Seoul, Republic of Korea, 1983.

127. **Thomann, R.V.,** Verification of water quality models, *J. Environ. Eng. Div.,* 108(EE5), 923, 1982.

128. **Thomann, R. V.,** *Systems Analysis and Water Quality Management,* McGraw-Hill, New York, 1972.

129. **Wlosinski, J. H.,** Flux use for calibrating and validating models, *J. Environ. Eng.,* 111(3), 272, 1984.

Chapter 2

RIVER TRANSPORT AND SURFACE EXCHANGE PROCESSES

I. MASS BALANCE AND MOVEMENT OF WATER

In Chapter 1, the governing equations were derived. In this chapter, the methods for measuring or predicting flow and time of travel are reviewed in relation to the governing equations. Where appropriate, techniques of measurement are compiled to assist in understanding how the methods are applied

A. USE OF VOLUME AND VELOCITY OF WATER IN WATER QUALITY MODELING

The understanding of how water moves in rivers and what volume of water is contained in various reaches is important because this factor alone can explain much of the observed variation in river quality. The variation of dissolved conservative substances can be adequately predicted if the movement and mixing in the water body is fully understood. For studies of the effect of conventional and toxic pollutants, understanding of water movement in a stream is necessary in almost all cases when tracing contamination to its source. Furthermore, if it is necessary to determine what water quality processes are important factors in determining whether a water quality problem occurs or does not occur, then it is crucial that the speed of water movement downstream be measured or predicted. In addition, the volume or the flow is a crucial component of the waste assimilative capacity of a stream (capacity to dilute or incorporate wastes into the natural geochemical cycle or food chains of aquatic plants and animals).

To solve the one-dimensional form of the mass balance equation for river water quality, the downstream advective velocity, U, must be measured or predicted. Recall that the mass balance or advective-dispersive equation for the concentration of a constituent, C, given earlier in Equation 35 clearly demonstrates the importance of knowing the velocity, U, to simulate water quality. Equation 35 was written as

$$\frac{\partial C}{\partial t} = \frac{\partial (UC)}{\partial x} + D_x \frac{\partial^2 C}{\partial x^2} + S \tag{35}$$

where x is the distance downstream, D_x is the longitudinal dispersion coefficient, and S is the sinks and sources of the constituent (i.e., inflows, decay, production, etc.). For descriptive studies or predictive studies where it is not necessary to predict water movement, time of travel or mean velocity is measured rather than simulated. Both measurement and simulation involve a water balance for the stream segment of interest. In Part 2, the steady-state water balance will be described. Part 3 will review methods of estimating time of travel. Part 4 will discuss methods for measuring discharge for steady-state and dynamic studies. As an alternative to measuring discharge and cross-sectional area, time of travel can be measured directly as described in Part 5. Where prediction of velocity in steady flow is important, the Manning equation, similar empirical equations, or a backwater solution can be used. These are reviewed in Parts 6 and 7. To predict velocities in unsteady flow, empirical equations and the equations derived from the conservation of momentum equation can be employed, and these are reviewed in Part 8. All predictive methods involve some empirical estimation of coefficients (generally the Manning or Chezy roughness coefficients — defined in

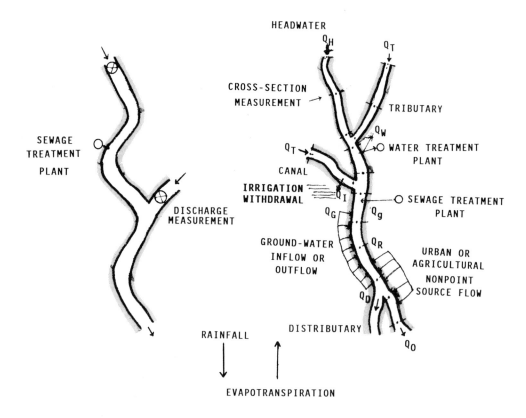

FIGURE 40. Components of a water balance for a river segment or network and minimum number of locations to measure discharge for a simple waste load allocation where nonpoint sources, withdrawals, diversions, rainfall, and evapo-transpiration are negligible.

French[1]). As a result, accurate prediction requires calibration of the flow equation, and methods useful to aid in this will be discussed as well.

B. STEADY-STATE WATER BALANCE

Performing a steady-state water balance for a stream involves measuring flows entering the stream segment or network and measuring the channel geometry at a sufficient number of locations to adequately characterize the water volume and distance along the stream (see Figure 40). At steady state, the sum of the flows entering the stream must equal the flows leaving. Typically, flows are measured entering at the head of the reach or network, entering from tributaries and canals, and entering from large sewage treatment plants. Flows withdrawn for irrigation and drinking water are measured, and usually flows leaving in distributaries can be measured. The flows that cannot be easily measured and can only rarely be accurately estimated (using groundwater or watershed models) include those associated with urban and agricultural run-off and the exchange of water with the adjacent groundwaters. In addition, it is very difficult to estimate water losses due to evapotranspiration. As a result, a number of measurements of flow throughout the stream network are made to define the unmeasurable losses or gains in flow rate.

In this section, some of the means of defining the flow balance and water volume are discussed. In addition, potential sources of the data necessary to make these calculations are given in this section.

The procedure for determining the flow balance can be as simple as measuring the flow into a stream segment and out of a segment to determine the net change in flow rate. For steady-state conditions, the net change in flow rate is zero. As an example, Figure 40 shows the absolute minimum number of discharge measurements necessary for a simple waste load allocation study (study to determine waste assimilative capacity and set discharge controls for sewage treatment design and operation). Recognizing that discharge measurements can be in error, however, it is advisable to at least measure the outflow to confirm the flow balance. Furthermore, for precise studies it is recommended that a series of localized flow balances be made in the vicinity of all inflows such as sewage treatment plant discharges to check the measurements (these tend to be difficult to make) and nonpoint sources to make estimates of the inflow when it is not possible to make such a measurement.

At present, the best method of detecting nonpoint source flows is to measure flow and pollutant loads throughout the reach and make localized flow and mass balance calculations to determine if all flows and loads are measured. Localized flow and mass balances that bracket important point sources are also important because it is usually not possible to be completely sure of the sampling frequency required to adequately measure treatment plant flows and loads.[2] Many treatment plants show significant variation in the flows and loads during daily operations, and these fluctuations are more noticeable in small and moderate-sized streams. Figure 41 shows a typical example of the variations that seem to be difficult to adequately measure in large scale river quality studies. Also shown are several examples where localized flow balances were made to assist in confirming critical flow measurements.

There is currently a lack of criteria to guide decisions about which inflows should be measured and which may be neglected beforehand. It is important to keep in mind that flow from sewage treatment plants (point sources), some polluted tributaries, and urban run-off may be small compared to the receiving flow, but that the total load of pollutants may be comparatively large. It is the total load (product of discharge and concentration) of important constituents that should govern when a tributary, point source, or nonpoint source is included in a modeling study. Because the relative load governs whether or not an inflow should be measured, the designer of a field study is frequently faced with a need to accurately measure tributary and river flows that may be different by several orders of magnitude.

For precise studies, loads entering the stream that are on the order of 1% of the load may be important. Loads on the order of 10% are almost certainly important.

With conservative substances, neglect of small loads results in an accumulation of error of prediction that usually remains negligible over short reaches. Errors in prediction from the neglect of small loads of nonconservative substances such as CBOD (carbonaceous biochemical oxygen demand) is less severe, but the error accumulates in calibration parameters such as the deoxygenation rate coefficient.

The usual practice is to measure all known loads in preliminary studies, use professional judgment to decide which loads to sample, and then check the decisions by performing sensitivity or error anlyses with the calibrated model (see Brown and Barnwell[4] for examples of the best error analyses currently available). Most studies are limited by the available resources that can be mustered for intensive sampling. Therefore, the decision regarding which tributaries should be sampled is an important component of planning a study. As a backup, it is possible to measure the river load at a number of points to be able to infer any neglected loads or unknown loads if the error analysis shows that the decisions on what loads should be sampled were not well taken.

C. TIME OF TRAVEL AT STEADY STATE

If water quality kinetics are to be accurately calibrated, the time over which the reactions occur is important. The time available for the water quality reactions to

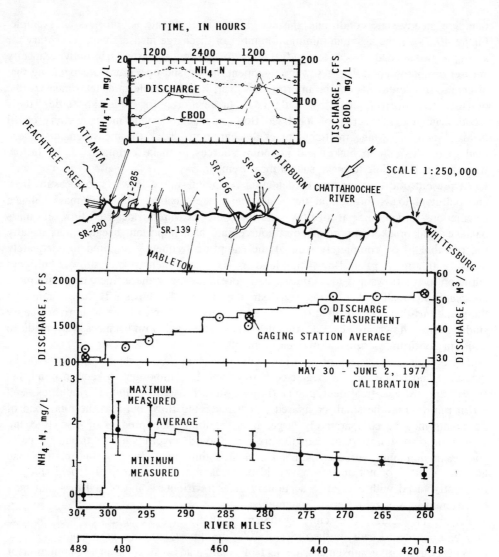

FIGURE 41. Distribution and variation of ammonia-nitrogen (NH$_3$-N) in the Chattahoochee River between Atlanta and Whitesburg, Georgia, May 31 to June 2, 1977. Note the variation of flow and ammonia entering from the R.M. Clayton Sewage Treatment Plant.[3] The single arrows entering the stream are tributaries, and the double arrows are treatment plant discharges or water withdrawals. Flow measurements are given for the inflows and instream flows in Table 7. Note the frequency of instream measurement and the measurement upstream and downstream of several inflows. Frequent measurement in the vicinity of Atlanta confirms that nonpoint flows are negligible. Time of travel in the reach of 1.5 d minimizes evapo-transpiration losses. The study was conducted during the dry season when rainfall and groundwater exchange are also negligible.

proceed is the time it takes water to move through the stream segment of interest. For steady-state conditions, the time of travel, t, can be derived from the continuity equation where t = AL/Q. Q is the stream discharge; A is cross-sectional area; and L is the length of the stream segment. Also note that Q/A = U. Therefore, to determine time of travel in a stream, channel geometry and flow must be predicted or measured at critical points (tracer studies are an alternative method to determine time of travel that will be discussed later).

It is obvious from the continuity equation that errors in discharge, area, or length lead directly to errors in time of travel. Therefore, errors of measurement in discharge, cross-sectional area, and distance along the stream can be important in calibrating kinetic coefficients for water quality that are governed by the estimated time of travel. As a result, cross-sections should be representative of the reach in which they are measured. For large rivers with approximately uniform channels, cross-section measurements can be spaced over several kilometers. In smaller pool-and-riffle streams, spacing on the order of meters to tens of meters to define the transition from pools to riffles may be necessary. Close spacing of cross-sections is necessary to adequately predict both the average depth and average velocity in reaches. This will be discussed further in the section on the Manning equation. That section presents one tentative method for determining how frequently cross sections should be measured.

The tentative criteria to guide how closely cross-section measurements should be spaced, given later in the section on the Manning equation, require *a priori* knowledge of the stream profile. Generalizations of stream geomorphology have not yet been fully pursued to the point where predictions can be made of channel nonuniformity.[5] Until this is possible, it will probably be necessary to chose cross-section spacing by trial and error and add additional sections if necessary to achieve greater precision in defining reaches.

Formerly, it was recommended[6] that cross-sections for rivers be measured at approximately every 150 m (500 ft), especially if it was possible to navigate the stream by boat and record the section shape with a depth sounder. For example, McKenzie et al.[7,8] followed this recommendation in a study of the Willamette River in Oregon and measured cross sections every 160 to 320 m (520 to 1050 ft), depending on the irregularity of the channel. More recent experience on the same river using cross-sections from flood studies spaced at greater distances (20 m [60 ft] at bridges to 4800 m [16,000 ft] in long uniform reaches) seems to indicate, however, that much greater spacing is normally more than adequate[2] in large rivers.

Presently, stream cross-section measurements are not recorded in readily available data bases. However, there are several sources that should be explored before planning surveys. First, the Federal Emergency Management Agency (FEMA) has commissioned surveys of most U.S. rivers that are near urban encroachments onto the floodplain. Since many pollution problems occur near urban areas, generally some useful information about channel morphology may be available from the regional FEMA office or the FEMA study contractor. Many such studies are on file with the state U.S. Geological Survey office or district U.S. Army Corps of Engineers office since they have performed a number of these studies for FEMA. These flood studies define high-flow sections. However, experience has shown[2] that these data can be more than adequate to define the cross sections of lage, stable, approximately uniform rivers and can be easily supplemented with additional data collected during field studies designed primarily to collect water quality data.[9] Second, the National Weather Service has collected data in flood-prone areas as well. These data may be more extensive and cover some of the same areas as the FEMA data. The National Weather Service collects cross-section data to make timely flash flood predictions in populated areas. Third, the Corps of Engineers conducts special studies on various rivers. These generally involve navigation studies on larger rivers where the navigation charts, like the coastal and estuarine charts of the National Ocean and Atmospheric Administration, may conservatively report depth (i.e., reported depth may be shallower than the actual depth). Fourth, the U.S. Geological Survey also conducts special studies and keeps extensive data on file. The U.S. Geological Survey also keeps on file extensive data collected at stream gaging sites. These data, however, should be used with strict caution because gaging stations are most often located at nonrepresentative sites where there exists a unique relation between river stage (depth) and flow.[10] Fifth, the Soil Conservative Service, regional agencies such as Tennessee Valley Authority and the Bureau of Reclamation, and state and local agencies may also keep on file morphological data. The Soil Conservation Service generally studies upland streams. The Tennessee Valley Authority and the Bureau study larger rivers and some small streams. State agencies such as the state geological survey compile morphological data from flood and water supply studies, and state pollution control agencies compile data from previous water quality studies. Cities and local agencies occasionally commission flooding, water supply, and pollution surveys and maintain the data on file. Finally, the U.S. Environmental Protection Agency regional offices occasionally conduct stream water-quality surveys from which data are available, and some screening level data on stream channels is available from the U.S. Environmental Protection Agency Headquarters in Washington, D.C. (U.S.) in what is known as the Reach File (a data base of information about a number of U.S. stream reaches).

If flow data are available for the reach of interest, it has almost certainly been collected by the U.S. Geological Survey. Unlike channel morphology data, flow data are readily available from the national data base, WATSTORE, that is maintained by the U.S. Geological Survey. Some, if not all, of the discharge data are transferred to the U.S. Environmental Protection Agency national data base, STORET. Edwards[11] describes the procedure for locating the nearest state office where the data can be obtained for a limited fee. The U.S. Geological Survey Office of Water Data Coordination may be able to provide information on the access to the data network from remote terminals. Limited flow measurements may be collected by other agencies such as the Tennessee Valley Authority, but these will be difficult to locate.

D. MEASUREMENT OF DISCHARGE OR VELOCITY
1. Discharge Measurements in Rivers, Streams, Channels, and Pipes

Measurement of average velocity or flow in a large stream or river is a straightforward task that requires a fairly complex field study to make point velocity measurements or to trace the movement of dye or a similar tracer along the stream. The U. S. Geological Survey[10,12] and the Bureau of Reclamation[13] have well established procedures to guide both types of measurements, and these procedures should be followed to insure accurate results. For smaller streams, the Bureau of Reclamation,[13] the U. S. Geological Survey,[10] the U.S. Environmental Protection Agency,[14] and the American Society for Testing and Materials[15] have well established methods for measuring flow. These methods are listed in Appendix II.

In addition, to the standard techniques, Appendix II also lists a number of less frequently used techniques that are compiled for the purpose of assembling in one place the flow measurement techniques that may be used in stream water quality studies. This is important to aid in preliminary studies where the infrequently used, imprecise methods may be important to estimate flows. It is also important to compile the methods that may be useful for nonpoint source studies. At present, it is very difficult to fully measure or even attempt to measure nonpoint sources.

In the remainder of this section, the most important flow measurement techniques will be reviewed. These methods include several techniques that are useful in determining discharge rates for a steady flow balance of a stream. Of the methods available, the standard current meter method is most widely used. The moving boat, acoustic, and electromagnetic methods are useful for defining inflow and outflow hydrographs for dynamic flow conditions. Dye dilution methods are useful for shallow and inaccessible flows. There are also several methods that are useful for measuring

STREAM CROSS SECTION

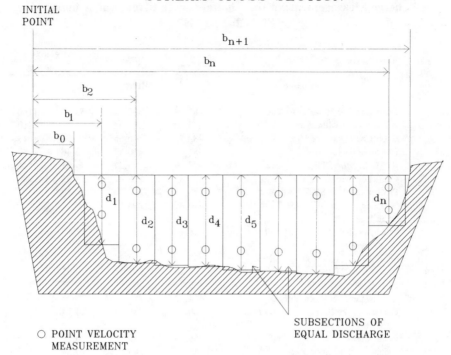

FIGURE 42.   Method to measure average velocity.

point source and tributary flows into a stream as well. Parshall flumes and weirs are frequently used to measure sewage treatment plant flows, and it is important to be able to determine if these flow devices are correctly measuring flow into a stream of interest. Furthermore, there are a number of flow measuring devices used to measure irrigation flows that must be understood for water studies in the arid western U.S.

2. Velocity-Area Methods

The direct measurement of average stream velocity by the most frequently used standard method involves point velocity measurements in at least 20 to 30 subsections in the cross-section as illustrated in Figure 42. Measurements in the subsections generally include at least two current meter measurements to characterize the subsection average. These 40 to 60 current meter measurements are appropriately averaged to determine the average velocity and discharge for the stream of interest. Table 8 summarizes the procedure involved.

Frequently, steps 3 and 5 are combined when the depth and velocity profile are measured simultaneously. In shallow rivers that can be waded, a current meter is attached to a sounding rod. In deeper rivers, the current meter is lowered into the water from a bridge or boat by a cable and held in place by a streamlined, bomb-shaped weight attached below the current meter.

Table 9 lists the current meters typically used to measure point velocities. While horizontal-axis meters are favored in Europe,[10] U.S. and Canadian engineers most frequently use the small Price (AA) current meter shown in Figure 44. In China both types are manufactured and used.[20] The vertical-axis meter better avoids bearing wear in sediment-laden flows that are more frequently encountered in the U.S. and China. Now with the introduction of solid plastic rotating cups, the Price meter better withstands

TABLE 7
Discharge Measurements of the Chattahoochee River near Atlanta, May 30 to June 2, 1977

Site no.	Station name	Location[a] River km	Location[a] River mi	Measured discharge (m³/s)	Measured discharge (ft³/s)
1	Chattahoochee River at Atlanta gage	487.58	302.97	32.56	1150
2	Atlanta water withdrawal	483.80	300.62	−3.11	−110
3	Cobb Chattahoochee STP[b]	483.70	300.56	0.37	13
4	Peachtree Creek	483.64	300.52	2.46	87
5	R. M. Clayton STP	483.19	300.24	4.25	150[c]
6	Chattahoochee River at SR[d] 280	480.82	298.77	34.55	1220
7	Proctor Creek at SR 280	478.78	297.50	0.20	7
8	Nickajack Creek at Cooper Lake Road near Mableton	474.97	295.13	0.59	21
9	Chattahoochee River at SR 139 near Mableton	474.19	294.65	37.10	1310
10	South Cobb Chattahoochee STP	473.60	294.28	0.42	15
11	Utoy Creek STP	469.28	291.60	0.51	18
12	Utoy Creek at SR 70	469.24	291.57	0.37	13
13	Sweetwater Creek near Austell	464.42	288.58	6.06	214
14	Chattahoochee River at SR 166 near Ben Hill	460.39	286.07	44.74	1580
15	Camp Creek STP	456.70	283.78	0.25	9
16	Camp Creek at Enon Road	456.31	283.54	0.54	19
17	Deep Creek at SR 70 near Tell	455.83	283.27	0.59	21
18	Chattahoochee River at SR 92 near Fairburn	453.64	281.88	45.31	1600[c]
19	Annewakee Creek at SR 166	452.98	281.47	0.65	23
20	Pea Creek at SR 70 near Palmetto	446.43	277.40	0.57	20
21	Bear Creek at SR 166 near Douglasville	444.10	275.95	1.30	46
22	Chattahoochee River above Bear Creek near Rico	443.87	275.81	48.14	1700
23	Bear Creek at SR 70 near Rico	441.75	274.49	0.45	16
24	Dog River at SR 166 near Fairplay	440.09	273.46	1.87	66
25	Chattahoochee River at Capps Ferry Bridge near Rico	436.44	271.19	51.25	1810
26	Wolf Creek at SR 5 near Banning	430.24	267.34	0.54	19
27	Chattahoochee River at Hutcheson's Ferry near Rico	427.54	265.66	51.54	1820
28	Snake Creek near Whitesburg	421.20	261.72	1.16	41
29	Cedar Creek at SR 70 near Roscoe	420.44	261.25	0.85	30
30	Chattahoochee River at US Alt. 27 near Whitesburg	418.18	259.85	52.10	1840

Note: Site numbers correspond to the arrows in Figure 41 and can be identified from the river miles given.

[a] Location upstream of the mouth of the Chattahoochee River at its intersection with the Flint River where discharge is measured or where a tributary or point source enters the stream.

[b] STP = Sewage treatment plant.

[c] Average flow. See Figure 41 for temporal variation over the period of interest.

[d] SR = State route.

[e] Average from gauging station records. One instantaneous measurement of discharge during the study was 43.04 m³/s (1520 ft³/s).

TABLE 8
Procedure for Measuring Average Velocity and Discharge (Velocity-Area Method)

Step	Procedure
1.	Select a cross-section from a straight, uniform reach with parallel streamlines and a relatively uniform bottom that is at least 0.15 m (0.5 ft) deep, that has velocities of at least 0.15 m/s (0.5 ft/s), and where there is easy access from cableways, bridges, or by wading (otherwise measure from boats). If possible the section should be free of large eddies with upstream circulation near the banks, areas of slack water, or excessive turbulence caused by upstream bends (Figure 43), radical changes in cross-section shape, and irregular obstructions such as boulders, trees, vegetation, and other debris in the vicinity.[10]
2.	Choose a time of measurement such that the discharge is steady or approximately steady during the period of measurement that usually ranges from 1 to 3 h depending on the size of the river. If flow changes rapidly, short-cuts in the method will be necessary.[10]
3.	Measure the cross-sectional area, A, by measuring depth with a sounding line or wading rod and width with hand lines or tapes.[10] In large rivers, electronic depth sounders and triangulation with transits or laser distance measuring equipment are used.
4.	Divide the section into at least ten subsections based on the expected distribution of discharge over the section. Typically, 20 to 30 sections are required for precise measurements.[10]
5.	Measure the vertical velocity profile in each subsection.
6.	Compute average velocity in each subsection, u_i, from the profile.
7.	Compute the subsection discharge, q:[10]

$$q_i = u_i \left[\frac{(b_i - b_{i-1})}{2} + \frac{(b_{i+1} - b_i)}{2} \right] d_i$$

8.	Compute the discharge from $Q = \Sigma q_i$.
9.	Compute the average velocity from Q and cross-sectional area, A, $U = Q/A$.

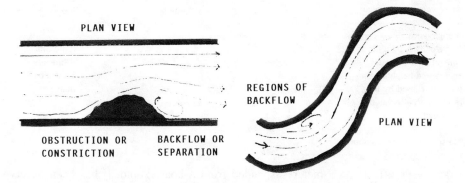

FIGURE 43. Flow separation areas to avoid in measuring flow.

damage and does not require recalibration. The inexpensive cups are precisely manufactured, and selected samples are tested to determine that the standard calibration is adequate for each batch. When the cups are damaged or worn, the cups are easily and quickly replaced in the field unless the yoke assembly is also damaged or worn.

French[1] estimates that Price current meters are accurate to within 2% for streams with a minimum of turbulence and currents parallel to the channel axis. However, Carter and Anderson[21] show that standard errors of 1% can be be expected for about two-thirds of the measurements. Similarly, A. Ott Kempten[16,17] indicates that measurements with an Ott meter are accurate to within 1%. Rantz et al.[10] note that the Price current meter under-registers when the meter is positioned near a vertical right bank (facing downstream), the water surface, and the stream bottom. The meter over-registers when it is positioned near the left bank. No other direct effect of turbulence

TABLE 9
Typical Methods of Measuring Velocity at a Point in a River

Physical principle involved	Current meter	Range of velocity		Minimum depth	
		m/s	ft/s	m	ft
Proportionality of water velocity and resulting angular velocity of the meter rotor.[10]	Horizontal Axis -Ott (Germany)				
	C 2 Small[16]	0.024—2.0	0.08—6.6	0.04	0.13
	C 31	0.024—10.1	0.08—33	0.49	1.6
	-Universal[17]	—	—	—	—
	-Neyrpic or Dumas (France)	—	—	—	—
	-Haskell (U.S.)	—	—	—	—
	-Hoff (U.S.)	—	—	—	—
	-Braystoke (UK)	—	—	—	—
	Vertical Axis (USGS, Can.)				
	-Small Price	0.0085—2.44	0.028—8.0	0.3	1.0
	-Pygmy Price	0.0085—0.91	0.028—3.0	0.18	0.6
	-Ice Meter	0.15—2.44	0.5—8.0	0.3	1.0
Voltage induced by water moving through a magnetic field perpendicular to the flow is proportional to water velocity.[1]	Electromagnetic -201 Marsh-McBirney	0.02—6.1	0.07—20	0.23[a]	0.75[a]
Floats with various levels of submergence move at the water velocity averaged over the depth of submergence.[18]	Floats	Does not seem to be any practical limitations that define a range.		Should be applicable to any flow.	
The velocity head $u^2/2g$ is converted to potential energy and height or the pressure caused by the resulting fluid column is measured.	Pitot tube	—		Approximately 4 times the Pitot tube diameter.	

[a]   Estimated by this author.

(surface wave effects on a meter suspended from a boat excluded) has been observed. One advantage of the Ott meter is that the measurement is less effected by oblique currents like those found next to channel walls. Based on the experience of the U. S. Geological Survey,[21] one may expect to measure flow with the Price, Ott, or electromagnetic current meters to within 2% in many cases.

Average velocities are determined from the vertical velocity profile by the vertical-velocity curve method, two-point method, 0.6 depth method, three-point method, and by less frequently used and less accurate methods.[10] The one-, two-, and three-point methods are less precise approximations of the vertical velocity profile method. The vertical-velocity curve method involves making a number of velocity measurements over the depth and averaging by the appropriate methods. The best estimates result when the measurements are closely spaced near the bottom where the velocity gradient is greatest. However, for convenience, velocity is usually measured at nine locations evenly spaced over the depth as shown in Figure 45. When the measurement locations are not equally spaced, the average is usually weighted by the relative distance between the measurements such that

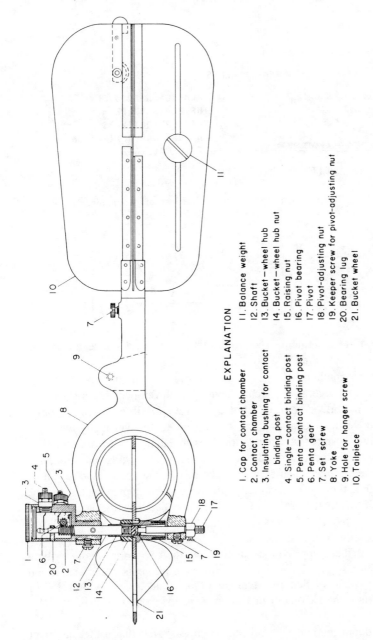

EXPLANATION

1. Cap for contact chamber
2. Contact chamber
3. Insulating bushing for contact binding post
4. Single—contact binding post
5. Penta—contact binding post
6. Penta gear
7. Set screw
8. Yoke
9. Hole for hanger screw
10. Tailpiece
11. Balance weight
12. Shaft
13. Bucket—wheel hub
14. Bucket—wheel hub nut
15. Raising nut
16. Pivot bearing
17. Pivot
18. Pivot-adjusting nut
19. Keeper screw for pivot-adjusting nut
20. Bearing lug
21. Bucket wheel

FIGURE 44. Diagram of type-AA price vertical axis current meter used by the U.S. Geological Survey and others.[19]

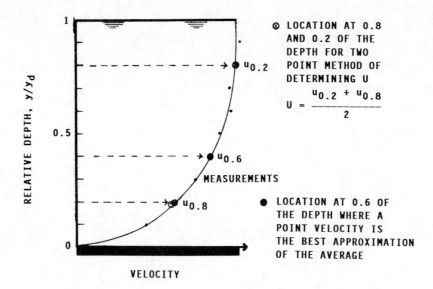

FIGURE 45.   Typical river vertical velocity profile and method of determining the vertically averaged velocity.

$$U = \frac{1}{y_d} \left\{ u_1 \left[\frac{y_2 - y_1}{2} + y_1 \right] + \sum_{i=2}^{n-1} u_i \left[\left(\frac{y_{i+1} - y_i}{2} + y_i \right) \right. \right.$$

$$\left. \left. - \left(\frac{y_i - y_{i-1}}{2} + y_{i-1} \right) \right] + u_n \left[y_d - \left(\frac{y_n - y_{n-1}}{2} + y_{n-1} \right) \right] \right\} \qquad (50)$$

where n is the number of measurements over the vertical, y_i is distance above the bottom, and u_i is the velocity measured at a point.

The vertical-velocity curve method has proven to be a time consuming procedure for most applications. As a result, several short cuts have been developed that are usually accurate[10] and highly useful. Accurate shortcuts are possible when velocity measurements at selected vertical locations can be related to the vertically-averaged velocity. Rantz et al.[10] give the most useful relationships in Table 10 that are compiled from the observational experience of the U. S. Geological Survey. In addition, the relationship between selected point velocities and the vertically averaged velocity can be confirmed by assuming that the vertical velocity profile can be approximated by the parabolic[22] or logarithmic[1] velocity profile. The difference between the accumulated observations of the U. S. Geological Survey and predictions based on the parabolic or logarithmic profiles are relatively minor.

The most widely used of these approximate methods is the two-point method where the vertically averaged velocity is the average of point measurements made at 0.2 and 0.8 of the depth. Table 10 shows that the average of $u_{0.2}$ and $u_{0.8}$ is within 1% of expected mean.[10] In addition, the two-point method also seems to be useful in partially full pipe flows.[23]

Also useful is the 0.6 depth method where it has been observed that a measurement at 0.6 of the depth is approximately equal to the average for most river and stream flows. The one-point approximation can be expected to be less precise (e.g., more error prone) than the two-point method. Table 10 indicates that this method is accurate to within 2%. Carter and Anderson[21] show that the standard error is approximately doubled for the 0.6 depth method compared to the two-point method as one would expect when an estimate is based on one half as many measurements. As a result, this

TABLE 10
Relationship between Point Velocities and
Vertically Averaged Mean Velocities

Relative depth (from surface)	Ratio of point velocity to vertically averaged velocity
0.05	1.16
0.1	1.16
0.2	1.15
0.3	1.13
0.4	1.11
0.5	1.07
0.6	1.02
0.7	0.95
0.8	0.87
0.9	0.75
0.95	0.65

From Rantz, S.E. et al., Measurement and Computation of Streamflow, Vols. 1 and 2, U.S. Geological Survey Water-Supply Paper 2175, U.S. Government Printing Office, Washington, D.C., 1982, 81.

method is only used when expediency is required, when the pygmy current meter is used in depths of flow of 0.09 to 0.46 m (0.3 to 1.5 ft) or the Price current meter is used in depths of 0.46 to 0.76 m (1.5 to 2.5 ft), and when velocity measurements at 0.2 ($u_{0.2}$) and 0.8 ($u_{0.8}$) of the depth are not possible. It may not be possible to accurately measure $u_{0.2}$ when waves, overhanging trees, or other obstructions affect the vertical velocity profile (see Figure 46). In the field, a check to see if the ratio $u_{0.2}/u_{0.8}$ is approximately 1.3 will show whether additional measurements are necessary. In some measurements from bridges, it may not be possible to lower a suspended meter and weight to 0.8 of the depth because the flow is too shallow or too swift.[10] It has also been noted that this method may be applicable in partially full pipe flow.[23]

The two-point method can be derived by assuming that the river flow is a uni-directional boundary-layer type that can be described by the logarithmic velocity profile[1] (see Figure 46):

$$u(y) = \frac{u_*}{\kappa} \ln\left[\frac{y_d}{y}\right] + \frac{u_*}{\kappa} + U \qquad (51)$$

where u_* is the shear velocity [$u_* = (gRS)^{1/2}$, g is gravitational acceleration, R is the hydraulic radius = area/wetted perimeter and is approximately equal to the depth, y_d, in wide rivers and streams, and S is the slope of the energy grade line that is equal to the slope of the water surface and stream bottom in uniform flows]; κ is the von Karman universal turbulence constant that has a value of 0.41;[36] and y is the distance above the bottom boundary. French uses Equation 51 to show that

$$U = \frac{u_{0.2} + u_{0.8}}{2} \qquad (52)$$

is approximately valid. This theoretically supports the extensive observations of the U.S. Geological Survey[10,22,37] that Equation 52 is valid. Equation 51 can also be used to show that the average velocity in the vertical is equal to the single measurement, $u_{0.56}$

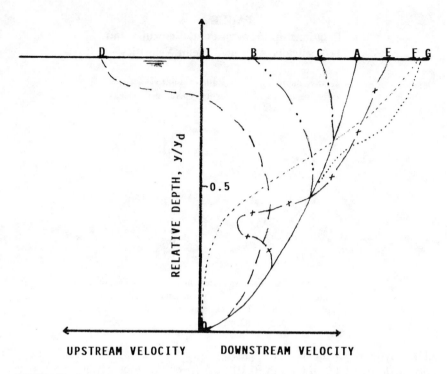

FIGURE 46. Regular and irregular velocity profiles observed in streams. (A) Logarithmic profile.[24] (B) Profile in a highly stratified flow where the boundary effect is magnified.[25,26,27] (C) Effect of weak secondary circulation on the profile.[5,10] (D) Profile caused by upstream flow of a cooling water discharge into subcritical flow or by a turbidity current flowing into a deep flow.[28,29] (E) Profile of a stratified flow in a bend.[30,31,32] (F) Profile upstream of an obstruction.[33,34] (G) Profile caused by incomplete mixing of a turbid tributary.[35]

(velocity at 0.56 of the depth), or for convenience in the field, approximately equal to $u_{0.6}$ (velocity at 0.6 of the depth), as assumed in the one-point method.

The practical result of being able to theoretically derive expressions for the one-point and two-point methods is that this provides guidance on when these time-saving methods may be inaccurate and indicates when the methods can be properly applied. First, it seems that weak secondary circulation and the irregularity in channel cross-section that contributes to these weak cross currents should not influence the choice of a measuring section for most field conditions except where sharp bends occur. This follows from the fact that Equation 52 is derived from field measurements that include some influence of secondary circulation and from theoretical velocity profiles (Equation 51) that do not. Therefore, reasonably accurate measurements do not necessarily require long straight approach flows with a uniform cross-section. Some irregularity and limited sinuosity may not affect velocity measurements.

Secondary circulation involves cells of spiraling flow in a channel in which there is a weak current acting perpendicular to the axis of the main downstream flow. These currents arise in straight channels because of the unequal shear along the wetted perimeter. Secondary circulation is most recognizable when the maximum longitudinal velocity in any section does not occur at the water surface. See Figure 47 and refer to French[1] for a fuller description that includes the distinction between the similar strong secondary circulation that occurs in channel bends.

Second, because the logarithmic velocity profile describes uni-directional boundary-layer flow, it can be noted that the flow should be reasonably straight and well behaved. Bends, constrictions, and obstructions can cause flow separation and significant

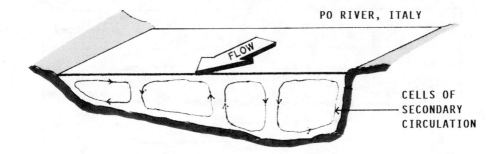

FIGURE 47. Typical secondary currents in rivers.[38]

eddies with excessive turbulence and upstream flow as shown in Figure 43 that is difficult to take into account. In addition, obstructions can cause backwater effects that modify the upstream velocity profile. Furthermore, highly irregular and turbulent flow can result which may require much longer measurement periods to adequately measure the true velocity at a point. Profiles B and C in Figure 46 can result when trees, debris, ice, and strong upstream wind currents affect the flow. Profiles like F result from boulders and obstructions on the bottom. In such cases, the two-point and 0.6 depth methods can be expected to be inaccurate.

Finally, the logarithmic profile does not include the stratifying effect of sediment, dissolved solids, or heat. McCutcheon[25] notes that stratification magnifies the effect of the side boundaries. Majeswski[27] observed the profile shape labeled B in Figure 46 downstream of the discharge point of a thermal power plant and in a model study as well. In addition, Parker and Krenkel[28] have observed upstream flows (like profile D) of heated water when the flow is subcritical $[U/(gy_d)^{1/2} < 1$, see French[1] for a definition of subcritical flow], and Wan[35] has observed a quite different effect when the flow is strongly stratified by suspended sediment (e.g., profile G). The most irregular profiles (e.g., profile E) seems to occur for stratified flows in bends. When any of these types of velocity profiles occur, more extensive velocity profile measurements are required to define the vertical average velocity.

When the measurement of $u_{0.2}$ seems inaccurate or the velocity profile is expected to be irregular because of channel alignment or other factors, it is useful to measure the average velocity by the three-point method. In this case, the 0.6 depth and two-point methods are combined to provide a more precise estimate of average velocity:

$$U = 0.25u_{0.2} + 0.25u_{0.8} + 0.5u_{0.6} \tag{53}$$

Marsh-McBirney[23] also recommends a three-point method for partially full pipe flow that is different in form in that

$$U = \left(\frac{u_{0.2} + u_{0.6} + u_{0.8}}{3} \right) \tag{54}$$

This formula seems to be based on observational experience involving partially full pipe flows where secondary circulation is more severe. At present, however, it is not clear why Equation 53 is not also applicable to partially full pipe flows.

In cases where irregular profiles like those in Figure 46 occur, Table 10 will not be valid. Irregular profiles must be defined by the vertical-velocity curve method. If necessary, a site-specific and flow-range-specific version of Table 10 can be constructed.

There are other less accurate methods that generally use Table 10 to relate single

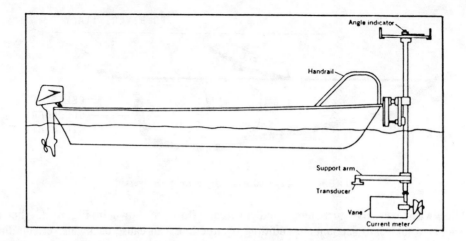

FIGURE 48.   Equipment used to measure average velocity by the moving boat method.[10]

point velocities at the surface, near the surface, or at 0.2 of the depth to the mean velocity. These methods are used when it is not possible to immerse the current meter to the proper depth during high flow or as part of special procedures like the moving boat method. In addition, Rantz et al.[10] review other methods in use in Europe.

The moving boat method involves continuously measuring velocity and depth across a large stream. A current meter is positioned at a constant depth below the surface. Depth is continuously measured with a depth sounder. Figure 48 illustrates the boat and equipment required. This method is ideal for flows that are too unsteady to permit a standard measurement over a 1 to 3 h period and when the standard method would be too tedious and costly.[10] This method should be particularly useful when a number of measurements are required to define an inflow hydrograph for the boundary condition for an unsteady flow model of a large river.

During the boat trip across the stream, about 30 to 40 velocity measurements of the combined water and boat velocity and the angle of the current meter with the current are recorded. To accurately measure velocity, the selection of a straight section of relatively uniform flow deep enough to permit the passage of a small boat is even more important than for the standard stream gaging method. The near-surface velocity measurements are more sensitive to secondary circulation in the channel than other types of measurements.

Velocities measured as part of the moving-boat procedure are best measured with a horizontal-axis type propeller meter, such as the Ott universal current meter. The horizontal-axis meters do not overregister velocity like the Price current meters (and other vertical axis meters as well) when surface waves cause the boat and attached meter to move up and down. The current meter is mounted at a depth of 0.9 to 1.2 m (3 to 4 ft) below the surface. The method assumes that the 30 to 40 velocity measurements are related to the vertical average in each section by the same constant. Measurements on the Mississippi River at Vicksburg and St. Louis, on the Hudson at Poughkeepsie, and on the Delaware at Delaware Memorial Bridge indicate that the proper correction to relate the velocity measured near the surface to the vertically averaged velocity is 0.90 to 0.92. For moderate rivers of depths exceeding 3 m (10 ft), Carter and Anderson[21] found from measurement at 100 sites that velocities recorded at a depth of 1.2 m (4 ft) should be multiplied by 0.9 to derive the vertically averaged mean.[10] Furthermore, these observations are somewhat consistent with the ratio of $1/1.15 = 0.87$ in Table 10 that is based on 40 measurements.

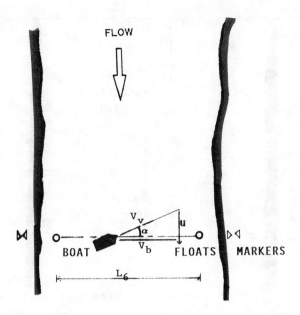

FIGURE 49. Moving boat method. Note the perpendicular boat path across the flow. The vector diagram relates the observed velocity vector, V_v, and current meter angle with the flow, α, to the river velocity at the location of the current meter. V_b is the velocity of the boat across the river.[10]

Discharge calculations are based on a modification of the method in Table 8. The point velocity measured at locations spaced across the stream is derived from observations of the velocity vector, V_v, and the angle of the current meter with the flow, α, according to the vector diagram in Figure 49. Point velocities are related to V_v and α as follows

$$u = V_v \sin(\alpha) \tag{55}$$

where u multiplied by 0.9 is the subsection average velocity used in step 7 of the procedure in Table 8 to compute discharge. Similarly, the distance across the stream, L_6, (required to compute cross-section area) is related to the recorded distance, L_v, multiplied by $\cos(\alpha)$. The total width is L_6 added to the distance from each float to the nearest shore.[10]

Acoustic discharge measurements, that are also useful for defining inflow hydrographs, are similar in concept to the moving-boat method but require a permanent installation. The advantage of this method is that rapid changes in discharge are more likely to be recorded.

The acoustic discharge method involves projection of sound waves across the flow and back. The difference in travel time of the sound waves along a diagonal path across the flow and back as shown in Figure 50 is related to the average velocity of the water crossing the acoustic path. Like the moving boat method, the average river velocity along the acoustic path must be related to the cross-sectional average. This is done by relating the acoustically measured velocity to the average measured by the standard current meter method given in Table 8 at times when the flow rate is not rapidly changing. Because this technique is not frequently applied in water quality studies, the reader should refer to Rantz et al.[10] and Laenen and Smith[39] to determine equipment setup and methods of calculation if it becomes necessary to consider its use.

Electromagnetic flow measurements are made in two ways (in addition to using an

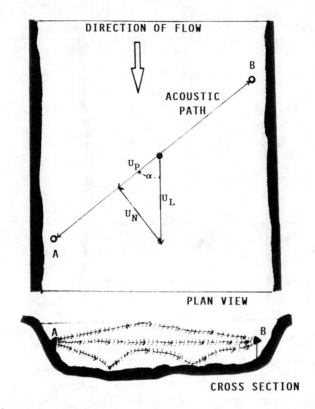

FIGURE 50. Diagonal relationship of the acoustic path between transducers and responders at locations A and B and the river flow. U_L is the average river velocity along the acoustic path; U_p is the velocity along the path; U_N is the velocity vector normal to the path; and α is the angle between the path and the river flow. The acoustic path may bend because of density gradients caused by temperature and salinity or may be reflected by the bottom and water surface. The most direct path will only be taken when vertical and lateral gradients do not exist in the flow.[39]

electromagnetic probe in place of the Price current meter for stream gaging). In highly irregular flows like those in tidally affected river reaches, an electromagnetic current meter is mounted in the flow and that point velocity is related to the cross-section average using the standard area-velocity method or moving-boat method.

For some small streams it is possible to bury electromagnetic coils in the stream bed. The movement of water through the magnetic fields around the probe or coil at 90° induces a voltage that is directly proportional to water velocity.[10] Probe measurements are accurate to the larger of 2 to 3% or 0.0002 m/s (0.007 ft/s). Measurements by the standard method given in Table 8 are used to relate velocity measurements to the cross-sectional average velocity. Electromagnetic probes and coils are especially useful when debris or heavy sediment loads would damage Price current meters, and occasionally it might be useful to substitute the probe for a price current meter when making standard gaging measurements.

Not only are standard gaging methods limited when the flow is varying rapidly, but these methods are also limited to depths of flow deep enough to allow current meter measurements. The information in Table 9 indicates that this means that flows without a suitable measurement section deeper than 0.15 m (0.5 ft, assuming that a pygmy current meter is available) require a different approach. The approach that has proven most successful for measuring discharge in shallow flows is the dye-dilution technique.

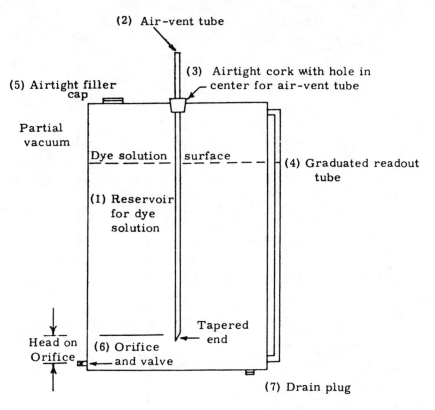

FIGURE 51. Mariotte vessel for injecting dye solution at a constant rate.[41]

3. Dye-Dilution Methods

The dye-dilution technique is most useful in flows that have good mixing but are too shallow or otherwise not easily accessible for current meter measurements. This may also include flows under ice where it is too difficult or too dangerous to go onto the ice and in pipes such as sewers and penstocks where access is severely limited.

There are at least two dye dilution techniques that have been developed. These include the constant injection method and the pulse or slug injection method. The constant injection method is more frequently used for stream measurements.

Of the two dye methods used to determine discharge described in Table 11, the constant injection method is the least error prone. Appendix II and Morgan et al.[45] indicate that errors of 2 to 3% can be expected. These errors arise from sample contamination, incorrectly prepared standards, incorrectly measuring the rate of dye injection, sampling before steady state concentrations are achieved, sampling too closely downstream where the dye is not laterally well mixed, losses of tracer due to sorption or photochemical decay, inadequate measurement or variation of the background fluorescence, recording the wrong readings, malfunction of equipment, failure to compensate for temperature effects, and computational mistakes. In addition to the errors that also affect the uncertainty in measuring discharge by the constant injection method, the slug injection method suffers from additional errors caused by errors in measuring the total volume injected, from infrequent sampling of the dye cloud to define the total mass of dye from the time-concentration curve for slug injections, and from inadequate sampling of the trailing edge of the dye cloud caused by diffusion and dispersion into nonflowing or very slowly moving dead zones and eddies in the channel and in the adjacent groundwaters.[40] Because the constant injection method minimizes

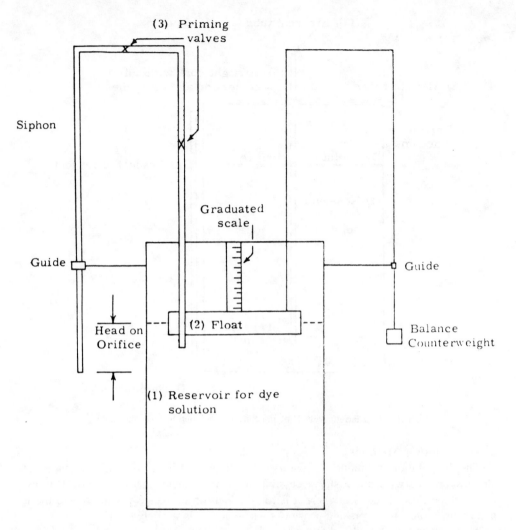

FIGURE 52. Floating siphon for injecting dye solution at a constant rate.[41]

potential for sampling errors and dye losses, it is more frequently used. Furthermore, the constant injection method is the only practical method for determining unsteady flow rates. Incomplete mixing can be more easily determined. In addition, the calculations for the constant injection method are less involved. However, more equipment and longer time in the field is required for the constant injection method.[40]

Losses of dye or interferences which cause measurement errors are caused by dye sorbing to benthic or suspended solids, failure to compensate for the decrease in fluorescence with increased temperature, failure to account for the decrease in fluorescence in highly acidic streams, photochemical decay, and chemical quenching.

Rhodamine WT is the best fluorescent dye for discharge measurements. Overall it best approximates an ideal tracer that is water soluble, easily detected, free of background interference, stable, harmless to man and wildlife, and inexpensive.

There are other methods as well for measuring river discharge that may occasionally be useful for stream surveys. These are reviewed in Appendix II. In addition to the acoustic and electromagnetic methods, there are several methods used to measure flows entering a stream from pipes and small channels.

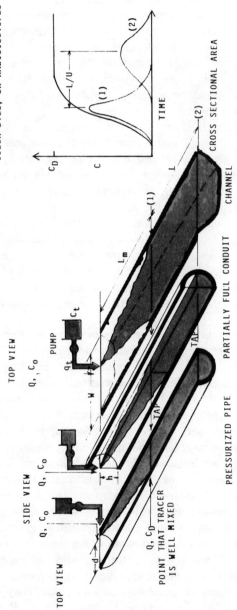

FIGURE 53. Dye injection into channels and pipes. Discharge is measured by two methods. For a slug injection [see dye curves (1) and (2)], discharge is determined by the time of travel over the distance, L, between stations (1) and (2) and the cross-sectional area of the flow. For a continuous injection, discharge is determined from a plateau concentration, C_D, at station (1) where the dye is well mixed, and the dye injection rate, q_t, the injected dye concentration, C_t, and the upstream dye concentration, C_o (usually zero).

4. Measuring Flows in Small Channels and Pipes

Flows from municipal and industrial wastewater treatment plants and polluted tributaries can be measured in the same manner as river flows, but usually it is better to use methods adapted to pipe flows and flows in small channels. Special methods are especially important to accurately measure small flows with relatively large pollution loads. Appendix II lists and illustrates the more useful methods.

The list of methods includes those used to determine flow in pipes flowing full and partially full and those useful in open channels. A number of these methods involve a permanent installation, and thus Appendix II is intended to aid in determining the accuracy and usefulness of existing flow records at wastewater treatment plants and at tributary gaging stations. Also included are a number of methods that are used during studies to determine flows where no other records are available. In addition, Appendix II includes methods that are not frequently used in water quality studies but may be encountered or may be useful in preliminary studies.

Almost all of the methods listed in Appendix II could be implemented in measuring flow in a treatment plant, but the simpler devices such as the Parshall flume and thin plate weirs are used more often because they are inexpensive to install and operate and the results are among the most accurate.[46] The velocity-area method for pipes or channels, tracer method, electromagnetic meters, and the volumetric method are most frequently used to check or establish the calibration of the device.

Checking the calibration of flow measuring devices is not only useful in operating a treatment plant and for determining if flow records for stream studies are accurate, but these methods are also used in National Pollutant Discharge Elimination System (NPDES) inspections to determine if permitted flows are being accurately measured. As a result, NPDES inspection procedures can be quite useful for checking the accuracy of records from wastewater treatment plants to avoid the need to make a redundant flow measurement.

Generally, for stream studies, checking plant flow records involves visiting the plant, viewing the measuring device, and discussing the operation and calibration of the device with the plant operator. Appendix II should prove very useful in describing the devices that may be encountered and in describing important operational characteristics. In cases where instream mass balances or plant visits indicate that flow records may be in error and the waste inflow is important in controlling stream water quality, Appendix II gives flow equations that can be used to check calibration curves. In cases like this, a more thorough inspection along the river bank can indicate if bypasses of raw sewage or leaks of partially treated sewage are involved. On rare occasions and during NPDES inspections, it may be useful to measure the volume of untreated sewage or confirm the calibration of flow meters at the entrance of the plant and do a water balance for the plant. Outflow measurements should also be confirmed or reproduced. Inspection of existing flow measuring devices should record important geometric dimensions such as throat and pipe diameter or width and depth of approach flows and tail waters at maximum flows; location of pressure taps or stilling basins and gages to determine water surface elevations; and length of meter, important elements, and approach flows. The most important attributes are given here and the reader may also wish to refer to ASME[47] and later editions, Rantz et al.,[10] and French.[1] Proper maintenance to keep the meters clear of debris and sediment should be evident or the flow should be checked during the inspection. The detailed inspection procedures to cover these and other concerns are compiled in Appendix II, from the NPDES procedures and from other sources. Coordination of plant inspections by state regulatory agencies to confirm the flow and water quality records during a stream water quality study may be useful if problems are anticipated with access to a particular plant.

When treatment plant records are unavailable or wastewater inflows and tributaries are ungaged (e.g., flow is unmeasured), there are a number of techniques that may be useful. The attachable acoustic meter of various types may be the most useful for attaching to a pipe at the river bank to record flow and check records, but it is one of the more expensive devices. When measuring flows at the river bank, the volumetric method is the most reliable but is the most difficult to apply especially to large flows where it is difficult to construct a large tank at the proper location. Many times the design of a sewage treatment plant makes the maximum use of the head available, and as a result, the outflow pipe is at or below the water surface of the receiving stream. In addition, the pipe is usually extended below the water surface for aesthetic reasons and to facilitate greater mixing. As a result, the electromagnetic meter that can be attached inside of a pipe would seem to be useful when concrete and PVC pipe are involved. While not frequently used, the parabolic flume would seem to be also useful for this purpose when the flow falls free at the end of the pipe and it is not feasible to use the volumetric method. The velocity-area method using electromagnetic probes, pitot tubes, small propeller-type current meters, and salt tracers are the most widely used methods to establish the calibration of pipe flow meters. But these methods are time consuming and tedious. When necessary, it may be feasible to use the Palmer-Bowlus flume in pipes of 0.61 m (2 ft) diameter and larger. Where the pipe ends before intersecting the stream, it may be possible to install a temporary flume or weir. Rantz et al.[10] describe the portable 3-in (0.076-m) Parshall flume and V-notch weir for flows of 0.00002 to 0.014 m^3/s (0.0007 to 0.5 ft^3/s) and 0.00057 to 0.057 m^3/s (0.02 to 2 ft^3/s), respectively, used by the U. S. Geological Survey. For larger flows the velocity-area method using the Price or pygmy current meter and tracer methods are the most useful for both measuring flow and calibrating weirs and flumes. Occasionally, it may be possible to use indirect methods where the open-channel flow goes through a culvert before entering the stream.

During preliminary visits to a stream or in streams involving a number of inflows that are difficult to measure, it may be possible to make flow estimates with approximate methods. These include (1) the velocity-area method where the velocity is observed at the surface with floats and related to the vertical average velocity with Table 10, (2) the slope-area method, (3) width contraction method, and (4) superelevation in bends. The slope-area method involves measuring or estimating the channel cross-sectional area and slope and estimating the Manning n. Matthai[48] shows how the drop in water-surface elevation through a channel contraction of known cross-section can be related to flow rate, but this requires a precise survey to relate the difference between upstream and downstream elevations. Usually the flow rate must be high such as during flooding or the contraction should be pronounced to produce a measurable head loss. This may occur during water quality surveys of dynamic conditions. Superelevation in bends also usually involves high flows but may occasionally be useful in high gradient streams found in mountainous areas.

E. TIME OF TRAVEL BY DYE TRACING

Given the frequency with which discharge gaging records are available from the U.S. Geological Survey and others, it is, on many occasions, more useful to use these data and other data available to define cross-sections to determine the width, depth, and travel time in the reach. In other cases, the necessary data are not available or are not easily measured. Shallow headwater streams are difficult to gage by the current meter methods. In large streams, it may occasionally be difficult to gain access, and excessive commercial navigation may disrupt field studies. Where highly dynamic flows are encountered, it may not be possible to measure the discharge quickly enough before the flow rate changes substantially. In addition, it may be generally more accurate to use dye tracers to determine average velocities.

Dye studies are conducted over the important reaches of interest by injecting dye and recording the time it takes the peak dye concentration or the centroid of the dye cloud to move from one location on the stream to another. Figure 54 shows that the travel time is determined by plotting measurements of dye concentration vs. time at two locations on the stream and measuring the difference in time that it takes the peak concentration to move from the upstream station to the downstream station. The points labeled A_C in Figure 54 are the locations of the centroid of the dye mass in the time vs. concentration curves. The time difference between the centroids is the exact travel time from one location to another, but the time difference between the arrival of the peak concentrations and the arrival of the centroids is rarely significant over short reaches on the order of kilometers or tens of kilometers (approximately miles to tens of miles). The difference in arrival time of the peak concentration vs. the centroid of the dye cloud may only rarely be significant in streams that exchange water with dead zones of backwater, stationary eddies, or macro-pore waters in the bed. The measurement is representative of the average water velocity if the dye is well mixed across the stream at the measurement sites as shown in Figure 55.

Important reaches that may require dye tracing to determine the time of travel are those where a sensitivity analysis indicates that the calibrated model is sensitive to estimates of travel time. These reaches generally involve steep concentration gradients like concentration gradients of dissolved oxygen and biochemical oxygen demand below major sewage treatment plant discharges. When resources permit, the dye study should cover the entire reach.

The procedure to measure time of travel in a stream reach is given in Table 12. The injection method should follow the guidance given in a previous section for dye-dilution methods of measuring discharge. For accurate studies, the dye should only be traced over short reaches. Dye tracing over longer reaches should be checked by measuring the complete dye curve to determine if significant quantities of dye are sorbed to bottom and suspended sediments. Hubbard et al.[12] shows how to determine if the dye is conserved and recommends correction methods for dye loss, but the dye loss corrections are conceptually incorrect if the dye is mainly lost by sorption.

Results are typically reported as shown in Figures 56 and 57. These data were collected in a study of the West Fork Trinity River in Fort Worth, TX (U.S.). In Figure 56, concentration vs. time curves are plotted to show the decrease in peak concentration at different stations downstream. This particular river reach is a pool-and-riffle type that mixes quite readily. Figure 57 shows the arrival time of leading edge, peak, centroid, and trailing edge (defined as 1% of the peak concentration) at stations downstream of the first measurement site at which the dye was well mixed. Figure 57 indicates that the time of travel based on the peak concentration and centroid is not significantly different and shows how the dye spreads out because of longitudinal dispersion over the reach. In this particular study, a very short reach was studied. Usually, dye studies extend over much longer reaches. These studies are described in detail in McCutcheon et al.[9]

For best results, time of travel studies should be conducted at the time of the water quality survey. Usually such a study can be combined with reaeration measurements. When it is not possible to combine the studies because of limited numbers of personnel, time of travel studies can be conducted at other times. In the event that river discharge at the time of the water quality study is different from the discharge during the dye study, then two or more studies may be necessary to bracket conditions of interest. Hubbard et al.[12] shows how time of travel studies can be extrapolated to different flow rates in Figure 58.

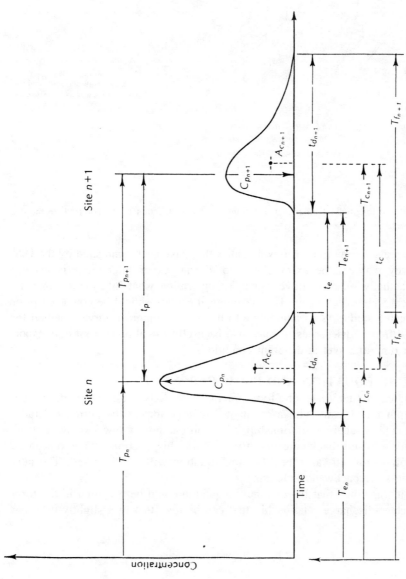

FIGURE 54. Typical time-concentration curves resulting from an instantaneous or slug injection. T_{pn} and T_{pn+1} are the times required for the peak dye concentrations to arrive at sites n and n + 1. t_p is approximately the time of travel between sites n and n + 1 based on peak concentrations, C_{pn} and C_{pn+1}. The exact time of travel between sites n and n + 1 is $t_C = T_{Cn+1} - T_{Cn}$ where T_{Cn} and T_{Cn+1} are times that the centroids of the concentration-time curves (A_{Cn} and A_{Cn+1}) arrive at stations n and n + 1. t_{dn} and t_{dn+1} are the duration times it takes the dye cloud to move past sites n and n + 1. T_{en} and T_{en+1} are the arrival times of the leading edge of the dye cloud. T_{fn} and T_{fn+1} are times at which the trailing edge of the dye cloud passes sites n and n + 1.[12]

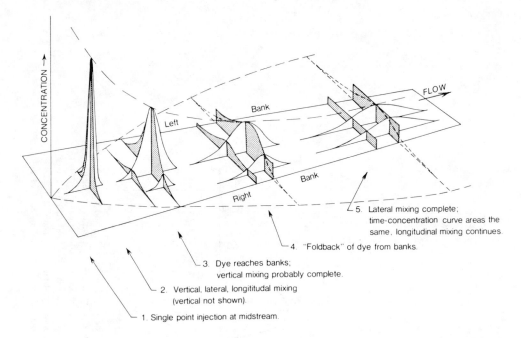

FIGURE 55. Lateral and longitudinal mixing of dye from a point of injection to the point where the dye is laterally well mixed.[12]

Because there are so many time of travel studies that have been conducted by the U.S. Geological Survey that can be extrapolated to discharges of interest, it is usually unnecessary to actually measure travel time. To determine what information may be available, the local state office of the U.S. Geological Survey should be contacted prior to undertaking such a study. As an alternative to the U.S. Geological Survey method for extrapolation to different discharges, it may also be useful to calibrate a mass transport model to predict time of travel at different conditions.

F. THE MANNING EQUATION AND FLOW EXPONENTS

Chapter 1 indicates that there are four practical methods for relating discharge, velocity, and depth for streams with a constant discharge during the period of study. These include stage discharge relationships,[49] empirical power law exponents,[4] the Manning equation,[4] and backwater solutions.[49] In this section, stage discharge relationships, flow exponents, and the Manning equation will be discussed. The next section will address the backwater solution.

In Chapter 1 it was noted that the continuity equation could be rewritten in the form of the standard stage-discharge relationship that can be rewritten in a slightly different form as

$$Q_n = KD^m \tag{56}$$

where Q_n is the discharge for uniform, steady flow in a channel, D is the depth of flow that is approximately equal to the hydraulic radius,[1] and K is a constant for a given location on the stream that can be expressed as

$$K = \Gamma A(S_o)^{1/2} \tag{57}$$

where Γ is an empirical friction coefficient, A is the cross-sectional area, and S_o is the channel slope. Generally, a stage-discharge relationship is written in the form:[10]

TABLE 11
Method of Measuring Stream Discharge by the Dye-Dilution Technique[a]

Step	Procedure

1. Select a sampling reach to provide lateral and vertical mixing in as short a distance as possible where easy access is available by bridge if possible. Mixing is enhance by injecting the dye at multiple points in the cross-section separated into areas of equal discharge or injecting a line source across the center of the flow. Injection in one half of the cross-section or at one bank increases the mixing length by a factor of approximately 2.2 to 4.0. Sampling reaches are normally chosen to avoid inflows and outflows. If a reach must be chosen that has an inflow, then the discharge measured will be the discharge at the measuring site if the dye is laterally mixed, implying that the inflow must be small enough to mix with the receiving stream and achieve a uniform concentration at the sampling site. Because most inflows enter at the side of a stream, there is usualy a longer distance required to allow the dye to mix across the stream. The effect of loss of flow between the injection point and sampling site depends on whether or not the dye is well mixed upstream of the diversion. If the dye is not well mixed, the discharge is indeterminate because a disproportionate amount of the dye may be diverted with the outflow. As a result, the injection point must be moved further upstream or moved downstream of the diversion. If the dye is well mixed before reaching the diversion, then the discharge measured is the discharge of the stream upstream of the diversion, and the only reason for locating the sampling site downstream of the diversion would be because more convenient access exists downstream (but measurement would still be of the total discharge above the diversion). Because diversions frequently involve water supply for irrigation and drinking, it would be advisable to move downstream of a diversion to avoid the perception that the study would affect water supplies if the objectives of the study permit this, or it would be better to sample upstream of the diversion to document the dye concentrations being diverted in the event that some concern arises.

2. Preliminary sampling or calculation should be used to define when adequate mixing has been achieved. Generally, 95% mixing is considered adequate when samples are collected from at least three points across the stream. The distance from the injection to the sampling site must be increased by a factor of 1.8 to achieve 99% mixing. Yotsukura and Cobb[43] amd Fischer et al.[44] give mixing length formulas of an approximate nature to guide sampling, but lateral sampling (across the stream) should be undertaken at the sampling site chosen and the station moved downstream if mixing is inadequate.

3. Determine the volume of dye needed in ml for a constant-rate injection as

$$V_d = 1.02 \times 10^8 (C_D/C_t) Q\ t_t$$

where C_D is the constant dye concentration (parts per billion) at the downstream sampling site; C_t is the concentration (parts per billion) of the injected dye (usually 20,000,000 ppb for a 20% rhodamine WT dye solution); Q is stream flow in ft^3/s; t_t is the time of the planned injection in hours; and 1.02×10^8 is a conversion factor. $C_D = 5$ ppb is generally adequate for most discharge measurements. Generally, the time of injection should allow for at least a 15 min plateau of constant concentration at the downstream sampling station. This requires at least a 15 min injection and usualy a much longer injection depending on the mixing in the stream between the injection and the measurement station. For a slug injection, the required volume in milliliters is as follows:

$$V_d = 3 \times 10^7 (C_p/C_t) Q\ t_p$$

where C_p is the peak concentration in parts per billion desired at the downstream sampling site; t_p is the estimated time required for the peak concentration to move from the injection site to the downstream sampling site; and 3×10^7 is a conversion factor. $C_p = 10$ ppb is recommended for rhodamine WT dye. In addition, the USGS recommends that C_D or C_p should not exceed 10 ppb or 10 μg/l at drinking water intakes. When possible, injections should be as dilute as possible to avoid public concern when injections are visible. To aid in this, rhodamine WT concentrations are easily detectable down to 1 μg/l and lower.

4. Choose the injection method for constant rate injections using small constant flow rate, battery driven pumps, or constant-head devices such as the Mariotte vessel (Figure 51) or floating siphon (Figure 52). The injection point should involve as much mixing as possible, such as that found in a riffle or waterfall, and the injection should not be directed to the bottom to avoid staining the sediments and losing some dye by localized sorption.

5. Sample background concentrations upstream of the injection at least three times or more frequently if the background is expected to change.

6. Before handling dye containers and equipment that has been in contact with dye, collect stream water that will be used to establish standards by diluting samples of the dye to be injected. A 5-gal (20-l) sample is usually sufficient.

TABLE 11 (continued)
Method of Measuring Stream Discharge by the Dye-Dilution Technique[a]

Step	Procedure
7.	Obtain at least three samples of the dye to be injected. These samples should be about 2 oz (60 ml) each.
8.	Precisely measure the volume injected during a slug injection or measure the injection rate of a constant rate injection using a graduated cylinder and stop watch at the beginning of the injection and at the end of the injection to ensure that the rate remains constant.
9.	Continue the constant rate injection long enough to produce a constant concentration for least 15 min at the sampling site. Depending on the longitudinal mixing between the injection and sampling points, longer injection times are needed to achieve the desired 15 min of sampling time. Once the length of time of the injection is chosen, the injection rate is computed from the estimated volume required divided by the injection time.
10.	Monitor the arrival of the dye at the sampling station and collect the necessary samples. The arrival of fluorescent dyes can be monitored with a fluorometer such as the Turner Designs Model 10 properly configured with the correct filters for the dye of interest. Fluorometers can be operated in the field by battery or generator and can be operated in a flow-through mode to continuously monitor fluorescence or can be used to measure the fluorescence of discrete samples. Samples can be collected at intervals of 5 to 15 min for the constant-injection method as the concentration plateau is approaching (see diagram in Figure 53). Five samples defining the plateau are returned to the laboratory. The remaining samples can be discarded. For the slug injection, samples are collected as frequently as necessary throughout the passage of the dye cloud to define the concentration time curve at the sampling site. This usually involves about 20 samples collected at intervals of about 5 to 15 min except on the trailing edge of the dye cloud where more infrequent sampling may be possible. As a rule of thumb, the sampling of the trailing edge of the dye cloud is usually discontinued when the concentration reaches 5% of the peak concentration. Where a long trailing edge exists because of exchange with dead zones in the channel (usually eddies and nonflowing pools) or with adjacent groundwaters, then a significant part of the total mass can be on the trailing edge, and sampling may need to continue longer. When significant exchange with the nonflowing areas is suspected, a selection of a different reach should be considered. More frequent sampling may be necessary if the peak is a sharp one. When possible, the person handling and injecting the concentrated dye should not be the one to collect samples because of the likelihood of sample contamination.
11.	Dye concentrations determined in the field can be used if temperature corrections are applied. To do this, water temperatures should be measured and the fluorometer maintained at a constant temperature. The difficultly in maintaining the fluorometer at constant temperature, however, generally makes it necessary to collect samples and analyze the samples in a laboratory where the samples and standards can be tested at a constant temperature.
12.	Samples should be collected below the water surface away from the bank to avoid low velocity dead zones. The best sampling site is at a constriction or where there is a reasonable distribution of velocity across the stream.
13.	Samples should be collected in small bottles (plastic is recommended for rhodamine WT dye) that are adequately washed in the laboratory and protected from contamination before use. Reuse of bottles in the field should not occur if possible, but if necessary, the bottles should be rinsed as thoroughly as possible in stream water that is free of dye. Samples should be well labeled.
14.	One the dye concentrations are determined, the discharge for the constant-injection method is computed as

$$Q = q_t \left[\frac{C_t - C_D}{C_D - C_o} \right]$$

where q_t is the dye injection rate; C_t is the concentration of the dye solution injected; C_D is the dye concentration measured downstream at the sampling location; and C_o is the background concentration upstream of the injection point. The discharge for the slug injection is computed as

$$Q = VC_t / \left[\sum_{i=1}^{n} (C_i - C_o)(t_{i+1} - t_{i-1})/2 \right]$$

where V is the volume of dye injected into the stream; C_i is the measured concentration in sequence of the first to the nth sample collected; and t is the time the sample was collected. t_{i-1} refers to the sample collected just prior to the sample of interest.

[a]   Adapted from References 40 to 42.

TABLE 12
Procedure for Conducting Time of Travel Dye and Dispersion Studies

Step	Procedure
1.	Define the extent of the study reach of interest and note the location of any drinking water withdrawals.
2.	Estimate river discharge rate at the time of the study and the dispersion coefficient for the reach.
3.	Back calculate the maximum amount of dye that can be injected without exceeding a standard for dye concentrations in the water intake. The U.S. Geological Survey uses a standard of 10 μg/l for rhodamine WT fluorescent dye. Hubbard et al.[12] give empirical formulas that are useful for estimating the amount of dye that should be injected.
4.	Choose the method of injection and determine the length of river required for complete mixing to occur. The point of injection should be chosen so that complete mixing has occurred before the dye flows into the reach of interest. See Section II.C on methods to estimate mixing lengths, or these distances can be more accurately determined in the field by measuring dye concentration at a number of points in the measurement cross-section chosen and determining if there is a significant difference in concentration.
5.	Prepare the solution of dye for transport to the site and injection at the point determined by the mixing length calculations.
6.	Sample water at predetermined sites downstream to measure the passage of the peak dye concentration at each site where the time of travel should be known. Since it may not be possible to track the dye peak as often as needed, sampling sites should be chosen so that velocity will be approximately constant over the reach between measuring sites. Sites should also be chosen to allow easy access to the river. At the first site, samples should be collected simultaneously at three or more sites across the stream to confirm that the dye is well mixed at the first location. Samples can be analyzed in the field with a portable fluorometer to determine how frequently to sample; otherwise, the flow should be collected in the appropriate type of bottle (plastic for rhodamine WT), and the samples defining the peak should be analyzed in the laboratory under controlled conditions. When a field fluorometer is available, sampling every 1 to 2 min to accurately define the peak may be more useful. Sampling to define the peak is all that is necessary for time of travel studies, but if the data may also be used for dispersion studies as well, then the entire dye distribution should be sampled at a site beginning with background concentrations and continuing until the concentration decreases to approximately 1% of the peak concentration or less.
7.	Discharge or reach averaged cross-sectional area should be measured over the reaches of interest. If discharge is measured, it should also be measured for any tributary entering the river where the flow rate is significantly increased. Where discharge cannot be measured, area should be measured over short enough intervals to accurately define the reach volume, and within the reaches, the flow should be constant. Therefore, sampling sites should be chosen upstream of tributaries and reaches should extend between tributaries. Knowledge of the flow rate is necessary if the results are to be extrapolated to different flow conditions.
8.	The length between measuring sites is typically determined from U.S. Geological Survey 7.5-min quadrangle maps or aerial photographs. These distances can be determined by field surveys, but such precision is rarely justified unless maps and photographs are not available.
9.	The reach averaged velocity, U, is computed from the length between measuring sites and the travel time. Discharge or reach averaged area is computed from the continuity equation, Q = UA.

$$Q = K(G - e)^m \qquad (58)$$

where G is gage height of the water surface elevation referenced to some datum (usually mean sea level) and e is the gage height of the water surface when the flow rate is zero. Therefore (G − e) is equal to the depth of flow in wide channels that are approximately shaped like a trapezoid or rectangle in the cross section. The accumulation of observational experience indicates that m has a value of 1.3 to 1.8 in uniform channels but can reach values of 2 and higher at gaging stations that are chosen because the control of the water surface by some irregular feature in the channel such as rapids, falls, or obstructions. Because typical gaging stations are purposefully located in very irregular sections, stage-discharge relationships and the flow exponents, to be discussed next, should never be based on U.S. Geological Survey gaging station relationships unless the stage-discharge relationship is controlled by channel friction loss representative of the entire reach.

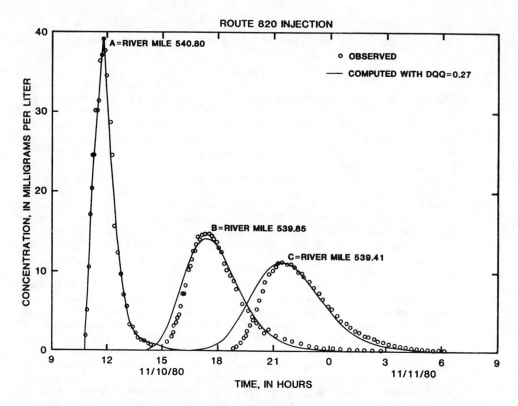

FIGURE 56. Dye distributions at three sites downstream of an injection made upstream of Interstate 820 across the West Fork Trinity River in Fort Worth, TX. River miles 540.80, 539.85, and 539.41 are also 870.33, 868.80, and 868.10 km upstream of the mouth of the Trinity River at the Gulf of Mexico. The solid curves are predictions of dye distributions using a Lagrangian routing method. DQQ is a dispersion factor. The solid curve shown for site A is the boundary condition used for the Lagrangian routing method.

It should also be noted that values of K and m may change for different ranges of flow if the nature of the channel in contact with the flow changes to become more or less rough. The coefficient K also changes with changes in cross-sectional area, and m varies with changes in channel shape that can occur as depth increases. For steady, low flow, however, these coefficients are usually constant.

Stage-discharge relationships are usually determined empirically by measuring discharge and depth of flow. For water-quality modeling, the relationship must be based on the reach-averaged depth for the reaches being modeled. USGS gaging sites relationships are usually only applicable to very short reaches and are not useful for this reason as well.

Stage-discharge relationships are useful for gradually varied flows in addition to uniform steady flows as long as the hysteresis in the rating curve is insignificant. Hysteresis occurs when two different flow rates are measured for the same water surface elevation as illustrated in Figure 59. The hysteresis is caused by changes in the energy slope and depends on whether the stage is rising or falling. Even if this occurs, a stage-discharge relationship can be used to properly describe flow and depth, but the calibrated values cannot be extrapolated to other flow conditions with great certainty.

The method of relating flow and velocity to depth using empirical exponential equations is very similar to developing stage discharge relationships. In fact the development of the empirical equations seemed to grow out of a desire to be able to predict stage-discharge relationships.[5] These methods have subsequently proven very useful for water quality modeling when flows are steady and have to be adapted to describe long reaches rather than very localized reaches in the vicinity of gaging sites.

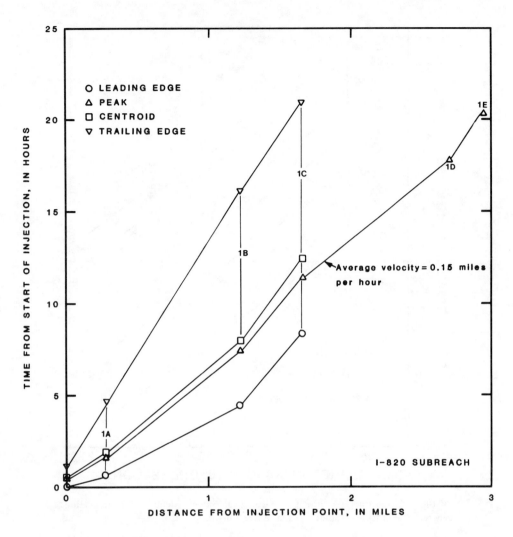

FIGURE 57. Arrival time of dye leading edge, peak concentration, centroid, and trailing edge of a dye cloud injected in the West Fork Trinity River just upstream of Interstate 820. Complete dye distributions were measured at sites A, B, and C. Only peak concentrations were measured at sites D and E. Note that 1 mi = 1.61 km. The average velocity of 0.15 mi/hr is also 0.24 km/hr. (Data courtesy of Ronald Rathbun, U.S. Geological Survey.)

The empirical flow-exponent equations are written as

$$U = aQ^b \tag{59}$$

$$D = cQ^d \tag{60}$$

and

$$W = eQ^f \tag{61}$$

where U is the average velocity in a reach, D is the average depth in a reach, and W is the average width of the channel. Because A = WD, the continuity equation, Q = AU, indicates that the empirical exponents and coefficients in the equations above are related as[50]

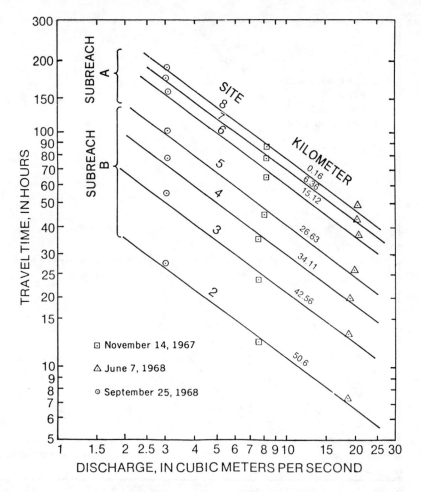

FIGURE 58. Illustration of the U.S. Geological Survey Method of extrapolating between time of travel measurements at different discharges.[12]

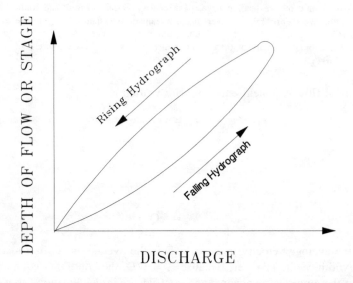

FIGURE 59. Hysteresis observed in stage discharge relationships for unsteady flow.

$$ace = 1 \tag{62}$$

and

$$b + d + f = 1 \tag{63}$$

from

$$Q = AU = UDW = (aQ^b)(cQ^d)(eQ^f) = (ace)Q^{b+d+f} \tag{64}$$

The form of the stage-discharge relationship given in Equation 56 is equivalent to Equation 60, where $K = (1/c)^{1/b}$ and $m = 1/b$. Therefore, it seems that both approaches have the same theoretical basis and are in effect, when properly applied, equivalent methods. The most significant difference seems to be that Equations 59 through 61 have proven easier to understand and apply.

By noting that the continuity equation for steady, uniform flow is the basis for the stage-discharge relationship and the equivalent flow-exponent equation, it should be clear that these methods are in fact limited to steady flow. Furthermore, these methods will have to be applied to larger streams where the channel is approximately uniform. The degree of unsteadiness in flow and nonuniformity of the channel that can be accommodated before these methods can be reasonably applied has not been explored. Theoretically, there is no such latitude. Practically, however, there has been extensive application beyond the theoretical limits, but the limits on use are at best only vaguely understood.

The stage-discharge relationship has been applied with some success to gradually varied flows in which the flow changes slowly and gradually. The degree of change possible has not been quantified. However, the next section on the Manning equation develops untested criteria to guide decisions on when a channel is not sufficiently uniform for the purpose of predicting depth and velocity in a channel.

Like the stage-discharge coefficients, the flow coefficients, a, b, c, d, e, and f, must be determined by calibration from reach averaged measurements of discharge, velocity, depth, and width. For screening-level studies where calibration is not possible, note that there is limited guidance for the choice of values and gives these in Table 13. Ambrose and Vandergrift,[51] however, give more extensive guidance for a number of different channels based on the work of Park[51,52] in Figure 60. Richards[53] provides more up-to-date guidance on coefficient selections.

In addition, the Manning equation can be used to express the coefficients in a more recognizable form. For example, in a wide rectangular channel the Manning equation in English units can be written in terms of the depth and width as

$$Q = \frac{1.49}{n} (S_o)^{1/2} W D^{5/3} \tag{65}$$

where 1.49 is a numerical constant that derives from the conversion of the Manning equation written for metric units to the English-unit version written here. If the Manning equation is used with width and depth in meters and U in m/s, then the numerical constant should be 1.0. See Chow[54] regarding the evolution of this manner of using English and metric units. n is the Manning roughness coefficient that is an empirically determined coefficient related to channel roughness. S_o is the slope of the channel. In a wide channel, the hydraulic radius (area/wetted perimeter) is approximately equal to the depth. This holds true as long as the width is at least 10

TABLE 13
Hydraulic Exponents for Various Types of Streams

Channel cross-section	Exponent for velocity (b)	Exponent for depth (d)	Exponent for width (f)
Rectangular	0.40	0.60	0.00
Average of 158 U.S. gauging stations	0.43	0.45	0.12
Average of 10 gaging stations on Rhine River	0.43	0.41	0.13
Ephemeral streams in semiarid U.S.	0.34	0.36	0.29

From Barnwell, T. O., Jr., Brown, L. C., and Marek, W., Development of a Prototype Expert Advisor for the Enhanced Stream Water Quality Model QUAL2E, Internal Report, Environmental Research Laboratory, U.S. Environmental Protection Agency, Athens, GA, 1987.

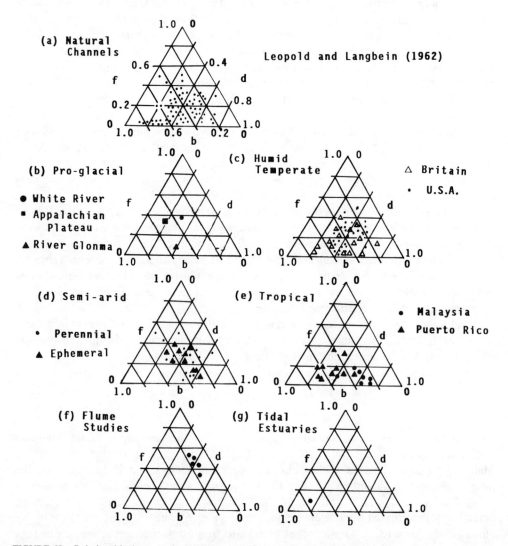

FIGURE 60. Relationship between hydraulic exponents for various types of channels in different regions. (From Park, C. C., *J. Hydrol.*, 33, 133, 1977. With permission.)

times the depth as is the case for almost all natural channels and many man-made channels. Solving for D yields

$$D = cQ^{3/5} \tag{66}$$

where

$$c = \left[\frac{n}{1.49WS_o^{1/2}} \right]^{3/5} \tag{67}$$

and $d = 3/5 = 0.60$. For the stage-discharge relationship, this translates into $m = 1.67$ because $m = 1/d$. In a similar, manner the coefficients in Equation 59 can be expressed as

$$a = \left[\frac{nW^{2/3}}{(1.49S_o^{1/2})} \right]^{-3/5} \tag{68}$$

and $b = 0.40$ as shown in Table 13. For a rectangular channel the width, W, does not change, and therefore, $f = 0$. As a result, $e = W$, which is constant in a rectangular channel. It also seems possible to repeat the same derivation for trapezoidal channels and irregular channels, but the derivation is more involved.

Equations 66 to 68, still involve an empirical coefficient, the Manning roughness coefficient, n. However, the revision of the equations represents a reduction from the original six empirical coefficients that were not well understood in terms of physical properties of the flow. At least for a rectangular channel, the coefficients can be written as constant numerical values derived from the Manning equation or written in terms of constant physical properties such as width and channel slope (constant for steady, uniform flow), plus the Manning roughness constant that is relatively well understood for stream flows.

The expression of the coefficients in the flow-exponent equations (Equations 59 to 61) in terms of slope, width, and n, implies that the Manning equation is also related to the governing equations. In fact, it can be confirmed in two ways. First, the equation of continuity was written in a stage-discharge relationship for steady, uniform flow (see Equation 20) that is similar to the Manning equation. Second, the Manning equation which can be used to express the flow coefficients can be derived from a simple force balance for steady, uniform flow in the exact same manner as the basic Navier-Stokes equations for fluid flow are derived.

If the Manning equation is written as it is traditionally expressed, it becomes immediately clear that this is a form of continuity equation as it is written in Equation 20 (see Chapter 1). The traditional form of the Manning equation is written as

$$U = \frac{c}{n} R^{2/3} S^{1/2} \tag{69}$$

where c is a constant equal to 1 when U is in m/s and R is in m and equal to 1.49 when U is in ft/s and R is in ft, n is the Manning friction coefficient, R is the hydraulic radius, and S is the slope of the energy grade line. The two equations are equivalent if the empirical coefficients in Equation 20 are written as $\Gamma = c/n$ and $m = 2/3$.

The Manning equation is derived from a force balance for a steady-flow in a uniform channel (i.e., the area and shape of the flow cross-section is constant, and the slopes of the energy grade line, water surface, and channel bottom are equal, $S_f = S_w = S_o$) as shown in Figure 61. The force balance involves equating the force due to the weight of

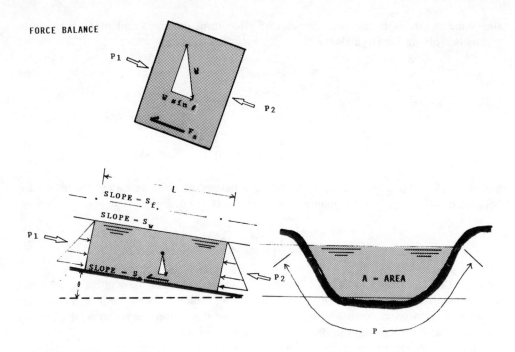

FIGURE 61. Force balance for a uniform steady channel flow. Note that $S_f = S_w = S_o$ and $p_1 = p_2$. The mean pressure forces, p_1 and p_2 at the beginning and end of a uniform reach are equal and act in opposite directions. P is the length of the wetted perimeter.

the fluid acting downslope, W sinθ, to the shear force on the channel bottom and sides resisting the water movement, $F_s = \tau LP$, where τ is the shear stress on the bottom and sides, L is the length of the channel over which the shear force acts, and P is the distance around the perimeter of the channel cross section along which the water is in contact with channel. Most natural and man-made channels have such shallow slopes that sinθ ≅ S_o, the channel slope. If it is assumed that the force resisting the flow is proportional to the velocity squared, U^2, then[1]

$$\gamma ALS_o = LPkU^2 \tag{70}$$

where γAL is the weight of water in a reach of length, L, and cross-sectional area, A. The specific weight of water is γ. kU^2 is assumed to equal τ, where k is a constant of proportionality. Solving for U yields the Chezy equation:

$$U = (\gamma/k)^{1/2}(RS_o)^{1/2} \tag{71}$$

where $(\gamma/k)^{1/2}$ is defined as the Chezy C, an empirical flow resistance factor that has the units length /time squared. The Chezy equation is widely used in Europe instead of the Manning equation. Henderson[55] gives representative values of C for various values of the Reynolds number (4UD/ν) and channel roughness. Chow[54] gives a few representative values and three formulas for the computation of C from channel characteristics and other empirical coefficients.

The assumption that the shear force is proportional to U^2 is consistent with observational evidence obtained for channels and pipes, and the same result can be derived by dimensional analysis.[55] The Darcy-Weisbach pipe resistance equation[56] and a number of other resistance equations that have proven useful in practice are of a form that indicates that this assumption is well justified.

The Manning equation was derived from Equation 25 independently by Gauckler in 1868[54,57] and Hagen in 1881.[55] Chow[54] indicates that the work first appeared in Hagen,[58] who developed simpler ways of expressing C. They noted that C was proportional to the sixth root of the hydraulic radius and defined a proportionality constant, n, which has become known as the Manning roughness coefficient. Even later, Strickler[54,59] reached the same conclusion by a different means. As a result of the work by Gauckler and Hagen, the simple expression for C is

$$C = \frac{R^{1/6}}{n} \tag{72}$$

Equation 69, that is called the Manning equation, results when Equation 72 is substituted into Equation 71.

Given that the theoretical basis for the Manning equation can be traced to a balance of forces for a steady, uniform flow, the similarity of the equation with Equation 20 is not coincidental. Equation 20 is derived from the application of Newton's second law to an incremental volume of fluid that was simplified for steady, uniform flows. The Manning equation is derived in a simpler, but no less rigorous fashion, from a balance of the forces acting on a reach of a uniform channel in which the flow is constant.

There is, evidently, some misconception about the theoretical basis of the Manning equation. In 1891, the Frenchman Flamant wrongly attributed the advances of Gauckler and Hagen to the Irishman, Robert Manning.[55] Manning had taken a different approach that was wholly empirical in nature. In a paper presented in 1889 to the Institute of Civil Engineers of Ireland that was published in 1891,[54,60] Manning fitted the observations of Bazin,[54,61] relating average velocity, U, hydraulic radius, R, and channel slope, S_o, for artificial channels and determined that Equation 20 was the appropriate form after some simplification.[54] Manning confirmed his choice of exponents in Equation 20, with an additional 170 observations[54] and evidently was aware of the consistency with theoretical form pointed out by Hagen.[54,55,62]

A rigorous derivation and reading of the work on the subject, therefore, indicates that the Manning equation does have a theoretical basis but that the theoretical basis was not derived by Manning. Manning's empirical approach to formulate the equation, after that same form had been theoretically defined, seems to have spawned the misconception that the Manning equation is completely empirical. Perhaps it is more accurate to view Manning's work as providing the experimental validation or confirmation of the theoretical form introduced earlier. A number of other studies have followed, such as the work by Strickler in 1923, that provide additional confirmation of the original observation represented by Equation 72.

As indicated earlier, it is difficult to determine the precise limitations of the Manning equation. The similarity in form between the Manning equation and stage-discharge relationships indicate that the Manning equation can be used in practice to describe friction losses in gradually varied flows. In addition, a number of models that simulate dynamic flows make use of the Manning equation to express friction losses in terms of the Manning roughness coefficient. In so far as this approach has proven to be quite useful, it is likely, but as of yet, not confirmed, that the application of the Manning friction coefficient is not sensitive to some unsteadiness of discharge. The effect of channel nonuniformity seems to be another matter, however.

Occasionally, it has been observed that it is difficult to predict both velocity and depth in highly nonuniform channels with methods that are similar to the Manning equation. In addition, it is difficult to calibrate models that rely on the Manning roughness coefficient as a calibration parameter to describe friction losses in highly

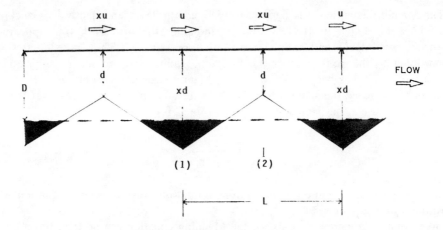

FIGURE 62.   Hypothetical pool and riffle stream at steady flow.

irregular channels. Finally, it can be noted there seems to be a lack of criteria to guide data collection efforts aimed at collection of channel cross-section data. Specifically, it is not clear how closely spaced should cross-sectional measurements be made to define short reaches that are approximately uniform to meet the theoretical limitation of the Manning equation and other uniform flow equations. To investigate the application of the Manning equation to nonuniform flows, the remainder of this subsection will analyze a simple case to show that the Manning equation is not exactly applicable to all stream flows. In addition, criteria will be developed to project how much nonuniformity may be permissible before it becomes difficult to predict both depth and velocity.

Most relationships between flow depth and velocity, including the Manning equation and equations solved in dynamic hydraulic routing models, assume that the reaches between or centered upon corresponding cross sections are approximately uniform. When it is not possible to adequately measure highly variable geometry, then predictions such as those for temperature and reaeration will suffer if cross-sectional areas and depths are reduced from the actual conditions, or else kinetic calibrations that depend on the average velocity or travel time in a reach will suffer. As a result, the calibrated model cannot adequately predict significantly different flow conditions. Therefore, models of highly irregular streams that are based on the assumption of uniform flow usually only describe the flow condition measured.

To illustrate the problem of predicting average velocity and depth in a nonuniform channel, consider the following simple but representative proof. Figure 62 illustrates a regular sawtooth representation of bottom elevation. The bottom varies linearly between points equally spaced along the stream where the depth is alternatively d and xd (x is assumed to be a constant). If the effect of acceleration on the water surface can be neglected (note that the average effect over reaches that are integer multiples of the length, L, should be negligible), the continuity equation then indicates that the average velocity at a cross-section varies linearly between xu above the high points of the bed and u where the depth is xd. From this, the reach-averaged depth, D, and velocity, U, of a reach of length, L, are

$$D = \left(\frac{x + 1}{2}\right)d \quad \text{and} \quad U = \left(\frac{x + 1}{2}\right)u \qquad (73)$$

If the expression given above for U is equated to the Manning equation, it can be noted that the resulting expression will only be valid when x = 1 or the channel is truly uniform. The resulting expression then must be written as an inequality that has the following form if the effect of channel nonuniformity is to be investigated. The inequality is written as

$$\left(\frac{x + 1}{2}\right)^{1/3} u \geq \frac{c}{n} d^{2/3} S_o^{1/2}$$ (74)

Note that this expression only reduces to the Manning equation if x = 1. Thus it is clear that the Manning equation and similar empirical equations used in steady-state models like QUAL2e[4] and others are not valid in a strict sense unless the flow is truly uniform. However, the Manning equation is frequently used to describe nonuniform flows because it is not possible to fully measure the nonuniformity and apply more precise methods. In many cases, especially those involving large rivers, $[(x + 1)/2]^{1/3} \cong$ 1. In fact, the cumulative observational knowledge for values of the Manning roughness coefficient[1,54,63] is frequently based on nonuniform channels. As a result, it seems that one should be able to confidently make screening-level calculations (preliminary estimate of stream conditions from information on file or easily observed) and easily calibrate a steady-flow model based on the Manning equation when the channel is approximately uniform.

If the channel is not very uniform, inordinately large values of n are required to calibrate a model. For the simple geometry in Figure 62, these inordinately large values of n are larger than the true value by the factor $[(x + 1)/2].^{1/3}$ This is explained by noting that the calibration value of n expresses the effect of friction losses due to channel form (or contraction and expansion losses) plus the skin friction effect using a term that was developed for the effects of skin friction alone.[56] Therefore, once the decision is made to apply the Manning equation to nonuniform flows where an expression equivalent to $[(x + 1)/2]^{1/3}$ for irregular channels is significantly >1, it should be realized that any calibration is severely limited to the range of conditions measured for calibration. This would, for example, be especially true in pool-and-riffle streams assumed on average to be uniform.

To relax the constraint related to assuming that the flow is uniform, several flow routing models do allow the user to express n as a function of depth. This is purely an empirical correction, however, because by definition the value of n is independent of depth for uniform flows. See Strickler's equation given by Henderson,[55] to confirm this. It should also be noted that the empirically-based practice of expressing n as a function of depth does represent the depth-varying effects of varied flow in constrictions and expansions of the channel, but it also represents other effects as well. The other effects may include unsteady flow and changing roughness due to vegetation. Therefore, these kinds of calibrations that employ a depth-variable n must be carefully confirmed for unsteady flow modeling and care must be used to not extrapolate low flow studies in nonuniform channels to significantly larger or smaller flow rates.

As an alternative, there is an approach that is occasionally used for calibration of the uniform flow equation in nonuniform channels. In this approach, the Manning n is limited to a restricted range of values that is representative of head losses due to the channel roughness but not the channel irregularity. Values of velocity and depth, or both, cannot be precisely predicted. For example, see McCutcheon et al.[9] where the pool-and-riffle stream in Figure 63 was described by keeping n in a reasonable range (less than 0.06) and changing the measured cross-sectional areas to reduce the depth. In the example study, the investigator originally assumed that cross-sections were not measured over short enough intervals and therefore assumed the latitude to amend the

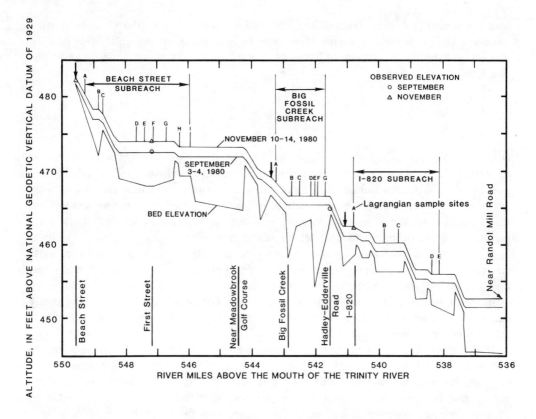

FIGURE 63. Bed profile and water-surface profiles of the West Fork Trinity River in Fort Worth, TX in the fall of 1980. Note that 1 mi = 1.61 km and 1 ft = 0.3048 m.

physical measurements of cross-sectional area (and thus depth), which is generally poor practice. The result was that the calibrated model used estimates of the Manning n that reasonably reflected the influence of skin friction losses to precisely predict travel time in the stream, but average depth was underpredicted because expansion and contraction losses were ignored. This result is consistent with the prediction from Equation 74 when the Manning n is held at the true value due to skin friction. In a highly nonuniform channel of a regular shape like that in Figure 62, predictions based on uniform flow equations would underpredict depth by a factor of $[(x + 1)/2]^{-1/2}$ if true values of velocity and Manning n are used in the flow equations. Velocity (or n) would be overpredicted by a factor of $[(x + 1)/2]^{1/3}$ if true values of n (or velocity) and d are used in the flow equations. Therefore, this analysis indicates that many flow calibrations probably account for errors in misusing the uniform flow equations by a combination of underpredicting depth, overpredicting velocity, and specifying n as a value larger than the true uniform flow value. McCutcheon[2] chose to use very large values of n (up to 0.12) in some reaches of the Chattahoochee River shown in Figure 64 so that depth and velocity could be accurately simulated. By contrast, the study of the West Fork Trinity River shown in Figure 63 limited the range of n and underpredicted depths (or cross-sectional area) on the order of 10 to 50% in some locations. In the case of the Chattahoochee River, the study of the flow condition was descriptive and intended to describe the measured conditions as accurately as possible and was not intended to predict the effect of different flows. The study of the West Fork Trinity River was originally intended to involve prediction of the effect of unsteady flows. Thus the purpose of the study dictates how approximate methods will be applied and when less approximate methods are required.

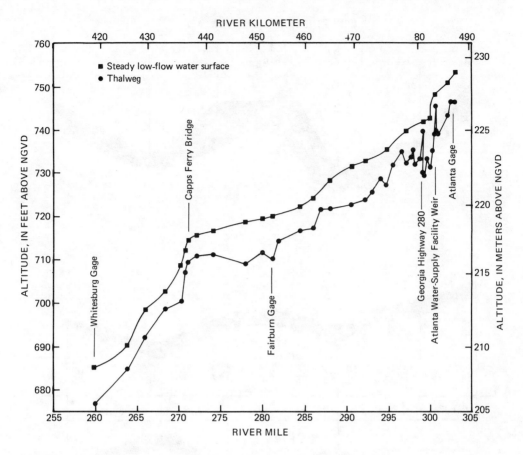

FIGURE 64. Bottom and water surface profile for the Chattahoochee River downstream of Atlanta, GA, in 1976. Note that NGVD is the U.S. National Geodetic Vertical Datum (mean sea level).

Recognizing that there are limitations to the use of uniform flow equations naturally leads to the next step of determining if criteria can be derived to specify the practical limits of applicability. A simple criterion can be derived from noting that one-half the sum of the velocities at stations 1 and 2 in Figure 62 is equal to the reach averaged velocity. Equating Equations 73 and 74 results in

$$\left(\frac{x + 1}{2}\right)u = \frac{1}{2}(u_1 + u_2) = \frac{S_o^{1/2}}{2n}(d^{2/3} + x^{2/3}d^{2/3}) = \frac{S_o^{1/2}}{n}D^{2/3} \qquad (75)$$

If it is noted that the reach averaged depth, D, is related to d in Equation 73 and d is relabeled as d_m (the adjusted depth obtained by shorting the true observed depth, d, so that the true nonuniform reach average velocity, U, can be described with a uniform flow equation), the following expression is the result

$$D = \left(\frac{x + 1}{2}\right)d_m \qquad (76)$$

From this equation, the fraction by which depth is underpredicted when the ratio of true velocity to predicted velocity is 1, can be written as

$$\frac{d}{d_m} = \frac{(x + 1)2^{1/2}}{(1 + x^{3/2})^{3/2}} \qquad (77)$$

LONGITUDINAL   SECTION A-A

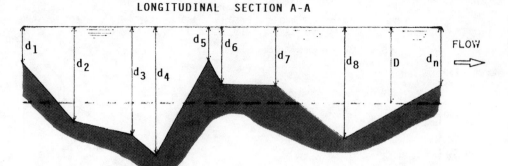

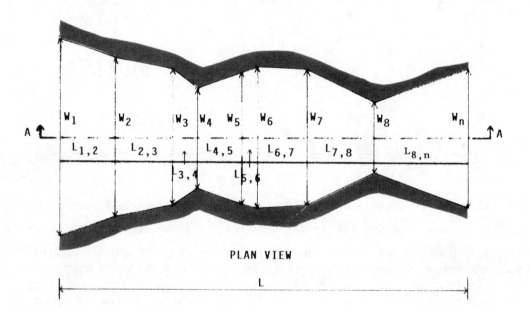

FIGURE 65.   Irregular stream geometry.

where it is important to note that $d/d_m = 1$ when $x = 1$. As x increases from a value greater than one, the depth becomes more difficult to represent with a uniform flow equation like the Manning equation where $d \geq d_m$. For example, if a stream has a profile similar to that in Figure 62 and the pools are twice as deep as the riffle sections, one would anticipate that the predicted depth would be approximately $1/1.26 = 0.79$ of the true depth.

It is also possible to derive a similar expression for $U \leq U_m$, but for this purpose it is useful to consider a more general case of irregular geometry. This is done by writing an expression for the average velocity, U, over the reach length, L, for the highly irregular reach shown in Figure 65. That equation is

$$U = \frac{\sum_{i=1}^{n-1}\left[\left(\dfrac{\dfrac{Q}{d_i w_i} + \dfrac{Q}{d_{i+1}w_{i+1}}}{2}\right)(L_{i,i+1})\right]}{\sum_{i=1}^{n-1}[L_{i,i+1}]} \tag{78}$$

If it is noted that $Q = U_m A$ where U_m is a measured or true value of the average velocity in the reach and A is the average cross-sectional area in the reach, then

$$\frac{U}{U_m} = \frac{\sum_{i=1}^{n-1}\left[\left(\dfrac{\dfrac{1}{(d_i)(w_i)} + \dfrac{1}{(d_{i+1})(w_{i+1})}}{2}\right)(L_{i,i+1})\right]\sum_{i=1}^{n-1}\left[\left(\dfrac{d_i + d_{i+1}}{2}\right)\left(\dfrac{w_i + w_{i+1}}{2}\right)(L_{i,i+1})\right]}{\left[\sum_{i=1}^{n-1}(L_{i,i+1})\right]^2} \tag{79}$$

where $L = \Sigma L_{i,i+1}$, d_i = depth, and w_i = width at section i. Note that Equation 79 should describe the overprediction of velocity or travel time in a nonuniform reach of an irregular shape.

Equations 77 and 79 have not been tested, but it seems probable that these equations (or a more rigorously derived form for depth) could be useful for determining if more cross-section measurements are necessary. Preliminary modeling should include sensitivity analysis to estimate the importance of predicting velocity and depth and the relative importance of each. If the study objective does not require prediction of velocity and depth for conditions different from the conditions used to calibrate the model (i.e., waste load allocation study at or near the critical low flow condition), then it may be possible to precisely describe the velocity and depth in the stream with values of the Manning n that may be less and less realistic as stream bed irregularity increases (see the hydraulic calibration for the Chattahoochee River in McCutcheon[2] for example). Where precise predictions are necessary and the calibrated model is not sensitive to the values of velocity or is not sensitive to the values of depth, it should be possible to chose realistic values of n and to accurately predict the parameter to which the model is sensitive but not both if the stream is irregular and nonuniform. When accurate predictions of velocity and depth would be useful, Equation 77 or 79 should be useful in defining when additional cross-sections should be measured if error criteria can be estimated from the sensitive analysis. For example, if it is determined that velocity should be predicted as accurately as possible and depth should be predicted to within 5%, then we can solve Equation 77 for $d/d_m = 0.95$ and determine that the change in average depth from one section to the next should not change by more than 15%. This estimate is limited by the assumption that the variation between sections is linear and this may not be true, especially for longer reaches or those reaches where morphological changes are expected. Examples of such changes include deposition of gravel bars at the confluence of tributaries, changes in geologic strata, and changes due to man, such as dredging. In addition, more precise flow routing models that incorporate fuller descriptions of flow accelerations in irregular channels and use higher order interpolation schemes between measured sections are not intrinsically error prone as is the application of the uniform flow equations to nonuniform conditions.

G. BACKWATER ANALYSIS

The backwater analysis is applicable when the flow rate is steady and the channel is nonuniform. Whether the backwater analysis is more accurate and easier to calibrate depends on how contraction and expansion losses due to channel irregularities are handled.

Typically, the backwater solution is used when the Manning equation is not applicable (i.e., the channel is too irregular to fit the uniform flow assumption). In addition, the backwater solution is used in dynamic modeling to provide the initial conditions that are tedious to measure, or to check cross-section measurements to be sure that they are consistent. Occasionally, benchmark errors occur in surveys that may be detectable with a backwater solution.

The equations for the backwater solution are derived from the St. Venant equation (Equation 21) by noting that the time-dependent terms are equal to zero:

$$\frac{\partial Q}{\partial t} = \frac{\partial A}{\partial t} = 0 \tag{80}$$

When the steady-state version of the St. Venant equation is solved for the change in water-surface elevation, y, with distance downstream, x, the resulting equation is written as[49]

$$\frac{\partial y}{\partial x} = \frac{-\dfrac{CQ}{gA^2}\dfrac{\partial Q}{\partial x} + \dfrac{Q^2}{gA^3}\dfrac{\partial A}{\partial x} - \dfrac{\partial a}{\partial x} - \dfrac{Q^2 n^2}{A^2 R^{4/3}}}{\left(1 - \dfrac{Q^2}{gA^3}\dfrac{\partial A}{\partial y}\right)} \tag{81}$$

where C is a momentum correction coefficient equal to 1 or 2 depending on whether discharge, Q, increases or decreases in the downstream direction due to tributary inflows, g is the acceleration of gravity, A is the cross-sectional area, n is the Manning roughness coefficient, and R is the hydraulic radius.

French,[1] Davidian,[64] and the U.S. Army Hydrologic Engineering Center[65] (HEC) derive a similar backwater equation, but they clearly identify the separate head losses due to contraction and expansion of the channel and losses due to roughness on the channel bed.

The limitations of the backwater analysis is best stated by French[1] who notes that solution technique is generally based on the following assumptions:

1. The head loss in each specified reach can be represented by the head loss in a uniform channel of the same hydraulic radius (R) and average velocity (U) or, as expressed in terms of the Manning equation:

$$S_f = \frac{n^2 U^2}{c^2 R^{4/3}} \tag{82}$$

   where c = 1 if metric (SI) units are used and 1.49 for use of English units.
2. The slope of the channel is small.
3. The shape of the vertical velocity profile is invariant over the reach such that the kinetic energy correction does not vary over the reach (see French[1] for a definition of kinetic energy correction factors).
4. The Manning roughness coefficient is independent of depth and constant throughout each reach of interest.

In addition, French[1] notes that the presence of entrained air may violate the assumptions under which the equations are derived.

Equation 81 is one flow simulation option in the WQRRS model,[49] and this is coupled to the water quality package of that model. In addition, backwater calculations can be easily performed by the HEC-2 Water Surface Profiles program[65] and the U. S. Geological Survey backwater program.[66] The HEC-2 model is the most widely tested model of the models available.

H. SIMULATION OF UNSTEADY FLOW

On some occasions, it may be necessary to model the effect of unsteady flow on water quality. Typically, dynamic modeling is useful in analyzing the effect of dam releases and flood run-off when these effects can be associated with water quality conditions that are more severe that those that occur during steady-state conditions.

The effect of dam releases and flooding is variable depending on the conditions that occur in a stream. If a stream does not accumulate settled material during periods of low flow, the effect of increased flow is to dilute pollutants entering the stream at other locations unless the flood water is more polluted. Faster velocities move phytoplankton out of the river reaches of interest faster and in some cases reaeration may be increased. If the stream accumulates organic material at lower flows, then high flow may resuspend the organic material and the reduced material in the bed as well. Whether dilution or reentrainment predominates depends on the increase in flow and the material that is suspended, plus the quality of the inflowing water. Stormwater may be too polluted to provide dilution.

Dynamic modeling seems to be taking on more importance as many point sources are cleaned up. In a number of cases, steady-state models have been used to design new wastewater treatment plants and upgrade old plants only to find that severe problems can still occur because of dynamic events that were not considered in the design. In addition, there is some interest in setting seasonal discharge limits that may be possible on a wider scale if dynamic modeling can be advanced. Furthermore, there are limits to the steady-state, critical flow analysis that can only be relaxed with a better understanding of dynamic water quality conditions. For example, some streams have drought flow conditions (usually measured by the 7-d average flow that occurs once every 10 years) that would require much longer that 7 d for the polluted waters to travel to the critical impact areas, such as dissolved oxygen sag point, before the flow would be expected to be increased naturally. A dynamic analysis may be useful in exploring the real effect of critical conditions on such an occasion. Finally, greater interest in the water quality effects of dams have been expressed over the last decade. More and more effort is being focused on determining the detrimental and advantageous effects of reservoir releases before projects are begun. There is also a need to determine the cause of water quality problems after completion. The review by Dortch and Martin[67] reports on the progress in applying models under these conditions.

There are two general categories for dynamic flow modeling. These include hydrologic and hydraulic flow routing. Hydrologic routing involves solving the continuity equation and an empirical relationship for the storage of water in a channel vs. discharge. Hydraulic routing involves solving the continuity equation and the conservation of momentum equation or an approximation of the momentum equation.

All methods for dynamic flow simulation have approximately the same data requirements. Cross-sectional shape must be measured at distance intervals that are spaced closely enough to adequately define the channel volume and the effect of irregularity on the head losses in the flow. As indicated in the section on steady-flow studies, little guidance exists in this regard. In addition, the purpose of the study

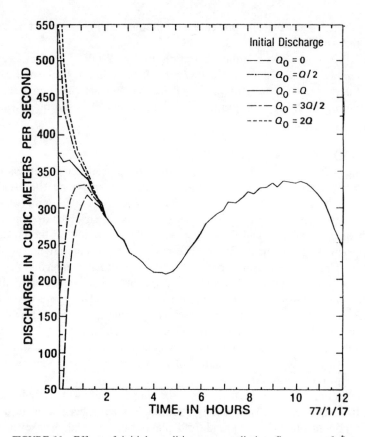

FIGURE 66. Effect of initial conditions on predicting flow rate of the Sacramento River near Freeport, CA, using the BRANCH flow routing model.[68]

dictates the precision necessary for measurement. However, little guidance is available relating modeling uncertainty to measurement intensity. This remains an art.

The cross-sections must be referenced to a common datum such as sea level (i.e., U.S. National Geodetic Vertical Datum, mean Yellow Sea level, etc.) or the slope between sections must be known. Alternatively, some hydrologic routing methods may require that channel volume vs. water-surface elevation be estimated.

Like any dynamic model, flow routing models require that boundary and initial conditions be specified. Boundary conditions consist of inflow hydrographs (discharge vs. time) for all significant tributaries, other inflows, and withdrawals, plus the inflow or outflow hydrograph (or select water surface elevations) must be specified. Initial conditions require that discharge and water surface elevations be specified at all internal computational nodes at the beginning of the simulation. However, if the simulation covers periods of time that are longer than the time it takes the dynamic wave (faster than mass transport) to move through the reach, then initial conditions are not as important. This is illustrated in Figure 66 where the BRANCH model[68] was used to simulate discharge at a station on the Sacramento River near Freeport, CA. Within 2 h, none of the assumed initial discharge rates at this station has any effect on the simulation of the tidally affected flow. In fact, many studies have relied on a backwater solution to estimate reasonable initial conditions rather than attempt measurements. Calibration data may consist of hydrographs at selected internal points in the model domain as the first choice or may consist of water surface elevations.

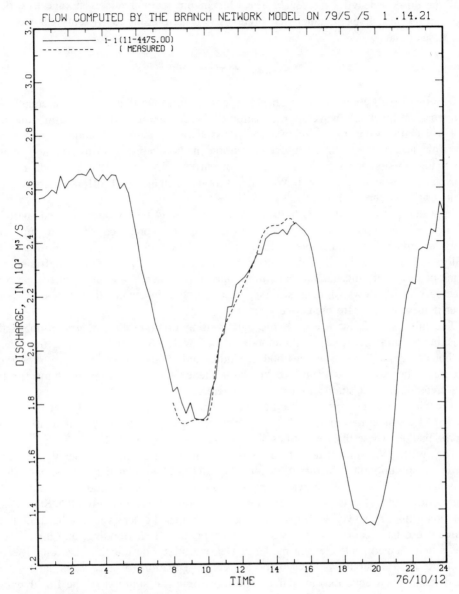

FIGURE 67. Predicted and measured discharge using the dynamic flow model, BRANCH, for the Sacramento River near Freeport, CA.[68]

Figure 67 illustrates the comparison of measurements and simulations used to calibrate values of the Manning roughness coefficient in the BRANCH model.[68] It may also be noted that the same types of cross-section data, boundary conditions, and calibration data are required for the backwater analysis except that a single steady-flow measurement is required in place of defining hydrographs.

Dortch and Martin[67] note that all hydrologic routing methods are based on the continuity equation written as

$$\frac{dS}{dt} = I - O \tag{83}$$

where dS/dt is the rate of change of storage in a reach, I is the inflow into the reach, and O is the outflow leaving the reach. The Muskingum method relates storage to inflow and outflow in a linear manner that involves two calibration coefficients per reach, k and x, such that

$$S = kO + xk(I - O) \tag{84}$$

The modified Puls method is somewhat less empirical in that the storage equation is derived from kinematic wave approximation to the conservation of momentum equation. The kinematic wave approximation is discussed later in this subsection. Dortch and Martin[67] note that the linear reservoir routing method (applicable to streams) relates the change in storage to the change of rate of outflow. There are other methods as well. Dortch and Martin note that Wurbs[67,69] reviews these and offers an annotated bibliography of the methods.

In water quality modeling, hydrological modeling is not frequently used, perhaps because of the empirical or semi-empirical basis of the storage equations. All methods seem to involve one or more empirical coefficients that must be determined by calibration. It is not clear if an effort has been made to express these coefficients in terms of physically measurable quantities. Therefore, much data collection seems to be necessary at all flows of interest. As a result of the data collection needs, it is not clear if these method are predictively valid.

The most significant use of hydrologic routing involves the incorporation of the Muskingum and modified Puls methods in the WQRRS model.[49] In addition, these methods do enjoy some usage in flooding studies and in studies associated with reservoir releases. Therefore it is useful to know of the existence of these methods for the interpretation of past studies in a reach of interest.

The three most common hydraulic routing methods involve the solution of (1) the full St. Venant equation, (2) diffusion analogy equation, and (3) kinematic wave approximation. These three equations are summarized in Equation 21.

The full St. Venant equation takes into account, in an explicit manner when solved correctly, all important factors that contribute to head loss in dynamic stream flow. This includes losses due to resistance to flow from channel roughness, expansion, contraction, and effects of acceleration of the flow. Unfortunately, the St. Venant equation cannot be solved with ease and has not been used extensively. Perhaps the most frequent use has occurred in the study of dam breaks, flash flooding, and mud flows. The undocumented finite element model of DeLong briefly described in DeLong[70] seems to be the most accurate solution of the St. Venant equation (or a very similar form of the dynamic flow equation) available. The most practical solution of the full dynamic equation seems to be that of Fread[71] incorporated in the DAMBRK model that has been widely applied. Bedford et al.[72] incorporate Fread's method in the CE-QUAL-R1 model. The WQRRS model[49] incorporates a finite element solution of the dynamic equation as well. In addition, there may be a number of other models based on the full equations that are useful in water quality modeling. It is difficult to determine this, however, because dynamic routing models can be quite complex. Due to the complexity of the code, few if any have received independent peer review of the computer code and documentation.

The diffusion analogy simplifies the full equation by one term that ignores the accelerative effects of the flow on head losses. The result is that occasionally the flow equation must be calibrated with atypical Manning n values, and the solution may be inaccurate and unstable for rapidly varying flows like dam breaks and flash floods. Evidently, however, the simplification does not lead to enough computational efficiency to facilitate widespread application. French[1] develops criteria to define the limits of use.

The kinematic wave approximation is perhaps the most widely used dynamic flow routing equation for water quality related studies. It is one of the flow routing options in the WQRRS model[49] and is the basis of flow routing in most advanced watershed models. Evidently, this popularity has arisen because of the ease with which the equations could be solved. With the recent introduction of almost unlimited computing resources (in the form of personal computing systems), however, the use of the kinematic wave approximation may decrease somewhat as solutions of the full dynamic equation become more practical.

The kinematic wave equation can be written as[49]

$$\frac{\partial A}{\partial t} + \frac{\partial Q}{\partial x} - s = O \qquad (85)$$

where s is the source and sinks of flow per length of channel. Comparison with the conservation of momentum equation (Equation 17), written in terms of similar state variables (Q and A), indicates that the kinematic wave approximation neglects those terms related to inertia or acceleration of the flow and pressure differences in the flow. The kinematic wave approximation retains only the effect of channel friction and gravity. In effect, the flow at any point in the stream is always uniquely related to depth, i.e., it is assumed that there is no hysteresis in the rating curve like that shown in Figure 59.

It has long been suspected that the kinematic wave approximation is quite limited for precise applications. Investigators have experienced difficulties in calibrating the kinematic wave equation in flat coastal areas such as Florida. Where channel slopes are flat, adequate calibrations have required increases in slope to values greater than actually measured. In most cases, such manipulation of physical data does not lead to confidence in the calibrated model. To avoid concerns about misuse of the approximation, French[1] reports criteria useful for defining the applicability of the kinematic wave approximation.

II. CONSERVATIVE TRANSPORT AND MIXING OF DISSOLVED SUBSTANCES

In the first section, it was implied that mixing may only be important in river water quality modeling under certain circumstances. Generally, it can be expected that mixing is important when temperature or concentration gradients are large. This tends to be the case under dynamic conditions when loads and flows may vary with time. Gradients also tend to be large in mixing zones where inflows mix with the receiving stream.

Generally, dispersive mixing is unimportant for steady-state modeling of conventional pollutants at the concentrations typically discharged into streams in the U.S. Bowie et al.[73] note that Thomann,[73,74] Li,[73,75] and Ruthven[73,76] have investigated the effect of mixing. There have been other studies as well. Ruthven developed a simple criterion to determine when dispersion may be important in the modeling of constituents released at a constant rate, into a constant flow, and subject to first-order transformation processes. That criterion indicates when dispersive mixing may cause calibration errors in first-order rate coefficients that are in excess of 10%. The criterion is written as

$$\frac{kD_x}{U^2} < 0.04 \qquad (86)$$

where U is average stream velocity, D_x is the longitudinal dispersion coefficient, and k is a first-order rate constant. As an example, Bowie et al.[73] investigated typical stream conditions for a decaying constituent such as biochemical oxygen demand where it was assumed that U = 0.3 m/s (1.0 ft/s); D_x = 46.5 m^2/s (500 ft^2/s); and k = 0.5 day^{-1} and found that kD_x/U^2 was 0.003. This is an order of magnitude less that the criterion (0.04). Therefore, Equation 86 indicates that dispersion may be important if the rate constants or dispersion coefficients are an order of magnitude higher than normally expected or if average velocity is extremely slow.

Although it has been shown that dispersion is relatively unimportant for many steady-state, one-dimensional flows, the effect of dispersive mixing is usually considered to be important when flows and loads are unsteady. The only exception seems to involve loads or flows that vary some but not enough to violate the vague definition of steady-state that is used for modeling purposes. In this regard, Bowie et al.[73] note that Thomann[73,74] indicates that dispersive mixing is important for small rivers with waste loads that vary with periods of less than 7 d. Because many wastewater treatment plants have a distinct diel variation in effluent water quality or at least a 5-d pattern due to the U.S. 5-d work week, Thomann's work seems to indicate that dispersion may be important in most dynamic water quality studies involving significant quantities of municipal or industrial wastewaters. In addition, Thomann indicates that dispersion is important whenever loads into large rivers are variable. Therefore, dynamic models seem to require that at least the longitudinal dispersion coefficient be specified as a matter of standard practice.

In Chapter 1, the derivation of the dynamic mass balance equation (advective-dispersive equation) shows that the dispersion coefficient is an empirical parameter derived from the eddy-diffusivity assumption. As a result of its empirical basis, the dispersion coefficient must be estimated or measured. The methods available for this will be discussed in this section.

Because tracer studies are necessary to measure dispersion coefficients, the use of fluorescent dyes is also reviewed in this section. The behavior of dyes in streams also has some influence on measurements of time of travel and dilution discharge measurements. However, it is expected that dye stability and losses have the greatest affect on dispersion measurements. Therefore, the use of fluorescent dye will be reviewed and the causes of instability and losses of dye described.

A second important mixing problem involves the typical assumption that a stream can be treated as a one-dimensional network of computational elements. As indicated in Figure 68, flows entering a stream do not mix immediately and knowledge of how quickly this mixing occurs is important in choosing and calibrating a model. In very large rivers like the lower Han in Korea and the lower Mississippi, reasonably complete mixing may never occur. If complete mixing does not occur, then one-dimensional models can be applied, but the results must be properly interpreted. For example, the calibrated model may require that unrealistic values of the longitudinal dispersion coefficient and kinetic parameters be specified to describe existing conditions. Such a calibration may not be useful for prediction or analysis of slightly different conditions. Even relatively quick mixing will cause the calibration to be slightly in error if the unmixed conditions cause the actual mass transport to be slower that the average that would occur if the inflow mixed instantaneously.

Calibration errors in the longitudinal dispersion coefficients arise because lateral and vertical mixing effects are lumped together with longitudinal mixing when a one-dimensional model is calibrated with laterally averaged data describing longitudinal gradients. This can be illustrated by comparing the mixing process in the mixing zone to mixing downstream of the mixing zone. Downstream of the mixing zone

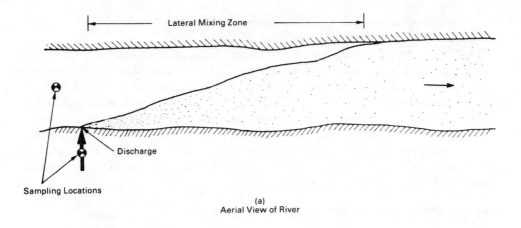

(a)
Aerial View of River

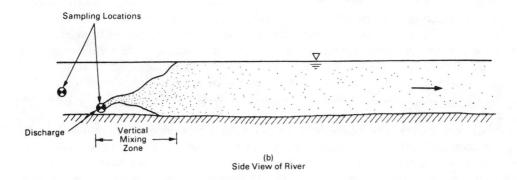

(b)
Side View of River

FIGURE 68. Zone of lateral and vertical mixing for discharges into a stream. (From Mills et al., EPA Rep. EPA/600/6-86/013, 1986.)

longitudinal mixing occurs because nonuniform lateral and vertical velocity profiles cause water in the center of the stream away from the bottom to move faster and overtake slower moving water downstream. This same process occurs in the mixing zone but, in addition, mixing due to vertical and lateral gradients is also superimposed. As a result, more mixing is occurring in the unmixed cross sections. In a uniform flow, the lumped one-dimensional coefficient must be increased in the unmixed zone compared to the mixed zone downstream.

In addition to overestimation of dispersion coefficients, kinetic parameters will be incorrectly estimated because the pollutant mass will generally not move downstream at the cross-sectional average velocity. If the inflow is introduced near the bank where velocities are slower than average, the pollutant mass will travel slower than average and vice-versa for discharges into the faster moving center flow. Thus a bank discharge will tend to cause kinetic decay coefficients to seem to be larger than actual values observed outside the mixing zone and vice-versa for center discharges moving faster than average.

Water quality kinetic calibrations do not not seem to be influenced by inhomogeneous concentrations in mixing zones unless composite samples are not representative of the cross-sectional average and the pollutant mass moves faster or slower than the water mass. This can be shown using a simple stream tube calculation[77] and comparing the total decay at the downstream end of a mixing zone to the decay determined by averaging concentrations over the cross-section throughout the mixing zone. When longitudinal mixing is unimportant (as normally expected), the one-dimensional and two-dimensional calculations are equal.

To aid in understanding the effect of mixing, the method to calculate the length required for mixing to occur will be reviewed. To be sure the one-dimensional approximation is accurate, the mixing distance can be computed to determine if calibration parameters are significantly different in the mixing zone compared to the calibration parameters determined for downstream reaches. It is also important to know the mixing distance to properly design calibration data collection studies. Composite samples representing laterally averaged concentrations must be collected in the mixing zone. Finally, simple mass balances will be reviewed in this section to determine methods of checking sampling programs.

A. ESTIMATION OF DISPERSION COEFFICIENTS

The derivation of the advective-dispersive equation makes it clear that any formulation to predict dispersion coefficients must be empirical in nature and based on observation under conditions similar to those being modeled. The extent to which a stream can be approximated by a one- or two-dimensional model also influences how well dispersion coefficients can be estimated. Furthermore, the degree of averaging in space and time in solving the advective-dispersive equation can influence the magnitude of the dispersion coefficient chosen by calibration. In some cases, numerical dispersion can be comparable to actual dispersion. When this occurs, the dispersion is specified as zero (i.e., $D_x = 0$), or better still, as the difference between the actual and the numerical dispersion.

Dispersion coefficients have been estimated for a number of rivers in the U.S. and elsewhere. Initially, these studies seemed to be aimed at quantifying the ability of streams to assimilate wastes.[78-81] Later dispersion studies were conducted to show that dispersion was not an important part of the steady-state waste assimilative capacity. More recently, interest seems to have returned to the dispersive capability of a stream as concerns have heightened regarding the mixing of hazardous waste spills and as efforts have intensified in the modeling of dynamic water quality conditions.

Table 14 from Bowie et al.[73] lists and reviews the longitudinal dispersion studies performed prior to 1985. This compilation shows that a number of empirical formulations are available to estimate dispersion coefficients.

Included in Table 14 are a number of important studies. Originally, Taylor[82] related longitudinal dispersion in pipes to the shear velocity and hydraulic radius (area/wetted perimeter) by assuming that the logarithmic velocity profile was valid. Elder[83] extended this analysis to infinitely wide channels that ignored the effect of lateral velocity gradients on one-dimensional dispersion. The work of Glover,[78] Godfrey and Frederick,[88] and Fischer[44] made it clear that the effect of the lateral velocity gradient was very important and in fact seems to be more important than the vertical velocity gradient. From studies by Elhadi and Davar,[85] it is not clear that the tendency to report measurements in terms of a dimensionless dispersion coefficient, D_x/Hu_*, derived from Elder's assumption about the importance of the vertical velocity gradient, is fully useful. Elhadi and Davar[73,85] found that values of D_x/Du_* do not remain constant as assumed.

A number of laboratory studies have been attempted, but these have done little to advance the understanding of the process. These efforts seemed to have been unsuccessful because of the lack of similarity between well-behaved and well-controlled flume flows and the flows occurring in streams.

Finally, several investigators have noted that the assumption regarding the Fickian diffusion definition of the dispersion coefficient may not be universally valid and have

TABLE 14
Studies Related to Longitudinal Dispersion

Ref.	Summary of results

82 $D_x = (10.1) R_p u_*$ (for pipe flow)

83 $D_x = (5.93) D u_*$ (lateral velocity variation not considered)

78 $D_x = (500) R u_*$ (for natural streams)

84 $D_x = 6.4(D^{1.24})E^{0.3}$ (for two-dimensional channel where $E = USg$)

79 $D_x = (14.3) R^{3/2} (2gS)^{1/2}$ (for open-channel flow)

80, 81   Method of moments:

$$D_x = \frac{U^2}{2} \left(\frac{\sigma_{t2}^2 - \sigma_{t1}^2}{t_2 - t_1} \right)$$

where the concentration variances, σ_{t2}^2 and σ_{t1}^2 are measured after the initial period when lateral and vertical mixing is complete. Long tails on the time vs. concentration curve at a site may introduce some error.

Integral equation:

$$D_x = - \frac{1}{A} \int_0^w q'(y)dz \int_0^y \frac{1}{N_y d(y)} dy \int_0^y q'(y)dz$$

where

$$q'(y) = \int_0^y [u(y,z) - U]dz$$

This formula considers the effects of lateral velocity changes across the stream.

Simplification of the integral equation above:

$$D_x = \frac{1}{2} \frac{d\sigma_x^2}{dt}$$

$$D_x = 0.3\overline{u'^2} \frac{l_y^2}{Ru_*}$$

Fischer also discusses another method for determining D_x called the routing procedure.

85   Reviewed many methods to predict D_x and found that $D_x/(Du_*)$ is not a constant as reported by many researchers.

86   Field measurements of D_x were made in the Green and Duwamish Rivers in Washington State.

87

$$\log\left(\frac{KU_s D_x}{U^2 D}\right) = 6.45 - 0.762\log\left(\frac{\rho U D}{\mu}\right)$$

$$\log\left(\frac{KU_s}{U} \frac{D_x}{u_* D}\right) = 6.467 - 0.714\log\left(\frac{\rho u_* D}{\mu}\right)$$

Formulas developed by curve fitting the data existing at that time. These equations have not been closely examined.

88   Dispersion tests were summarized for five natural streams. The measured dispersion coefficients were from 4 to 35 times greater than predicted by Taylor's[82] method.

89 $D_x = (7.25) D u_*$ (for two-dimensional channels)

90   The limitations of dispersion equations which do not consider lateral velocity variations are discussed. Site-specific measurements of D_x are recommended.

91   In laboratory experiments, D_x varied from 0.056 m to 6.13 m²/s (0.6 to 66 ft²/s).

92   Dispersion coefficient studies and the associated hydraulic data were reviewed for 17 rivers and the following equations were developed.

$$D_x = 0.66 \frac{U^3}{C_w} \frac{Q_o}{2S_o W_o}$$

TABLE 14 (continued)
Studies Related to Longitudinal Dispersion

Ref.	Summary of results

$$D_x = 0.058 \frac{Q_o}{S_o W_o}$$

93
94

Dispersion tests performed in the Mississippi River are summarized.

$$D_x = \frac{\beta Q^2}{u_* R^3}$$

where

$$\beta \simeq 0.18 \left[\frac{(gRS)^{1/2}}{U} \right]^{1.5}$$

Summary of D_x values also reported.

95

$$D_x = \frac{0.011 U^2 W^2}{D_u}$$

Liu[94] shows this is a special case of his formula when $\beta = 0.011$.

96   Several conceptual models of mass exchange with dead zones are presented, and the Fickian equation is modified to include mass transfer to and from dead zones.

97   Application of Hays et al.[96] dead zone model to TVA stream data.

98   Longitudinal dispersion of fluid particles in small mountain streams in New Zealand was investigated. It was shown that the dispersion coefficient increased with distance and never approached an asymptotic value.

99   Longitudinal dispersion of fluid particles in the Missouri River and in a small mountain stream was investigated. The dispersing particles were shown to behave differently from the Taylor type model. A method to predict dispersion was developed.

100   A non-Fickian model is presented to predict stream dispersion.

101   A modified model of stream dispersion is presented that includes the effects of storage along the bed and banks.

102,103   Effects of dead zones on longitudinal dispersion are demonstrated, and it is shown how dead zones modify longitudinal dispersion.

104   A hybrid method is discussed to predict dispersion in the Waikato River, New Zealand.

105   Dispersion processes in streams are reviewed, and it is shown that many experimental results do not agree very well with the Fickian dispersion theory. A non-Fickian dispersion model is proposed.

106   Dispersion in steep mountain streams is examined.

107   Fischer's methods are successfully applied to predict dispersion in mountainous streams.

108   Methods are developed to predict dispersion in rivers including the effects of dead zones, using a (j, n, m) type model. See reference for further explanation.

109   The Fickian equation is solved with a Lagrangian scheme to avoid lumping numerical dispersion with actual physical dispersion.[110]

111,112   Determined that D_x and coefficients for nonconservative water quality constituents could be determined simultaneously during calibration. D_x determined by this method is in good agreement with literature values (Jobson) or match D_x values determined from dye studies.

113   Numerical dispersion minimized with a Lagrangian routing procedure that provides more consistent estimates of D_x for pool and riffle streams than the methods of moments. Applying this procedure to peak dye concentrations yielded D_x to within 10% of estimates based on the entire concentration-time curves.

Notation:

D_x	= longitudinal dispersion coefficient
A	= cross-sectional area
W, W_o	= channel width; the subscript o refers to width at steady, base flow

TABLE 14 (continued)
Studies Related to Longitudinal Dispersion

C_w	= wave velocity
$d(y)$	= depth of water at y
E	= rate of energy dissipation per unit mass of fluid = USg
N_y	= lateral turbulent mixing coefficient (lateral eddy-diffusivity)
D	= stream depth
K	= dispersion factor
l_y	= lateral distance from location of maximum velocity
σ_x^2	= variance of distance — concentration curves
$\sigma_{t1}^2, \sigma_{t2}^2$	= variance of time — concentration curves at stations 1 and 2
t_1, t_2	= mean times of passage for the dye clouds to past stations 1 and 2
ρ	= mass density of water
Q, Q_o	= discharge; the subscript o refers to discharge at steady, base flow
$q'(y)$	= integral of velocity deviation over the depth
R	= hydraulic radius
R_p	= pipe radius
S, S_o	= slope of energy gradient; the subscript o refers to slope at steady, base flow
U	= average velocity in a stream reach, or at a section at an instance in time
u'	= deviation of velocity from cross-sectional mean velocity
U_s	= mean velocity of flow at a sampling point
u_*	= shear velocity
g	= gravitational constant
$u(y,z)$	= point velocity at cross-sectional coordinate
y	= depth above bottom of stream
z	= coordinate for lateral distance across a stream flow
t	= time
μ	= coefficient of water viscosity

From Bowie, G. L., Mills, W. B., Porcella, D. B., Campbell, C. L., Pagenkopf, J. R., Rupp, G. L., Johnson, K. M., Chan, P. W. H., and Gherini, S. A., Rates, Constants, and Kinetics Formulations in Surface Water Quality Modeling, 2nd ed., EPA/600/3-85/040, U.S. Environmental Protection Agency, Athens, GA, 1985.

offered supplemental corrections for other effects that are neglected or poorly represented. Fischer[44] notes that Fickian dispersion is not valid during an initial mixing period. Hays et al.[96] and other investigators have developed dead zone corrections that assume that mixing occurs into adjacent quiescent areas. In addition, a number of different conceptual approaches have been proposed to explain the non-Fickian behavior observed. Other hybrid methods and alternative simple turbulence closure methods have also been proposed.

Despite the impressive efforts to discover appropriate phenomelogical descriptions to cover a wide range of stream conditions or to cover narrow and important classes of conditions, there is little real guidance presently available. A number of investigators seem to imply that well-defined guidance cannot be reasonably derived by suggesting that the best approach is to measure dispersion coefficients in the field.

The most practical approach to estimating dispersion coefficients seems to be the approach employed in the QUAL 2e model.[4] Following the work of Elder, the dispersion coefficient is assumed to be proportional to shear velocity and depth

$$D_x = K_d D u_* \tag{87}$$

where K_d is a constant of proportionality. If the Manning equation is used to express the shear velocity $[u_* = (gDS_o)^{1/2}]$ in terms of more easily measured and estimated parameters, the expression above can be rewritten as

$$D_x = CK_d nUD^{5/6} \tag{88}$$

TABLE 15
Representative Values of the Manning Roughness
Coefficient, n[a]

Typical value	Range	Type of channel and condition
		Artificial Channels, Flumes, and Canals
0.010	0.008—0.013	Glass, plastic, machined metal (i.e., bronze)
0.011	0.010—0.015	Dressed timber, joints flush
0.014	0.010—0.018	Sawn timber, joints uneven
0.011	0.010—0.013	Cement plaster
0.013	0.011—0.015	Concrete, steel troweled
0.014	—	Concrete, timber forms, unfinished
0.016	0.015—0.017	Untreated gunite
0.014	0.011—0.018	Brickwork or dressed masonry
0.017	—	Rubble set in cement
0.020	0.016—0.033	Earth, smooth, no weeds
0.025	0.022—0.030	Earth, some stones, and weeds
		Natural Stream and River Channels
0.030	0.025—0.040	Clean and straight
0.040	0.033—0.050	Winding with pools and shoals
—	0.075—0.150	Very weedy, winding and overgrown
—	0.025—0.160	Floodplains
—	$0.031\ d^{1/6b}$	Clean, straight alluvial channels

[a] Table adapted from References 4, 54, and 55.
[b] d = D-75 size in feet = diameter that 75% of particles are smaller than.

where C = 3.82 if English units are used to express numerical values of velocity, U (ft/s), and depth of flow, D (ft), or 1.00 if the metric units of meters per second and meters are used. n is the Manning roughness coefficient (representative values are given in Table 15 from Brown and Barnwell[4] but comprehensive compilations are given in French,[1] Chow,[54] and Barnes[63]), and K_d is the dispersion constant. Representative values of K_d (dispersion constant) are given in Table 16 from Brown and Barnwell.[4] In addition, there may be other determinations of K_d published elsewhere.

Equation 88 has proven adequate because it has been applied in steady-state or quasi-dynamic conditions where dispersion is rarely of importance. Where dispersive mixing may be important for steady-state conditions, the QUAL2e model is sufficiently flexible to allow the user to change the value of K_d in different reaches to compensate for any basic flaw in the form of the equation (i.e., that D_*/Du_* is not constant[85]).

Many, if not all, of the studies reviewed in Table 9 seem to be aimed at deriving laws for describing longitudinal dispersion coefficients for stream. There are, however, a number of studies by the U. S. Geological Survey and other agencies where dispersion coefficients have been measured in streams. These studies are intended to provide estimates for dispersion of spills of hazardous materials. In addition, time of travel has been measured by dye studies in a number of stream reaches in the U.S. that can also be used to estimate dispersion coefficients. Other studies of reaeration using dye tracers provide similar information. These studies are too numerous to list but should be readily available from state offices of the U.S. Geological Survey and other federal and state agencies. Furthermore, many state pollution control offices make similar measurements.

Data for estimating dispersion coefficients are collected using the method of Hubbard

TABLE 16
Measurements of Longitudinal Dispersion in Open Channels[a]

Channel	Depth (ft)	Width (ft)	Average velocity (ft/s)	Shear velocity (ft/s)	Dispersion coefficient (ft^2/s)	Dispersion constant (unitless)
Chicago Ship Channel	26.5	160	0.89	0.063	32	20
Sacramento River	13.1	—	1.74	0.17	161	74
River Derwent	0.82	—	1.25	0.46	50	131
South Platte River	1.5	—	2.17	0.23	174	510
Yuma Mesa A Canal	11.3	—	2.23	1.13	8.2	8.6
Trapezoidal laboratory	0.115	1.31	0.82	0.066	1.3	174
channel with roughened	0.154	1.41	1.48	0.118	2.7	150
sides	0.115	1.31	1.48	0.115	4.5	338
	0.115	1.12	1.44	0.114	0.8	205
	0.069	1.08	1.48	0.108	4.3	392
	0.069	0.62	1.51	0.127	2.4	270
Green-Duwamish River	3.61	66	—	0.16	70—92	120—160
Missouri River	8.86	660	5.09	0.24	16,000	7500
Copper Creek (below gauge)	1.61	52	0.89	0.26	215	500
Clinch River	2.79	154	1.05	0.22	151	235
	6.89	197	3.08	0.34	581	245
	6.89	174	2.62	0.35	506	210
Copper Creek (above gauge)	1.31	62	0.52	0.38	97	220
Powell River	2.79	112	0.49	0.18	102	200
Clinch River	1.90	118	0.69	0.16	87	280
Coachella Canal	5.12	79	2.33	0.14	103	140
Bayou Anacoco	3.08	85	1.12	0.22	335	524
	2.98	121	1.31	0.22	420	640
Nooksack River	2.49	210	2.20	0.89	377	170
Wind/Bighorn Rivers	3.61	194	2.89	0.39	452	318
	7.09	226	5.09	0.56	1722	436
John Day River	1.90	82	3.31	0.46	151	172
	8.10	112	2.69	0.59	700	146
Comite River	1.41	52	1.21	0.16	151	650
Sabine River	6.69	341	1.90	0.16	3390	3100
	15.6	417	2.10	0.26	7200	1800
Yadkin River	7.71	230	1.41	0.33	1200	470
	12.6	236	2.49	0.43	2800	520

Note: 1 ft = 0.3048 m, 1 ft/s = 0.3048 m/s, and 1 ft^2/s = 0.0929 m^2/s.

[a] Adapted from References 4 and 44.

et al.[12] The procedure is given in Table 12 and is the same as that for time of travel studies except the complete dye curve passing a station is defined. In essence, the method consists of collecting concentration vs. time data like those defining the curves shown in Figure 69.

Dispersion coefficients are estimated from dye measurements by at least five distinctly different methods. All methods match mass transport predictions (like those represented by solid lines in Figure 69) with measured tracer concentrations by calibrating the dispersion coefficient or a similar factor.

The first method is known as the method of moments. This approach assumes that only Fickian dispersion occurs and thus cannot treat some aspects of natural dispersion too well. The primary manifestation of non-Fickian dispersion is the long tails frequently associated with concentration-time curves. The long tails are normally due

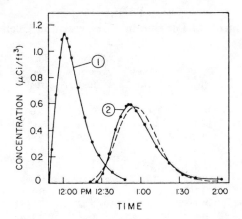

FIGURE 69. Time-concentration curves for an instantaneous or slug release of a radioactive tracer and illustration of predictions using Fischer's routing method[44] (dashed curve). The data defining curves (1) and (2) are from field experiment 1-60 of Godfrey and Fredrick.[88] The injection was made at 11:08 A.M. Curve (1) was measured 2399 meters (7870 ft) downstream and curve (2) at 4130 m (13,550 ft) downstream. The routing curve was predicted by choosing the dispersion factor, K, as 21 m^2/s or 230 ft^2/s. (From Fischer, H. B., List, E. J., Koh, R. C. Y., Imberger, J., and Brooks, N. H., *Mixing in Inland and Coastal Waters*, Academic Press, New York, 1979. With permission.)

to exchange with "dead zones". The long tails on the concentration-time curves are important because they distort estimates of D_x determined by the method of moments.

The dispersion coefficient is computed by the method of moments[1] as

$$D_x = \frac{U^2}{2}\left[\frac{\sigma_{t_2}^2 - \sigma_{t_1}^2}{t_2 - t_1}\right] \tag{89}$$

where U is average stream velocity, t_1 and t_2 are the mean times of passage for the dye cloud to move past stations 1 and 2, and σ_{t2}^2 and σ_{t1}^2 are the variances of the time vs. concentration curves.[44] French[1] gives an example calculation using this method and notes that the dye tracing data must be collected in long prismatic channels free of dead zones (i.e., zones of limited water movement like large eddies, adjacent backwater areas, and bed macro-pore waters). In addition, the sampling stations must be located far enough downstream from the injection point to avoid problems of mixing in the initial period. This distance, x, is defined such that

$$\frac{x\epsilon_y}{UW^2} > 0.4 \tag{90}$$

where W is the width of the stream, $\epsilon_y = 0.6Du_*$, and x is the mixing length from the point of discharge to the point where the inflow is well mixed across the stream.

Fischer et al.[44] propose a more accurate technique for natural streams called the routing method. The concentration vs. time curve at a downstream station, x_2, is predicted by

$$C(x_2,t) = \int_{-\infty}^{\infty} C(x_1,\tau) \frac{\exp\left[\dfrac{-[U(t_2 - t_1 - t - \tau)]^2}{4D_x(t_2 - t_1)}\right]}{[4\pi D_x(t_2 - t_1)]^{1/2}} U d\tau \tag{91}$$

where $C(x_1,\tau)$ represents the time-concentration curve at an upstream station. D_x is determined by trial and error until $C(x_2,t)$ matches the observed curve as shown in Figure 69. French recommends the following procedure:

1. Estimate D_x from Equation 89
2. Compute a trial distribution of $C(x_2,t)$
3. Calculate the root mean square error between the measured and computed values of $C(x_2,t)$
4. Use a trial and error search to see if other trials of D_x result in lower values of the root mean square error

The data defining the time vs. concentration curves should be collected far enough downstream to meet the distance criterion in Equation 90 (i.e., the inflow should be well mixed across the stream).

A third method was proposed by Krenkel[1] that is based on the solution for $C(x,t)$ resulting from a pulse input of tracer, or

$$C(x,t) = \frac{M}{A(4\pi D_x t)^{1/2}} \exp\left[-\frac{(x - Ut)^2}{4D_x t}\right] \tag{92}$$

where M is the weight of tracer added and A is the cross-sectional flow area of the stream. This method requires only one downstream time vs. concentration curve, but mixing must occur very rapidly to avoid the initial mixing effects.

For modeling, it may be best to calibrate the mass balance equation over the time and distance scales that will be simulated to compensate for any effect of numerical dispersion. This would involve modeling the dispersion of a dye cloud from an upstream to a downstream measurement location and changing the dispersion coefficient until predictions and observations agree. This would involve solving Equation 35 using an upstream concentration vs. time curve as a boundary condition and predicting downstream concentrations that could be compared to measurements. For example, in Figure 69, data collected at station 1 could be the boundary condition for predictions at station 2. In Figure 56, station A could be used as a boundary condition to predict concentrations at stations B and C, or station B could be used to predict concentrations at station C.

This same procedure can be used for Lagrangian models and is illustrated in Figure 56. In this case, Equation 44 is solved by changing the dispersion factor DQQ (defined in Equation 44 as D_Q) until the predicted curve best matches the measurements. The investigator who obtained these results[113] also noted that simply fitting the predicted peak concentrations to the measured concentrations was equally accurate compared to matching the entire time concentration curves. The procedure matching predicted to measured peak concentrations obviously would be much less tedious and could use the extensive number of dye studies performed to determine time of travel.

Finally, McBride and Rutherford[112] show that the nature of the dispersion process is sufficiently different from other processes that affect the distribution of mass in streams so that dispersion coefficients can be predicted from slugs of nonconservative substances. They were able to calibrate a Lagrangian model with data for biochemical oxygen demand and dissolved oxygen by changing the dispersion coefficient and the deoxygenation coefficient (reaeration coefficient was estimated by formula) until predictions matched measurements. In this case, McBride and Rutherford documented the result of a large spill of milk into the lower Waipa River (New Zealand) where very large gradients of biochemical oxygen demand and dissolved oxygen evidently resulted in significant dispersive mixing. This procedure was successful because the effect of the dispersion coefficient on concentration vs. time curves of nonconservative materials could be easily distinguished from the effects of the kinetic parameters defining deoxygenation and reaeration.

The method of McBride and Rutherford has not been widely tested and the limits of applicability are not well understood. Nevertheless, it is the only method that does not require an organized dye tracing study.

B. USE OF FLUORESCENT WATER TRACING DYES

Although a number of tracers such as salt, color dyes, and radionucluides can be used for water tracing, fluorescent dyes have proven most useful. Fluorescent dyes are more useful because they have most of the ideal traits of a water tracer. These ideal traits are that the water tracer be (1) water soluble, (2) easily detected, (3) not subject to background interference, (4) stable in streams, (5) harmless to aquatic wildlife and nontoxic to man, and (6) inexpensive. Of the more practical dyes listed in Table 17, rhodamine WT seems to have the best combination of traits that approximate an ideal tracer.[114] Occasionally, however, other dyes are used because of availability or because they sorb less to solids in the water.

Rhodamine WT is readily dissolved in water, has one of the lowest detectable thresholds for fluorescence, is not subject to significant background interference, and is relatively inexpensive. Only rhodamine B seems to have lower detection limits and is apparently less expensive. Rhodamine B, however, strongly sorbs onto solids in the water. Rhodamine WT can be toxic under extreme circumstances and does suffer some loss under typical conditions. To avoid problems with dye loss, procedures have been developed to limit dye losses and avoid unstable conditions in streams.

1. Filter Fluorometers

In general, fluorescent dyes are more easily detected than most other tracers because fluorometers have been designed that detect small amounts of dye. Fluorometers detect dye concentrations as low as on the order of 0.01 to 0.5 µg/l. Table 18 indicates that rhodamine WT can be detected at levels as small as 0.013 µg/l above background concentrations. Of the fluorescent dyes, only rhodamine B is slightly more detectable. Approximately 4 or 5 times more pontacyl pink dye (sulphorhodamine B) is needed for minimum detection. The blue and green fluorescent dyes are 7 to 40 times less detectable than rhodamine WT.

The typical filter fluorometer that makes these low detections possible involves six relatively simple elements shown in Figure 70. Experience has shown that it is possible to use the relatively simple filter fluorometer with the proper calibration procedure to determine dye concentration relative to calibration standards without having to use more elaborate fluorescence spectrometers to determine the absolute magnitude of the sample fluorescence.

The choice of light source and combination of filters has to be tailored to the excitation spectra of each individual dye. The light source for excitation of a fluorescent response in a sample is usually a low pressure mercury vapor lamp or a green T-5 envelope lamp.[114,115] The low pressure mercury vapor lamp is a general purpose ultraviolet or far ultraviolet lamp emitting light at a wavelength of 546 nm — the mercury "green line". The T-5 lamp emits a band of light from less than 520 to more

TABLE 17
The Most Useful Fluorescent Water Tracing Dyes[a]

Dye	Color index no.[b]	Generic name	Alternative names
Blue Fluorescent Dyes			
Amino G acid	—	—	7-amino 1,3 naphthalene di-sulphonic acid
Photine CU	—	CI fluorescent brighter 15	—
Green Fluorescent Dyes			
Fluorescein	45350	CI acid yellow 73 ($C_{20}H_{12}O_5$)	Fluorescein LT Uranine Sodium fluorescence D&C yellow nos. 10 and 11
Lissamine FF	56205	CI acid yellow 7	Lissamine yellow FF or FP (Overacid) Brilliant sulpho flavine FF Brilliant acid yellow 8G
Pyranine	59040	CI solvent green 7	D&C green 8
Orange Fluorescent Dyes			
Rhodamine B	45170	CI basic violent 10	D&C red no.21 Rhodamine BA (?)
Rhodamine WT	—	—	Intraacid rhodamine WT
Sulpho rhoda-mine B	45100	CI acid red 52 ($C_{27}H_{29}N_2O_4Na$)	Pontacyl brilliant pink B Pontacyl pink Lissamine red 4B Kiton rhodamine B Acid rhodamine B Intraacid rhodamine B

[a] Adapted from References 114 and 115.
[b] From the 1971, 3rd ed. of the *Colour Index* of the Society of Dyers and Colourists (UK). Also see the *Technical Manual of the American Association of Textile Chemists and Colorist* and Grove.[116]

than 560 nm with a peak at 546 nm. Typically, mercury vapor lamps are used in Turner fluorometers when tracing the orange fluorescent dyes.

Better sensitivity and accuracy are expected if the light source does not overlap and interfere with the emission spectra of each individual dye. Where overlap of the excitation and emission spectra occur, unreasonably high fluorescence readings may occur if the excitation light passes directly through the sample or is reflected into the photomultiplier. For this reason, the sample chamber is usually designed so that the detector is at 90° with the pathway of the excitation light as shown in Figure 70. In addition, blackened surfaces and special sample-containing cuvettes are used in the sample chamber to reduce reflection. As a result, the light that reaches the photomultiplier is usually only that due to fluorescence. In fact, Wilson et al.[115] indicate some fluorometers are designed well enough that overlap of the excitation and emission bands does not present a problem.

TABLE 18
Comparison of Different Characteristics of Fluorescent Dyes and Recommended Primary and Secondary Filters[a]

Dye	Maximum excitation wavelength (nm)	Maximum emission wavelength (nm)	Background fluorescence of distilled water (relative scale units)	Minimum detection limit ($\mu g/l$[b])	Relative costs[c]	Recommended filters[d] Primary	Recommended filters[d] Secondary
Blue Fluorescent Dyes							
Amino G acid	355(310)[e]	445	19.0	0.51	6	(C) 7-37	(W) 98
Photine CU	345	435(455)[e]	19.0	0.36	7	(C) 7-37	(W) 98
Green Fluorescent Dyes							
Fluorescein	490	520	26.5	0.29	5	(W) 98	(W) 55 & 44
Lissamine FF	420	515	26.5	0.29	8[c]	(W) 98 or 2A & 47b	(W) 55 & 44 / 2A-12
Pyranine	455(405)[e]	515	26.5	0.087	4	(W) 98	(W) 55 & 44
Orange Fluorescent Dyes							
Rhodamine B	555 ± 1[f]	580 ± 1[f]	1.5	0.010	1	2×(C) 1-60 & (W) 61[g]	(C)4-97 & 3-66[g]
Rhodamine WT	555 ± 4[f]	580 ± 2[f]	1.5	0.013	2	2×(C) 1-60 & (W) 61[g]	(C)4-97 & 3-66[g]
Pontacyl pink	565 ± 1[f]	590	1.5	0.061	3	2×(C) 1-60 & (W) 61[g]	(C)4-97 & 3-66[g]

Note: Relative scale units are from the Turner 111 fluorometer scale.

a   Adapted from References 41, 114, and 115.

b   Based on the use of the Turner 111 fluorometer with high sensitivity door, recommended filters, and lamp at 21°C.

c   Ranking from out-of-date information in Smart and Laidlaw[114] and Wilson et al.[115] The relative costs should be further investigated when new studies are planned. The relative cost of Lissamine FF dye seems to be especially out-of-date.

d   (C) is a Corning filter and (W) is a Kodak Wratten filter.
e   Secondary maxima in parentheses.
f   The maxima are those measured by Smart and Laidlaw[114] plus or minus any differences in the maxima measured by Turner and Associates and reported by Cobb and Baily.[41]
g   For greater sensitivity, the (C)1-60 & (W)58 combination is used for the primary filter and the (W) 23A is recommended as the secondary filter.

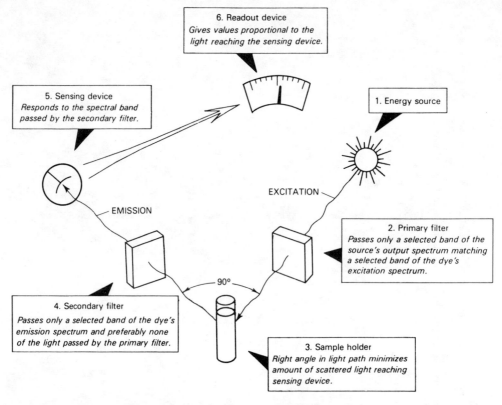

FIGURE 70.   Basic structure of typical filter fluorometers.[115]

The optimum or maximum excitation light wavelengths for the most practical fluorescent dyes are given in Table 18. Figure 71 gives the excitation and emission spectra for the three most useful dyes used in tracing surface waters. For rhodamine WT, pontacyl pink, and acid yellow 7 dyes, the mercury vapor lamps found in most fluorometers[115] are quite adequate to excite the florescence at wavelengths of 405, 436, and 546 nm. Different bulbs may be necessary to excite the blue fluorescent dyes. The fact that mercury vapor lamps do not excite dye fluorescence at the maximum in the emission spectrum for rhodamine WT, pontacyl pink, and acid yellow 7 is not a disadvantage because the less than optimum excitation frequencies are adequate and more importantly do not fall within the emission spectra.

If the excitation frequency or band does overlap the emission spectrum of a dye, some filtering of the light is useful to fully exclude all of the excitation light and parts of the emission spectra. The primary filter is chosen to exclude as much light as possible from the sample chamber that falls within the emission spectrum in the event that reflection into the detector is possible. The secondary filter is designed to exclude light outside the emission spectra that may stray into the sample chamber. The commercially available filters that can be used for primary and secondary filters are listed in Table 18.[114,115]

As an example, Figure 72 illustrates the spectral-transmittance characteristics for the filter combinations recommended when detecting rhodamine WT and pontacyl pink dyes.[115] For the orange fluorescent dyes excited with a mercury vapor lamp (546 nm, mercury green line excitation), the best primary filter consists of a combination of two Corning 1-60 filters and one Kodak Wratten 61 filter. This combination has a primary maximum transmittance corresponding to the green line of mercury. As a result, this

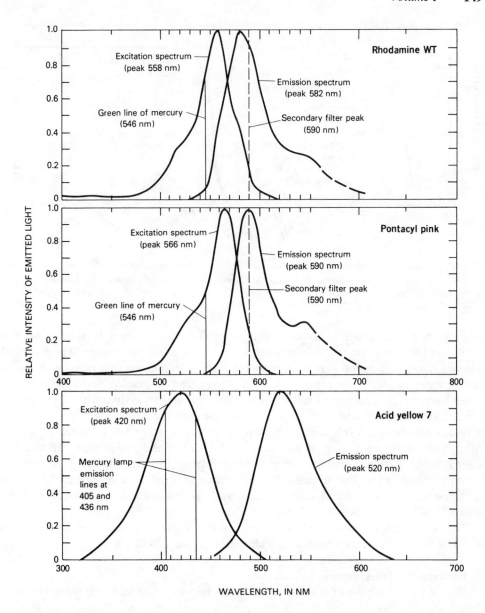

FIGURE 71.   Excitation and emission spectra of rhodamine WT, pontacyl pink, and acid yellow 7 dyes. Spectrofluorometric analysis for rhodamine WT and pontacyl pink by G.K. Turner Associates. Spectra for acid yellow 7 adapted from Smart and Laidlaw.[114,115] (From Wilson et al.[115])

primary filter combination is as efficient as any other combination with the same transmission losses.

The secondary filter combination of a sharp-cut (Corning 3-66) filter (placed closest to the photomultiplier) and blue-green band-pass (Corning 4-97) filter has a peak transmittance at 590 nm. This corresponds closely with the maximum emission wavelengths of the orange dyes. The relatively high transmission through the 590 filter allows greater sensitivity. However, when high dye concentrations cause fluorescence greater than the maximum detection level for the photomultiplier in use, neutral density filters must be used with secondary filters to further reduce the transmission of light.

The transmittance spectra of the 590- and 546-nm filter combination overlap

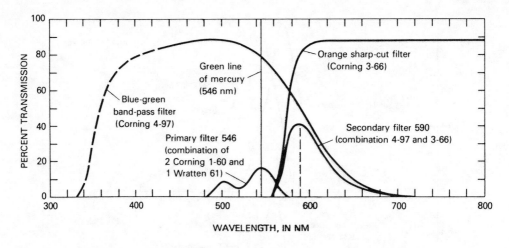

FIGURE 72.   Spectral-transmittance characteristics of preferred filters for rhodamine WT and pontacyl pink dyes. Sources: Corning Filter Catalog for curves for filters 4-97 and 3-66; Feuerstein and Selleck,[119] Figure 3, for curve for primary filter combination 546. Separate curves for Corning 1-60 and Wratten 61 not shown.[115] (From Wilson et al.[115])

slightly, but this is not a problem unless the T-5 Green lamp is used. The T-5 excitation band overlaps the 590-nm filter spectrum by a few nanometers. The 590- and 546-nm filter combination is especially useful when detecting the fluorescence of dye in turbid water excited with a mercury vapor lamp. The suspended particles in turbid water have a high tendency to reflect light. However, any reflection of the light at 546 nm is completely screened by the 590-nm filter combination.

When background levels are absent, sensitivity can be increased by using a primary filter combination consisting of one Corning 1-60 filter and one Wratten 58 filter to screen the light from a mercury vapor lamp. The green T-5 lamp is not recommended for this combination. The Wratten 23A filter can be used as a secondary filter, but background fluorescence and scattered light such as that caused by turbidity must be minor[115] or errors can be significant.

The recommended primary filter for acid yellow 7 dye is a combination of the 2A and 47B filters. The recommended secondary filter is the 2A-12.

2. Background Interference

One of the ideal traits of a tracer is that it not be subject to background interference. Smart and Laidlaw[114] indicate that background interference is important for at least two reasons: (1) background interference masks low tracer concentrations and (2) the interference may indicate that the stream may contain more dye mass than was injected (i.e., apparent dye recovery exceeds 100 percent). Suspended sediment and natural fluorescence such as that from organic material are the two main sources of background interference.

Suspended sediment absorbs and scatters the excitation light.[114] In addition, fluorescence is reduced when some of the dye is adsorbed to the particles. At a concentration of 1000 mg/l, sediment may sorb only limited amounts of dye, but fluorescence in distilled water can be reduced to as much as 65 to 30% after suspended sediment is added. The reduction, however, depends on the nature of the sediments. Lighter colored, silica type sediments may reflect and scatter more light than darker sediments. Therefore, the effect of sediments on the fluorescence of a sample cannot be reasonably predicted, especially if the concentration and mineral composition change from one sampling station to another during a stream tracing study. An extreme example might occur when tracing a sediment laden tributary into another river where the sediments are very different in composition and concentration if some settling of the original sediments occur or the original sediments are measurably diluted in the receiving stream. Nevertheless, it is possible to compensate for the effect by calibrating to determine the exact effect of sediments at each site or, more practically, allow the sediments to settle or centrifuge the samples and then determine the fluorescence. In addition, the effects of fine suspended sediments can be minimized by diluting the samples before determining the fluorescence if sorption to the sediments is limited.[114]

Natural fluorescence at emission bands similar to those of fluorescent dyes occurs in quantities that primarily effect blue and green fluorescent dyes. Natural interference with the orange dyes is limited. The interference arises from natural organic material. Green algae actually causes very little interference, despite the general perception in this regard. The chlorophyll *a* emission spectrum and the spectrum of other algae cellular components exceeds 600 nm except for some red algae components.[114] As shown in Figure 71, there is some opportunity to overlap the emission spectra of rhodamine WT and pontacyl pink dyes, but evidently the 590 secondary filter combination reduces the interference enough so that there are no practical effects of algae on background fluorescence.

Of the more useful dyes, interference by natural organic fluorescence has the most significant effect on acid yellow 7 or lissamine FF dye. Smart and Laidlaw[114] find that interference to other green fluorescent and blue fluorescent dyes can be on the order of 50 µg/l for photine CU and 20 µg/l for acid yellow 7, compared with 0.07 µg/l for rhodamine WT for the surface run-off in an agricultural catchment.(The concentrations are equivalent concentrations of the interferences derived from the calibration curve for each dye and the fluorescence.) Furthermore, the background fluorescence increases linearly over the wavelengths of 400 to 600 nm as the organic carbon content increases. Unfortunately, the background can be quite variable or it would be possible to compensate by using larger quantities of dye or subtracting the interference. For example, Smart and Laidlaw[114] found that the coefficient of variation for the background levels reported above were approximately 45% for the blue and green dyes. As a result, flows that are heavily influenced by run-off, swamp drainage (water has a yellow-brown color characteristic of humic and fluvic acids), and sewage (especially effluents from pulp and paper mills) will present significant background fluorescence in the emission spectra of blue and green dyes such as acid yellow 7. That background fluorescence is likely to be so variable as to make compensation difficult and impractical in many cases. Wilson et al.[115] indicate that the variable organic content in storm sewers make it especially difficult to use acid yellow 7 for nonpoint source pollution studies.

Another characteristic that determines how well a dye approximates an ideal tracer involves dye losses. Real or apparent losses or instabilities of dyes are caused by sorption, failure to compensate for the decrease in fluorescence with increased temperature, failure to account for the decrease in fluorescence in highly acidic streams, photochemical decay, and chemical quenching. Each of these processes has a different effect on each individual dye.

3. Sorption to Sediments

Dyes sorb to mineral and organic solids that are suspended or are a part of the bed. As a result, dyes are not conserved over long distances for extended periods of sampling. Because some sorption is unavoidable, the effect must be minimized for dye tracing to be useful. At this time the sorption effect has not been quantified. Guidance on how to minimize sorption is limited, but Smart and Laidlaw[114] indicate that contact

TABLE 19
Qualitative Description of the Tendency of
Fluorescent Dyes to Sorb to Solids[a]

| Dye | Relative sorption tendency to sorbents | |
	Organic	Inorganic
Blue Fluorescent Dyes		
Amino G acid	Moderate	Low
Photine CU	Moderate	Moderate
Green Fluorescent Dyes		
Fluorescein	Moderate	Low
Lissamine FF	Low	Low
Pyranine	Moderate	Low
Orange Fluorescent Dyes		
Rhodamine B	High	High
Rhodamine WT	Moderate	Moderate
Pontacyl pink	Low	Moderate

[a] Adapted from References 41, 114, and 115.

with less than 100 mg/l of solids does not cause significant dye losses except for rhodamine B dye. It is not clear how contact with bed sediments should be treated. At least a small part of the dye is in contact with the bed at all times after the initial mixing takes place. In the bed, the concentration of solids is about four orders of magnitude higher than the 100 mg/l suggested by Smart and Laidlaw. Experience of this writer and others indicates that dye losses to bottom sediments tend to be small for rhodamine WT. However, if a significant amount of the flow moves through the bed, sorption losses should be investigated.

Primarily, sorption is expected to be a potential problem in small streams with cobble and gravel beds in which the volume of bed pore space is a significant portion of the volume of stream flow. Kenneth Bencala (personal communication, 1987, U.S. Geological Survey, Menlo Park, California) indicates that it is difficult to use rhodamine WT for quantitative dye tracing in mountain streams.

It is especially important to avoid direct contact of concentrated dye with bottom sediments at the time of injection. The staining of bottom sediments at the injection can occur for any dye considered here. The occurrence of staining indicates that significant dye loss may have occurred.

It has not been possible to correct for dye losses due to sorption or to predict such dye losses. The reason for this is that the sorption mechanisms have not been fully explored. It is known that for different fluorescent dyes, there are different tendencies to sorb to mineral and organic surfaces. Organic content, pH, sediment concentration, and chemical characteristics of the dye are known influences on sorption.

For the eight dyes listed in Table 19, there is a stronger tendency to sorb to organic materials. The data given in Table 20 are from Smart and Laidlaw[114] who investigated sorption to seven types of solids. They found that sawdust, humus, and heather generally sorbed more than clays, limestone, and sand. However, mineral solids in streams are generally coated with organic films.

The one dye that is distinctly different from the other seven is rhodamine B. On the

TABLE 20
Percent Dye Sorption for Mineral and Organic Solids

Dye	Sediment concentration (g/l)	Mineral				Organic			Rank
		Kaolinite	Bentonite	Limestone	Orthoquartzite	Sawdust	Humus	Heather	
Amino G acid	2	99	—	95	—	66	75	—	1
	20	97	—	96	—	17	39	—	
Photine CU	2	93	80	93	58	48	60	57	7
	20	90	38	40	83	—	14	23	
Fluorescein	2	98	98	98	98	86	83	41	5
	20	93	87	94	98	11	17	0	
Lissamine FF	2	97	96	96	99	83	90	88	3
	20	96	92	88	95	70	68	54	
Pyranine	2	95	100	96	100	70	76	74	4
	20	95	98	85	87	30	31	18	
Rhodamine B	2	1	4	8	10	12	3	4	8
	20	4	8	2	8	4	2	1	
Rhodamine WT	2	89	92	93	98	81	82	81	6
	20	67	79	99	90	42	11	18	
Pontacyl pink	2	88	98	97	—	92	92	—	2
	20	51	—	76	—	—	63	—	
Rank		4	2	3	1	5	6	7	

Note: Rank increases with sorption tendency, and an initial dye concentration of 100 µg/l was used.

From Smart, P. L. and Laidlaw, I. M. S., *Water Resour. Res.*, 13(1), 15, 1977. With permission.

basis of their experiments reported in Table 20, Smart and Laidlaw[114] recommend that rhodamine B not be used for quantitative dye studies. Unlike the other seven dyes, rhodamine B is a cation that readily attaches to anionic sites prevalent on mineral and organic surfaces. In addition, the rhodamine B cation is more readily attracted to glassware surfaces. This accounts for a significant loss of rhodamine B dye.[114]

Smart and Laidlaw[114] observed no significant loss of the anionic dyes in contact with hard and soft glass (Pyrex) containers for up to 10 weeks. Acid yellow 7 and rhodamine WT did not show losses during 10 weeks of storage in polythene (plastic) bottles or while in contact with rubber stoppers and Parafilm sealing film. Interference with the blue fluorescent dyes does occur if samples come into contact with paper, cotton wool, textiles, and other products treated with blue fluorescent optical brighteners. Domestic detergents and plastic tubing may also introduce optical brighteners that should be avoided. For best results, Smart and Laidlaw recommend that potential contamination be investigated with a hand-held ultraviolet lamp and by testing sample blanks. Contamination of samples containing green and orange fluorescent dyes is minimal under conditions frequently encountered in the laboratory.

The four dyes that have sulphonic acid functional groups, (1) pyranine, (2) pontacyl pink, (3) amino G acid, and (4) acid yellow 7, are the most resistant to sorption. Rhodamine WT, photine CU, and fluorescein are less resistant to sorption, but when suspended solids are less than 100 mg/l, the difference seems to be insignificant. The influence of pH is important in distinguishing between these two groups and will be discussed later.

In the absence of better guidance, the qualitative information in Table 19 and the selective information for sediment concentrations of 2000 mg/l and 20,000 mg/l in Table 20 should prove useful. Generally, it seems that the anionic dyes are useful when sediment concentrations are less than 100 mg/l. For the conditions normally encountered in streams, sorption losses of rhodamine WT are insignificant. If large sediment concentrations are encountered, site specific investigations of the seven anionic dyes should be considered. In sewage and other organic-rich waters, pontacyl pink dye may prove more useful than rhodamine WT. The background interference should be constant if acid yellow 7 is to be used under the same conditions. Sorption to mineral sediments present in large concentrations can be minimized by using pontacyl pink or acid yellow 7 dyes instead of rhodamine WT dye.

The ranking of sorption tendencies in Table 19 and the data of sorption in Table 20 may not be applicable to all conditions that may be encountered. In addition, there are some differences of opinion about relative sorption capacity. Cobb and Bailey[41] note that the original manufacturers found that rhodamine WT sorbs less that pontacyl pink. As a result, the best use of the results reported here may involve using the studies of Smart and Laidlaw as guidance for when additional study of a chosen dye and sediment system is necessary.

4. Temperature Effects

The amount of dye being followed in a stream appears to increase or decrease if the effects of temperature are not taken into account. As water temperature increases, the intensity of fluorescence decreases.[115] The change in fluorescence with changing temperature for a given concentration of dye can be compensated for in at least one of two ways.

First, temperature effects can be avoided if samples and calibration standards are analyzed at a standard temperature such as 20°C or a constant laboratory temperature. For this method, standards made up with stream water and samples from the stream are carried back to the laboratory for analysis. The samples are placed in a water bath or allowed to adjust to the temperature maintained in the laboratory. Analysis of samples in the field without correction for the effect temperature are conducted only to provide a qualitative indication of dye concentrations to guide sampling.

Second, the effects of temperature can be taken into consideration by using the compensation method developed by Smart and Laidlaw.[114] As illustrated in Figure 73, Smart and Laidlaw and Wilson et al. determined temperature correction factors for fluorescence. The temperature corrections (curves shown in Figure 73), can be approximated by

$$F = F_o exp(n\Delta T) \tag{93}$$

where F is fluorescence (dial reading from the fluorometer) at temperature, T, F_o is the fluorescence at a standard temperature (Smart and Laidlaw used 0°C), n is a constant for each dye, and ΔT is the temperature difference between the reference temperature and the temperature of the sample at the time of measurement. The values of the temperature correction coefficient, n, are given in Table 22.

Equation 93 can be used to make corrections in the field when precise tracking of the tracer dye is required. For example, sampling at the arrival of the peak dye concentration in a Lagrangian water quality study is difficult to do in shallow streams without correcting for temperature differences that occur during the day at different locations. It should be noted as well that care must be exercised in the field to maintain constant temperatures in the sample chamber to be able to make quantitative measurements. The field fluorometer must be insulated from direct sunlight and be maintained at a reasonably constant temperature. The fluorometer should be operated continuously to avoid warming effects on the sample chamber temperature. Due to the fact that it is difficult to maintain relatively constant temperatures in the sample chamber under field conditions, most studies rely on the first method of temperature compensation — samples are analyzed in the laboratory.

TEMPERATURE CORRECTIONS FOR VARIOUS DYES

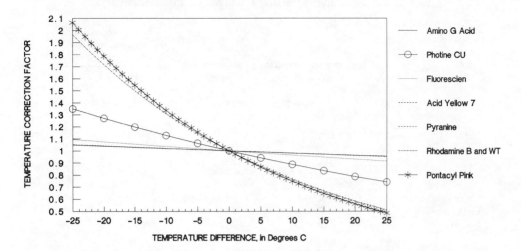

FIGURE 73. Temperature correction curves for amino G acid, photine CU, fluorescein, acid yellow 7, pyranine rhodamine B, rhodamine WT, and pontacyl pink dyes. Curves for pontacyl pink, rhodamine WT, and acid yellow 7 dyes are from Wilson et al.[115] Curves for amino G acid, photine CU, fluorescein, pyranine, and rhodamine B dyes are from Smart and Laidlaw.[114] Wilson et al.[115] illustrats how a fluorescence reading in the field at 62°F (16.7°C) can be corrected to compare with laboratory standards or other measurements made at 72°F (22.2°C).

The data plotted in Figure 73 indicate that the fluorescence of rhodamine WT and pontacyl pink dyes are most affected by temperature variations in samples. The effect of temperature changes is quite limited for fluorescein, acid yellow 7, pyranine, and amino G acid dyes. There is approximately a 70% change in fluorescence of photine CU dye for a 50°C temperature change.

5. pH Effects

The effect of pH also makes it difficult to obtain consistent measurements that indicate that dye is conserved. The effects of pH on fluorescence are reversible by adjustment of the samples in the laboratory, but the effects of a dye reaction with compounds in acidic water are not always readily reversible.[115] Therefore, alternative dye tracing methods may need to be considered for streams impacted by mining and other acidic wastes.

Although the effect of sample pH can be compensated for by using native stream water for calibration standards, the reversible effects of pH cannot be readily quantified. Smart and Laidlaw[114] show that four different acids ($KHPO_4$, HCl, HNO_3, and HSO_4) have a significantly different effect on fluorescence when the pH is less than 7.0. As a result, they recommend that natural waters from the stream under study be used to formulate calibration standards in the laboratory when waters with high or low pH are involved. This should work unless pH changes significantly over the stream reach being sampled.

Smart and Laidlaw[114] investigated the effect of pH over the range of 3 to 11 to determine when some correction should be considered. In using hydrochloric acid (HCl) and sodium hydroxide (NaOH) to control the pH, they determined that there were fairly extensive ranges over which pH had no effect. The one exception was pyranine dye where the fluorescence of a constant concentration varied continuously over the pH range of 3 to 11. Table 21 gives ranges of pH over which fluorescence of a constant

TABLE 21
Effect of Various Environmental Changes on Fluorescent Dyes[a]

Dye	Temperature correction coefficient, n	pH range for no effect	Percent of maximum fluorescence for chlorosities of	
			3.6 g/l	17.8 g/l
Blue Fluorescent Dyes				
Amino G acid	−0.0019	6.0 to 9.0	100	100
Photine CU	−0.012	6.5 to 11.0	100	100
Green Fluorescent Dyes				
Fluorescein	−0.0036	6.5 to 11.0	100	100
Lissamine FF	−0.0020	4.0 to 10.0	100	100
Pyranine	−0.0019	none	100	100
Orange Fluorescent Dyes				
Rhodamine B	−0.027	5.0 to 9.0	100	98
Rhodamine WT	−0.027	5.0 to 11.0	97	92
Pontacyl pink	−0.029	4.0 to 10.0	100	96

[a]   Adapted from References 41, 114, and 115.

concentration is constant. The ranges given indicate that acid yellow 7 (Lissamine FF) and pontacyl pink dyes are the most stable when pH changes. For these dyes there are no significant changes in fluorescence of a constant concentration solution in the range of pH 4 to 10. The fluorescence of rhodamine WT dye is stable over the range 5 to 11. Rhodamine B dye is stable over the range of 5 to 9. Amino G acid dye is stable over the pH range of 6 to 9. Photine CU and fluorescein dyes are stable over the pH range 6.5 to 11. As a result, rhodamine WT is expected to be stable in most streams. Only streams impacted by acid mine drainage may represent difficulties and then only if pH changes significantly in the impacted stream.

Of the dyes that are affected by ionization, fluorescence changes because of the effect on degree of molecular resonance when acid functional groups become protonated.[114] Acid yellow 7 and pontacyl pink dyes are stable at lower pH values because the sulphonate acid groups do not dissociate until the pH approaches 3. Dyes containing carboxylic groups begin to dissociate pH values between 4 to 6 compared to 6 to 7.5 for phenolic groups. Pyranine dye undergoes rapid changes in fluorescence near pH 7. This occurs because of the ionization of a phenolic OH group.

Fluorescein dye changes fluorescence with pH because of a structural change. As a result, fluorescence decreases more rapidly as pH is decreased below pH 7 compared with the other dyes considered here that change fluorescence because of ionization.

If the pH of the stream is outside the range of 5 to 11, then it seems more practical to use an alternative dye (pontacyl pink or acid yellow 7) rather than attempt corrections for measured fluorescence of rhodamine WT dye. If the pH is outside the ranges given in Table 21 and does not change in the reach of interest, then the use of natural stream water to make calibration standards will correct for the reduction in fluorescence. In cases where the pH is less than 3.0 and changes over the reach, it will be possible to adjust the pH of the sample in the laboratory if the effect of pH is to cause ionization or protonation, or readily reversible structural changes in dye molecules. A practical limit of the pH minimum for acid yellow 7 and pontacyl pink dyes where the fluorescence is reduced no more than 3% is 3.0. Because natural waters rarely drop below pH 3,[117] it is unlikely that pH adjustments will be required.

6. Photochemical Decay

Photodecomposition of dye in water causes a permanent loss of fluorescence that must be minimized by keeping sampling reaches as short as possible since the effect increases gradually with time. Photochemical decay is minimized in deeper, more turbid, shaded streams sampled on cloudy days.

Smart and Laidlaw[114] indicate that photodecomposition is often expected as dye molecules are raised by the excitation light source to a higher energy level at which they fluoresce. Residence in higher energy levels increases the likelihood of photochemical reactions that change the nature of the dye molecules. Decomposition depends on both light intensity and wavelength. Ultraviolet light causes more rapid decomposition than longer wavelengths of light.

Table 22 is a compilation of the photochemical decay rate coefficients measured for the decomposition of dyes by Smart and Laidlaw.[114] The rate coefficient, k, for photochemical decay given in Table 22 is defined by

$$F = F_o \exp(-kt) \tag{94}$$

where F is fluorescence at time t, and F_o is the initial fluorescence at t = 0. The rate coefficients given in Table 22 do not give quantitative information that can be used to make dye loss corrections It is very difficult to extrapolate from the rates compiled to the unique conditions in a stream. Photodecomposition is directly influenced by light intensities that cannot be precisely measured nor estimated at stream sites influenced by shading, reflection, cloudiness, turbidity, and algae concentrations. However, the compilation of rate coefficients in Table 22 do offer a relative comparison between dyes of the various types. In addition, these rates can be used for order of magnitude estimates when no other data are available.

Acid yellow 7 and the orange fluorescent dyes have the lowest photodecomposition rates. The rapid decay of photine CU, pyranine, and fluorescein dyes preclude use in streams.[114] The decay rate of amino G acid is large but not as large as those dyes that should not be used in streams. Wilson et al.[115] indicate that acid yellow 7 dye seems to have an order of magnitude less decay than pontacyl pink dye. The behavior of rhodamine WT dye has not been fully examined, but accumulated field experience seems to indicate that significant dye losses do not occur for several days.[115] Smart and Laidlaw[114] offer a more quantitative assessment from the compilation of decay rate coefficients. In sunlight, acid yellow 7 and rhodamine WT dyes are at least 3 times more resistant to decomposition than pontacyl pink dye. For the photodecomposition rates given, dye tracing studies using acid yellow 7 and the orange fluorescence dyes can extend for at least a week before significant losses occur.

7. Chemical Quenching

A significant amount of real or apparent dye loss is caused by chemical quenching. Chemical quenching is caused by agents or impurities in the water that absorb exciting light, absorb light emitted by the dye as it fluoresces, degrade the excited-state energy of the dye, or chemically react to change fluorescence.[115] Examples of such agents include dissolved solids, chlorine, high concentrations of dissolved oxygen, high levels of turbidity or suspended solids that have already been discussed, and organic material that has also been previously discussed. Large concentrations of turbidity or suspended solids and large dye concentrations cause a screening effect similar to true chemical

TABLE 22
Photodecomposition Rate Coefficients in Per Day for Commonly Used Fluorescent Dye Tracers

Investigator and environmental conditions	Amino G acid	Photine CU	Fluorescein	Acid yellow 7	Pyranine	Rhodamine B	Rhodamine WT	Pontacyl pink
Pritchard and Carpenter:[118]								
Artifical light						1.7×10^{-5}		
Sunny			1.3×10^{-1}			1.7×10^{-5}		
Feuerstein and Selleck:[119]								
Cloudy			5.1×10^{-2}			4.5×10^{-3}		2.0×10^{-3}
Sunny			2.6×10^{-1}			2.2×10^{-2}		1.0×10^{-2}
Yates and Akesson:[120a]								
Minimum rate			4.5×10^{-2}	0		3.6×10^{-2}		4.4×10^{-2}
Maximum rate			3.9×10^{-1}	4.6×10^{-1}		1.2		6.4×10^{-1}
Watt:[121]								
Sunny (10 μg/l)								
Sunny (100 μg/l)								
Von Moser and Sagl:[122]								
Cloudy			1.5×10^{-2}	8.0×10^{-4}		5.6×10^{-4}		1.8×10^{-3}
Sunny			2.6×10^{-1}	7.4×10^{-3}		3.4×10^{-4}		1.0×10^{-2}
Abood et al.:[123]								
Sunny						8.3×10^{-3}	1.5×10^{-3}	5.6×10^{-3}
Smart and Laidlaw:[114b]								
Sunny	3.2×10^{-2}	>1.28	1.9×10^{-1}	$<2.0 \times 10^{-4}$	2.4×10^{-1}	1.1×10^{-3}	$<2.0 \times 10^{-4}$	6.6×10^{-4}
Artificial light	7.4×10^{-4}	1.1×10^{-1}	2.6×10^{-2}	$<2.0 \times 10^{-4}$	3.2×10^{-4}	3.0×10^{-4}	$<2.0 \times 10^{-4}$	$<2.0 \times 10^{-4}$

a Decomposition rate coefficients of a dry dye expressed in terms of a solution.

b 6-h experiments conducted for dye solutions of 100 μg/l. Assuming that the sun shines 12 h/d, these rate coefficients would be one half the values shown to express the daily photodecomposition.

From Smart, P. L. and Laidlaw, I. M. S., *Water Resour. Res.*, 13(1), 15, 1977. With permission.

quenching that can be controlled by dilution if necessary.[115] The fluorescence of the red components of algae[41,114] and organic materials[115] is an interference that is somewhat similar to that of quenching agents.

The effect of dissolved solids or salinity on dye fluorescence does not seem to have been fully investigated. Evidently, the dissolved solids in seawater cause quenching like other agents.[41,119]

Smart and Laidlaw[114] note that only the orange fluorescent dyes seem to be affected by dissolved solids in concentrations typically encountered in marine waters. The relative effect of chlorosity on the most frequently used dyes are shown in Table 21. Rhodamine WT dye is the most affected, but only about an 8% reduction of fluorescence occurs if chlorosity is increased to 17.8 g/l (sea water has a chlorosity of 19 g/l[117]). Smart and Laidlaw also indicate that the nature of the dissolved solids in water have some influence. Potassium chloride causes a greater reduction in fluorescence than sodium chloride at equivalent concentrations. Furthermore, the effect of salinity is slow acting. There is some indication that equilibrium levels of fluorescence are obtained only after 300 h in some cases.[114] Smart and Laidlaw indicate that past observations of dye losses in saline environments are probably due to the effects of salinity but offer little guidance on how corrections may be obtained or the effects avoided. Therefore, it may be best to consider acid yellow 7 as a substitute when quantitatively tracing fresh waters into coastal areas with significant increases in salinity.

There is only limited information on the effect of chlorine. The studies that are readily available involve effects on rhodamine B and WT dyes in chlorine contact chambers of sewage treatment plants. Smart and Laidlaw[114] note that Deaner found that initial chlorine residuals of 22 mg/l were much more effective in reducing the fluorescence of rhodamine B dye than rhodamine WT dye. After 5 h, rhodamine B fluorescence was reduced 31%, whereas rhodamine WT fluorescence was reduced 8%. Some difference in solids concentrations leading to greater sorption of the rhodamine B influences these differences, however. Therefore, the results cannot be quantitatively compared.

Smart and Laidlaw[114] found that the reduction of fluorescence is independent of dye concentration but is dependent on the chlorine concentration. This seems to clearly indicate that the quenching effect is responsible for the reduced fluorescence.

In streams, free chlorine residuals, if they occur, are on the order of 0.1 to 1.1 mg/l.[124] Typically, these concentrations are only present in narrow plumes that only extend partially across the stream before the chlorine dissipates. Combined residuals can be present in concentrations of 10 mg/l or more (from data collected by the U.S. Geological Survey, 1980 to 1981, West Fork Trinity River, Dallas, TX). However, these concentrations do not seem to have the same effect as free residuals.

Where the residual is approximately 1 mg/l, there is some indication from extrapolation of the results of Smart and Laidlaw[114] that these levels would not have an effect on rhodamine WT dye and only a limited affect (less than 10%) on rhodamine B dye. As a result, effects of chlorine are probably only of concern when the dye is injected into a chlorinated effluent and remains in contact with the wastewater for several hours before mixing with the receiving river water. Mixing of dyed river water into a wastewater plume in a river does not seem likely to affect rhodamine WT concentrations.

Of greater concern is the use of chlorinated tap water to make up standards. Hem[117] indicates that chlorine concentrations are typically 0.1 to 0.5 mg/l in chlorinated tap water. These concentrations are of such concern (quenching of the low-concentration standards is the primary concern it seems) that Wilson et al.[115] recommend such chlorinated tap water be allowed to sit for approximately 12 h to dissipate the chlorine.

The effect of high dissolved oxygen levels seems to be of two types. First, it is suspected that the effect of dissolved oxygen is similar to the quenching caused by chlorine.[115] The quenching effect of dissolved oxygen must be very similar to the effect of chlorine if chlorine reduces fluorescence because of an oxidation reaction. Dissolved oxygen is also an oxidant even though it is not as strong as chlorine. Second, samples supersaturated with oxygen or samples that are highly agitated form minute bubbles that reflect and absorb both excitation and fluoresced light. Typically, minute air bubbles enchance fluorescence[45] whereas the quenching effect (chemical reaction of the oxidation type?) decreases fluorescence.[114]

As a result, it is not recommended that dye studies be conducted in highly turbulent, white-water reaches that experience intense aeration or highly productive reaches that are supersaturated during periods when sunlight is the most intense. At this time, there are no means of correction for these effects. Highly aerated wastewaters should also be closely checked for supersaturation. Regardless of stream conditions, samples should be checked for bubbles before a cuvette is placed in the sample chamber. It may be necessary to allow supersaturated or highly agitated samples to sit on the laboratory bench until excess dissolved oxygen is dissipated, but when bubbles are observed, they should be noted and the results treated with some caution since the effects are not fully understood.

Relative effects on different dyes have not been fully defined. Smart and Laidlaw[114] indicate that Watt finds rhodamine WT and pontacyl pink dyes are less prone to the effects of vigorous agitation than rhodamine B dye.

In addition, other chemical reactions are possible. G.K. Turner and associates indicate that contact with metals reduces fluorescence.[114] Any effects that seem to occur because of chemical reactions may possibly be compensated for by using stream water from the site for making up standards and avoiding dilution of samples unless all samples are diluted by the same amount with the same water.

8. Toxicity

The most important toxicity issue evolves around the formation of a human carcinogen when rhodamine WT dye is present with high concentrations of nitrite. The carcinogen has been observed in the laboratory but not in streams. In addition, there are a number of other studies of dyes in extremely high concentrations where various effects on wildlife have been observed.

When extremely high concentrations of nitrite are present (in excess of 1 mg/l) with rhodamine WT dye, the formation of diethylnitrosamines (DENA), a carcinogen, has been observed in controlled laboratory tests.[115,125] As of yet, DENA has not been observed in stream waters treated with rhodamine WT dye. Wilson et al.[115] note that the U.S. Geological Survey has conducted preliminary studies with water from four streams and could not detect DENA when the water was treated with dye. In addition, the Geological Survey has conducted a risk analysis that indicates that the risk to human health seems sufficiently small to justify continued use as long as dye concentrations remain less than 10 μg/l[115] at points where water is withdrawn for consumption.

One reason that DENA may not be detectable in streams is that nitrite concentrations are typically much smaller than 1 mg/l. Generally, nitrite concentrations are 0.1 mg/l or less.[2,117] Occasionally, values of 0.8 mg/l have been observed downstream of sewage discharges where ammonia is rapidly oxidized,[2] but these seem to be limited to small streams where the physical and biological kinetics are rapid. Fortunately, drinking water withdrawals are not usually located just downstream of sewage inflows where the highest nitrite concentrations seem to occur. Given the remote likelihood of large nitrite concentrations occurring in surface waters, the inability to detect DENA in four streams injected with rhodamine WT,[115] and the results of the Geological Survey risk analysis, the use of rhodamine WT dye seems safe. Nevertheless, if high concentrations of nitrite are suspected in the vicinity of water withdrawals, an alternative dye such as acid yellow 7 (Lissamine FF) should be considered.

Dye concentrations that are known to affect fish and other wildlife are several orders of magnitude larger than concentrations that are needed to trace water movements in streams. The readily available studies for aquatic wildlife toxicity are reported in Table 23.

C. MIXING LENGTHS

The prediction of mixing lengths is important to determine how useful one-dimension models may be, to determine where to sample dye when measuring time of travel and dispersion in streams and when measuring discharge by the dye dilution method, to determine how to sample for stream water quality surveys, and to define mixing zones. The mixing length is the distance downstream from a source that is required for the source to enter the river and mix with the receiving waters until concentrations are essentially constant over the cross-section (distance from location 1 to location 5 in Figure 55 or the zone of lateral mixing in Figure 68).

Despite the importance of the mixing length, the available predictive methods are not completely reliable. Typically, estimated mixing lengths should be confirmed by measurement. This seems to be especially necessary for the dye dilution method of measuring discharge and precise time of travel studies.

The best predictive equation is the empirical one developed by Fischer et al.[44] Fisher et al. indicate that mixing occurs at different rates depending on where the inflow is introduced in the receiving stream. Sources introduced at the center of the stream mix most rapidly while sources located at the bank mix the slowest. The predictive equations for these source locations can be written as

$$L = C \frac{UW^2}{\epsilon_y} \qquad (95)$$

where C is 0.4 for sources located at the bank and 0.1 for sources located at the center of the flow. Note the similarity with the criterion given in Equation 90 when C = 0.4. This is the same equation defined in a slightly different context. ϵ_y is the transverse mixing coefficient defined previously that depends on the channel characteristics. In the most general form, the transverse mixing coefficient can be written as

$$\epsilon_y = C_m D u_* \qquad (96)$$

where C_m is a coefficient that has a value of approximately 0.1 to 0.2 for straight, uniform channels and 0.4 to 0.8 for slowly meandering channels with moderate irregularity. Even higher values have been observed for highly irregular, rapidly meandering streams. At the present time, this qualitative guidance about the values of the transverse mixing coefficient (ϵ_y) is the best available.[44] See Fischer et al.[44] for example calculations.

There has been additional work on the mixing of line sources across the stream (continuous injection across the stream) and multiple point inflows. However, predictive equations for these sources do not seem to be developed to a fully practical stage.

TABLE 23
Studies of Wildlife Toxicity to Fluorescent Dyes

Investigator	Dye	Concentration (mg/l)	6 h	24 h	48 h	96 h	Species	Effect
			LC$_{50}$ Values (12°C)					
Brandt[126]	Fluorescein	100					Trout and roach	Not toxic.
Sowards[127]	Fluorescein	Visible concentrations					Longnose dace (*Rhinichthys cataractae*)	Did not affect the toxicity of the fish toxicant, Pronoxfish.
Marking[128]	Rhodamine B		—	736	306	217	Rainbow trout (*Salmo gairdnerii*)	LC$_{50}$ is the concentration at which 50% of the test species are killed or affected. The 6 h data are estimated from other sources.
			—	962	647	526	Channel catfish (*Ictalurus punctatus*)	
			1176	754	700	379	Bluegill (*Lepomis machrochirus*)	
	Fluorescein		6410	4198	3420	1372	Rainbow trout (*Salmo gairdnerii*)	
			—	3828	2826	2267	Channel catfish (*Ictalurus punctatus*)	
			—	5000	4898	3422	Bluegill (*Lepomis machrochirus*)	
Pritchard and Carpenter[118]	Rhodamine B	100					Unnamed	No ill effects for at least 2 months.
Panciera[129]	Rhodamine B	100					Oyster larvae and eggs (*Crassostrea virginica*)	Larvae died within 2 d and no eggs developed. Temporary retardation of larvae growth and 27% of eggs were abnormal.
		1						No effect.

Ref.	Dye	Concentration	Test organism	Effects
Woelke[130]	Rhodamine B	32	Sea urchin eggs (species *hemicentrotus*)	Observable effect.
		10	Embroys of bay mussel (species *mytilus*)	No observable effect. Observable effect. No observable effect.
		3.2		
		1.0		
Parker[131]	Rhodamine WT	10 (for 17.5 h at 22°C) 375 (for 3.5 h)	Smolt of silver salmon and Donaldson trout	No mortality or respiratory effects noted. The total experiment was 21 h long and involved a step increase in concentration from 10 to 375 mg/l
	Rhodamine B	10 (48 h at 24°C)	Pacific oyster (*Crassostrea gigas*)	Eggs and larvae continued to develop normally.
		>0.09	Quahog clams (*Mercenaria mercenaria*)	Flesh staining occurred but cleared rapidly in dye-free water.
		>8.4		Clams show avoidance reaction.
Wortley and Atkinson[132]	Rhodamine WT	2000	Water flea (*Daphnia magna*), shrimp (*Gammarus zaddachi*), log house (*Asellus aquaticus*), may fly (*Cleon diperum*), and pea mussel (species *pisidium*)	No mortality for 48-h and 1-week tests at 10°C compared to control tests.
Akamatsu and Matsuo[133]	Stilbene triazine optical brighteners	10 and 20	Goldfish (*Carassius auratus*)	No abnormalities in body weight or length were noted.
		LC$_{50}$ = 2000 mg/l	No details given.	
		LC$_{50}$ (96 h)		
Keplinger et al.[134] and Strum and Williams[135]	Optical brighteners	32 — 474	Bluegill (*Lepomis machrochirus*)	
		108 — 1780	Rainbow trout (*Salmo gairdnerii*)	
		86 — 1060	Channel catfish (*Ictalurus punctatus*)	

From Smart, P. L. and Laidlaw, I. M. S., *Water Resour. Res.*, 13(1), 15, 1977. With permission.

D. SIMPLE MASS BALANCES

A steady-state mass balance for a conservative substance such as dissolved solids is as simple as the water balance reviewed earlier. In fact, the method is so simple that a dissolved solids balance should be conducted in conjunction with any water balance to assist in measuring any nonpoint source flows that may have been difficult to identify during water quality surveys of a stream. Most experienced modeling experts automatically include a conservative mass balance in the procedure to verify and calibrate a water quality model.

Figures 74 to 76 are used to illustrate a steady-state mass balance for dissolved solids in the Arkansas River downstream of Pueblo, CO. Figure 74 is a map of the stream reach that shows locations at which flow and specific conductivity were measured. Specific conductivity, C_s, is linearly related to dissolved solids or salinity concentrations, C, over short reaches in a stream when the nature of the dissolved solids does not shift (because of changes in the relative amounts of ions present due to inflows of waters with different ionic compositions). The relationship is written as

$$C_s = bC \tag{97}$$

where b is an empirical constant. See Hem[117] for the basis and limitations of such relationships.

Flow and dissolved solids balances in the stream are given in Figures 75 and 76. In Figure 75, it can be noted that most of the inflows into the stream flow seemed to be identified for the 2-d periods in April 1976 and September 1979 when the stream was intensively sampled. If it is noted that discharge measurements are accurate to approximately 2 to 5%, then one would conclude that there is a very consistent relationship between flow measurements in the stream and the measurements of flow at the head of the reach and in the tributaries and waste inflows (i.e., the boundary conditions). From the April 1976 results one could reasonably conclude all boundary conditions for flow were measured unless there was some extraordinary exchange of flow of approximately equal amounts between gaging locations that was not documented. During the September 1979 survey, the instream measurements of flow are less consistent with measurements of flow for the boundary conditions. In the lower 30 km (19 mi) of the reach during the September 1979 survey measurements were consistently higher than predictions from the flow balance.

The discrepancies of the water balance predictions with the September 1979 measurements could arise from two causes. First, and most likely for a small stream of this type, is that the flow rates in the downstream section of the reach were influenced by higher flow rates entering the stream just before the survey began. Second, and most disconcerting for a study of this type is that some unmeasured nonpoint source flows may have entered the stream in 1979 but were not present in April 1976. Such a quandary regarding what actual occurred could best be addressed by the field crew who worked at the site (a good argument for having the modeler assist in, or direct from on-site, all field studies) or by reliance on other types of measurements such as discharge records at the head of the reach or dissolved solids data. In this case, the field crew notes made no mention of any other flows, and they traveled the entire reach measuring cross-sections. Furthermore, there did not seem to be any data collected just prior to the study for discharge at the reservoir upstream of the study reach or from the sewage treatment plant that supplies approximately 50% of the flow in the stream. Therefore, lacking definitive boundary condition data, the information available on the dissolved solids shown in Figure 76 were investigated.

The information about the dissolved solids balance indicate that the higher than

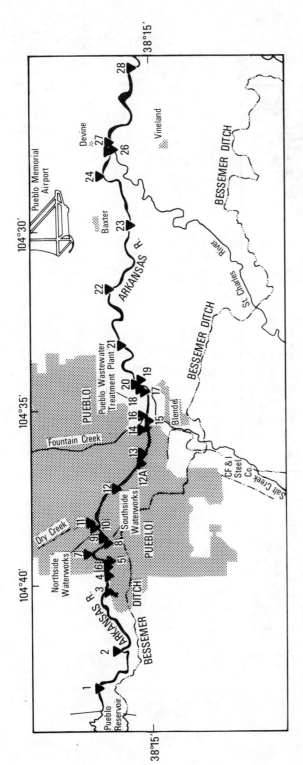

FIGURE 74. Arkansas river between Pueblo Reservoir and Nepesta Gage near Pueblo, CO, showing tributaries, points of discharge, points of water withdrawal, and sampling locations.

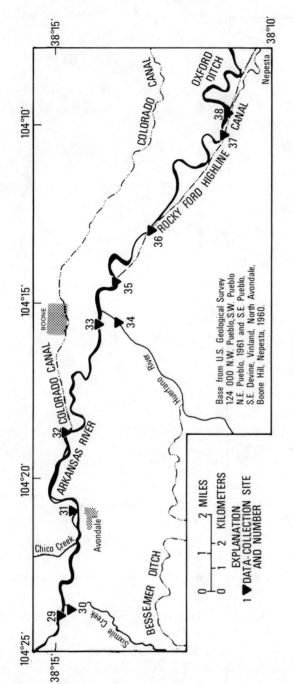

FIGURE 74. (continued)

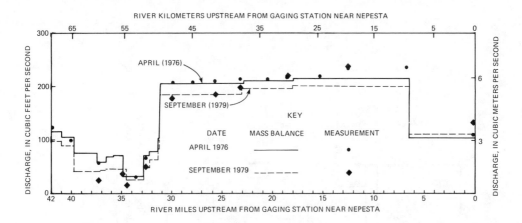

FIGURE 75. Predicted and measured flow in the Arkansas River near Pueblo, CO.

expected flow conditions observed for the September 1979 data may have been due to inconsistency between boundary conditions and instream measurements that should be investigated further. Also of interest is the discrepancy between observed and predicted dissolved solids in the downstream section of the reach during the April 1976 study. The travel time in the reach was 2 d and the studies only lasted 2 d. Therefore, the data in the downstream reaches should be examined to see if samples collected on the first day are different from those collected on the second day to determine if the steady-state assumption is valid. These same types of discrepancies between observed and predicted data are evident in similar measurements of ammonia and dissolved oxygen for the same surveys (see Figures 77 and 78). The prediction of ammonia and oxygen for this stream is described in detail in McCutcheon.[2]

These simple mass balances may not indicate exactly what problem exists with the data, such as missing some inflows or inconsistencies between inflow and instream measurements because of unsteady flow in the stream, but these are good indicators of when data may be substandard and should be de-emphasized during calibration. As of yet, there are few criteria for confirming that a steady-state model is appropriate when minor unsteadiness occurs. A simple mass balance and the appropriate statistical analysis of measurement and modeling error may provide a much needed quantitative framework.

It may also be noted that a simple mass balance is useful in determining the adequacy of sampling location in addition to providing valuable information about timing of data collection. Figure 76 also shows that there are several sampled locations in the upstream section of the reach at which the specific conductivity measurements do not agree with measurements made upstream and downstream of the site. In these cases, it turns out that the samples were grab samples taken from a single location. In the upstream reach, inflows are closely spaced and it is obvious that the samples were collected before the inflows were fully mixed. The inflows upstream of these sampling sites were drains introducing irrigation return flows that are typically high in nitrates. When one observes the nitrate measurements present in Figure 79, it becomes clear that the samples were in fact incorrectly collected at these sites. The sites that were poorly mixed are designated with a question mark in Figures 77, 78, and 79. Samples from these mixing zones also show the greatest amount of data scattered at a point on the river. This occurs because of turbulent mixing of the plumes entering the flow.

Simple mass balances assuming no decay and no reaction for a number of constituents have been found to be equally useful. For example, Figures 80 and 81 show that simple

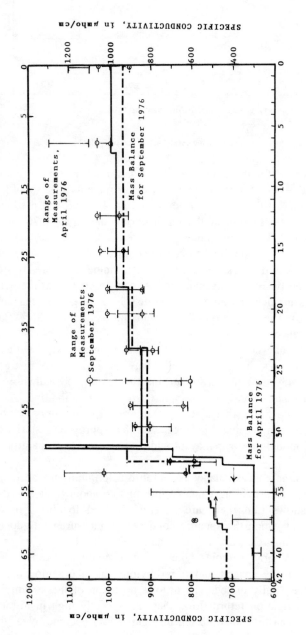

FIGURE 76. Predicted and measured dissolved solids in the Arkansas River near Pueblo, CO. Note that dissolved solids are directly proportional to specific conductivity. The solid curve for the April 1976 mass balance corresponds to the left specific conductivity axis. The dashed curve and the vertical ranges of data defined by circular points for the September 1976 mass balance correspond to the specific conductivity axis on the right.

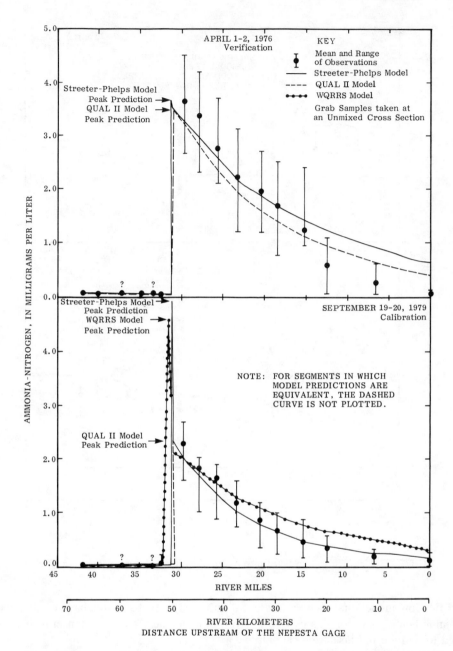

FIGURE 77. Ammonia-nitrogen measurements and predictions in the Arkansas River near Pueblo, CO.

mass balances of total nitrogen and orthophosphate may be useful in determining the overall fate of nutrients in a river. Figure 81 is a plot of measured and predicted orthophosphate based on the assumption that phytoplankton uptake and other reactions are negligible. The data are quite scattered at a few locations below the sewage treatment plant inflow, but overall seem to be in adequate agreement with the simple mass balance that assumes that orthophosphate is conserved.

The total nitrogen balance in Figure 80 indicates that nitrogen may not be conserved or that inflows may not be adequately measured. A simple mass balance does not make the exact form of nitrogen kinetics obvious, but it does point out that a simple

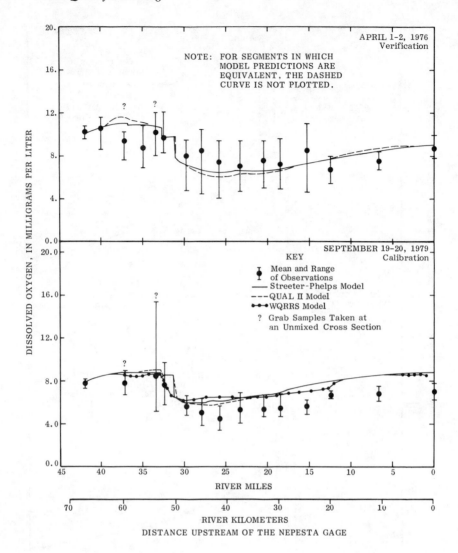

FIGURE 78.   Measured and predicted dissolved oxygen in the Arkansas River near Pueblo, CO.

nitrification model that ignores losses of nitrogen may be inadequate or that data defining boundary conditions may be in error. In either case, these are concerns that may need to be followed up during calibration.

These simple mass balances may also be performed in a Lagrangian manner as well. Figure 82 shows two sets of measurements made at the same time at two stations that are 3.9 km (2.4 mi) apart. When the sets of measurements are lagged by the travel time between sites (18.6 h), the result shows what happens to a parcel of water that leaves the upstream site and arrives at the downstream site 18.6 h later. In this particular steady flow case, mid-afternoon peaks of dissolved oxygen due to intense photosynthetic activity in the upstream, treeless section, arrived at the shaded, downstream section to cause abnormal mid-morning peak concentrations of dissolved oxygen. As a result, advection along with limited dispersion and volatilization were processes that contributed to the interesting dual peaks in diel dissolved oxygen at the downstream site.

Therefore, simple mass balances are very useful diagnostic methods that can be used

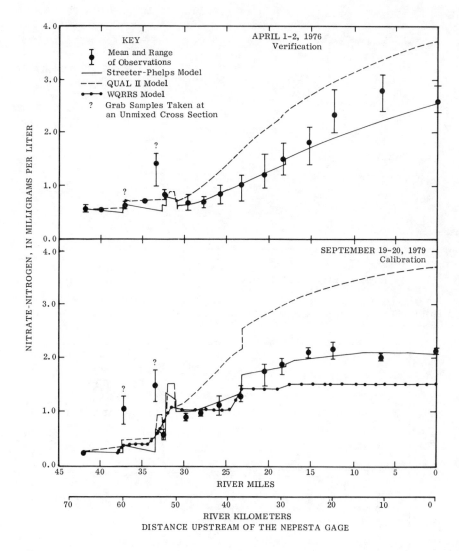

FIGURE 79. Nitrate-nitrogen measurements and predictions in the Arkansas River near Pueblo, CO.

to check the adequacy of data and model assumptions and can be used to indicate what form of the water quality kinetics may be appropriate. Other similar mass balances to track slug injections and continuous injections of tracers have also proven very useful in measuring discharge, time of travel, dispersion, and gas transfer. These methods have been discussed in other sections.

III. HEAT BALANCE AND TEMPERATURE MODELING

A. IMPORTANCE OF TEMPERATURE IN DETERMINING WATER QUALITY

Many water quality models include the simulation of water temperature not only to simulate that parameter for exposure calculations but to also predict the effect of changes in temperature on the biochemical kinetics of a water body. In addition, temperature is an important determinate of chemical reactions. It is equally important to know water temperature as it is to know pH and other geochemical properties when

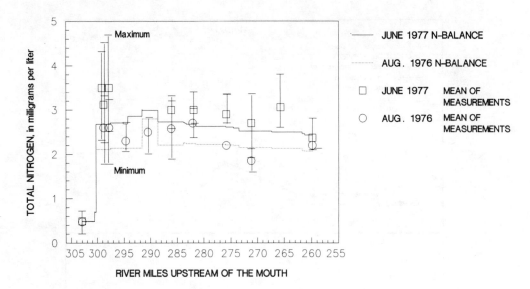

FIGURE 80.   Total nitrogen balance for the Chattahoochee River downstream of Atlanta, GA.

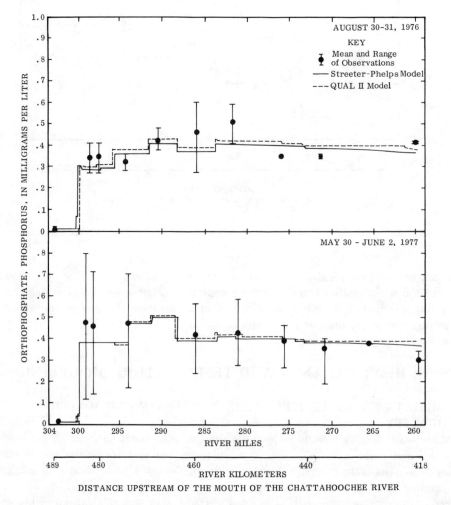

FIGURE 81.   Orthophosphate balance for the Chattahoochee River downstream of Atlanta, GA.

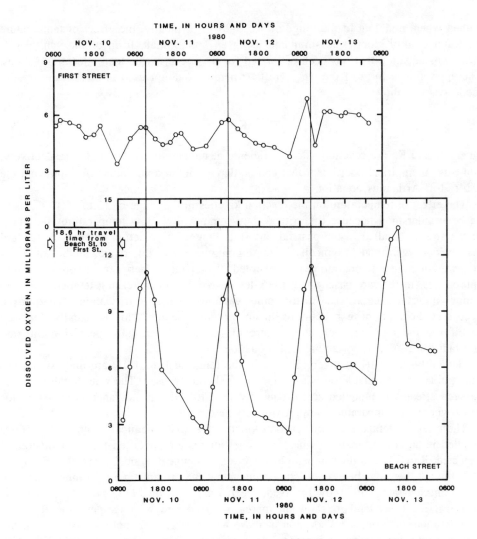

FIGURE 82. Convective effects on dissolved oxygen in the West Fork Trinity River in Fort Worth, TX. Beach Street is the upstream end of the reach and First Street is the downstream end. A crude mass balance results when measurements at the upstream end are lagged after the measurements at the downstream end by the time of travel.

forecasting the effect of physical-chemical processes. Finally, aquatic organisms have different growth responses at different temperatures that can be simulated if temperature changes are simulated.

Given the importance of temperature effects on reactions, only screening level analyses and simple computerized models neglect temperature modeling. Because temperature in a water body may not vary significantly and may not be outside ranges where reactions or growth is impeded or unduly increased, there are a few cases where temperature modeling is not important in describing (i.e., calibrating a model) or predicting stream water quality. Given the ease with which temperature can be modeled, however, there do not seem to be any important reasons for neglecting temperature in the formulation of a computerized model code or calibration of a model. It is very easy to measure the necessary boundary conditions and calibration data for temperature modeling.

Generally it is possible to incorporate the effects of temperature into water quality models over a range from approximately 0 to 35°C. However, the best results seem to gained when modeling in the range of 15 to 25°C. Typically, the effect of temperature on water quality kinetics is simulated using an approximate form of the van't Hoff-Arrhenius equation that has been calibrated for the use in waste load allocation studies. The approximate form of the van't Hoff-Arrhenius equation used in most water-quality models is as follows:

$$K = K_0\theta^{(T - T_0)} \tag{98}$$

where K and K_0 are reaction rate coefficients at temperatures, T and T_0, respectively, and θ is an empirical coefficient that can be derived in an approximate fashion from the van't Hoff-Arrhenius equation.

The effect of temperature on the reaction kinetics in the range of 15 to 25°C seems to be best understood because the critical, late-summer, low-flow temperatures for many of the streams studied in the past fall into this range. The reaction kinetics in standard dissolved oxygen models typically increase continuously in this range. Studies of rivers at lower or higher temperatures[2] have noted discrepancies between various modeling approaches. In the rare case where the rates decrease with increasing temperature (i.e., temperatures in excess of 28°C for some reactions such as nitrification), it would be necessary to treat these exceptions in an *ad hoc* manner. This is usually done by specifying the rate for each temperature of interest rather than specifying the rate constant for 20°C and correcting with Equation 98.

Generally, the effects of temperature in the range of 0 to 15°C are understood, but little guidance is available on what values of θ should be used. The water quality of ice-covered streams is modeled so infrequently that little if any guidance is available for cases where the temperature is approximately zero.

The effect of temperature on reaction kinetics for toxic chemical models is generally handled in an *ad hoc* manner where rate coefficients are specified for the temperature expected. Rarely are water temperatures and toxic chemicals modeled simultaneously.

Although there has been conflicting evidence at the present time regarding reaeration at low mixing rates[136,137] (now resolved, see *J. Environ. Eng.*, 114 (4), 1989), physical-chemical reactions tend to show an increase in the rate with temperature. Rates of volatilization, diffusion, and sorption increase with temperature increases because of the higher molecular energy at increased temperatures. Decreasing water viscosity with increasing temperature causes many sediment particles to settle faster and increases the potential mass flux of constituents through viscous and laminar regions around particles.

The general effect of increasing temperature on biologically mediated rates is to cause rates to increase until a maximum rate is achieved. Thereafter, rates are constant or immediately begin to decrease. Typically the decrease in rate occurs more rapidly than the increase. Evidently, enhanced diffusion and chemical reaction contribute to the increasing rate. Organism toxicity to higher temperatures contributes to the decreasing rate. High temperatures cause quick death as cell cytoplasm suffers irreversible damage.[138]

In the typical range of stream temperatures (0 to 35°C), nitrification is one reaction that achieves a maximum rate and begins to decrease with continued increasing temperature. Other reactions are less sensitive to temperature and may not achieve a maximum rate until the temperature exceeds that normally found in streams. Biochemical chemical decay is such an example where rates continually increase in a limited range of interest. Some responses to changing temperature are illustrated in Figure 83.

Those reactions that continually increase over the typical range of 15 to 25°C are the easier to model (using Equation 98). Those reactions that achieve a maximum rate in the range of temperatures normally encountered are more difficult to simulate.

The physics of heat transfer between water and the atmosphere have been extensively studied and are well known. In fact, the physical processes that influence heat transfer are so well known that temperature has been used as a calibration parameter for mass transport calculations. In addition, the exchange of heat with the bottom is understood well enough to provide adequate models of all types of streams. Data collection techniques are refined and well developed as well. Our understanding of the physical process is so advanced, however, that data collection techniques seem to be the limiting factor. This is almost unique among process models for water-quality constituents of interest.

In this section, the standard method of correcting reaction rates and constants will be derived. In addition, various means of reporting the effect of temperatures on reactions will be reviewed to assist in interpretation of studies of the effect.

Also in this section, the heat balance will be described along with the measurements necessary to support modeling studies. Generally, data from the National Weather Service is will be sufficient to allow studies to forego extensive meteorological data collection except in remote mountainous areas.

Modeling techniques that will be described are diverse and differ primarily in degree of approximation of the complete heat balance. Like the mass balance expressions derived earlier, practical equations for the heat balance rely on averaging. In most cases, the eddy-diffusivity approach has proven useful. Exceptions involve the modeling of thermal discharges of cooling water from power plants that have already been discussed in general terms. In addition to the semi-empirical closure method required to solve the heat balance equation, it is necessary to introduce other approximations to represent evaporation and other heat transfer processes. Approximation of the evaporation process introduces the only other empirical coefficients besides eddy-diffusivity that must be determined by calibration. These are the wind speed coefficients for which reasonable but not extensive guidance is available.

The modeling approaches range from representation of the full heat balance to simple first order approximations. Useful and accurate approximations involve equilibrium temperature approaches where the equilibrium temperature has been estimated in various ways, including by assuming that equilibrium temperature can be represented by the air temperature. Each of the more useful techniques will be reviewed. In addition, at least one fully empirical technique will be reviewed as well.

B. RELATIONSHIP BETWEEN REACTION RATES AND TEMPERATURE

The general practice is to specify reaction constants for a standard temperature that is usually taken as 20°C. Although 25°C is sometimes used for reaeration coefficients.[141] A form of the van't Hoff-Arrhenius equation,[142] i.e., Equation 98, or an equivalent equation is used to predict the change in reaction coefficients caused by changes in temperature.

Svante Arrhenius, Jacobus H. van't Hoff, and others observed that reaction constants, K, were related to activation energy, E, absolute temperature, T, and the gas constant, R, in the following way:

$$\frac{\partial \ln(K)}{\partial T} = \frac{E}{RT^2} \tag{99}$$

This equation may also be derived from the Gibbs-Helmholtz equation and ultimately can be derived rigorously from basic thermodynamics.[143,144] Where the activation energy

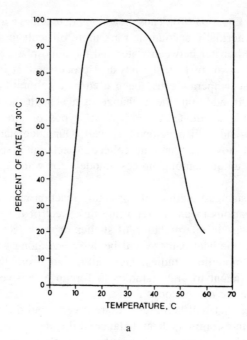

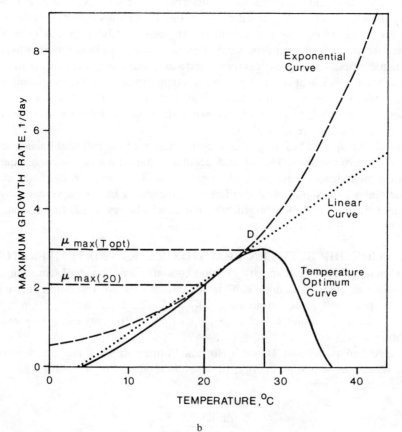

FIGURE 83. Typical response of growth or chemical reactions to temperature change. (a) Effect on nitrification as reported by Borchardt.[73,139] (b) Major types of temperature response curves for algae growth from Bowie et al.[73] (c) Respiration rate multiplier for (1) brown bullhead, (2) yellow perch, and (3) rainbow trout during active swimming (from Thornton and Lessum[140]).

Brown Bullhead

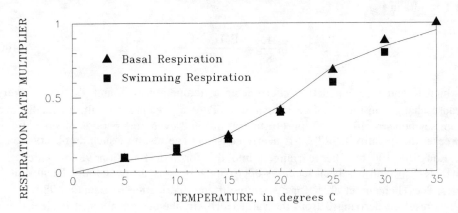

FIGURE 83(c)1

YELLOW PERCH

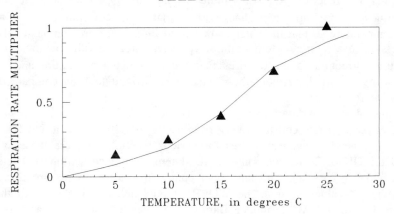

FIGURE 83(c)2

RAINBOW TROUT

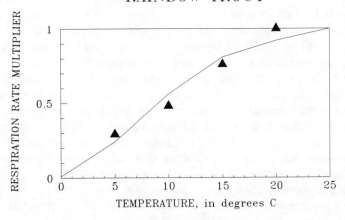

FIGURE 83(c)3

is constant, Rich[142] notes that the integration of Equation 99 between the limits of T_0 and T results in

$$\ln \frac{K}{K_0} = \frac{E(T - T_0)}{RTT_0} \tag{100}$$

in which K and K_0 are reaction coefficients at temperatures T and T_0, respectively. Because changes in the absolute temperature, T, of a stream are small, especially over stream segments of 10 to 100 km (16 to 160 mi) and over time periods of several days or weeks, the quantity $E/(RTT_0)$ is nearly constant. The result of such an observation is that Equation 100 can be rearranged into the form of Equation 98 in which $\theta = \exp[E/(RTT_0)]$ and has been taken as approximately constant over limited temperature ranges. See Thomann et al.[145] for a slightly different derivation of Equation 98.

As a result of deriving θ from the van't Hoff-Arrenhenius equation, it is clear that θ is only approximately constant for limited ranges of temperature and is mildly affected by changes in temperature. This explains some of the divergence in reported values for θ as well because past studies do not seem to be based on consistent temperature ranges. The fact that θ is function of $1/(TT_0)$ indicates that any experimental value should be mildly influenced by temperature changes (note that these temperatures are in terms of absolute temperature so that the changes that occur in streams are comparatively small) and is related to a reference temperature, T_0. Therefore, the full specification of the temperature correction factor requires that the reference temperature and range of application be specified. Without these data, it is difficult to fully compare different experimental results.

Estimates of θ from laboratory and field studies for the important physical and biochemical reaction rate constants vary from 1.008 to 1.20.[73] Brown and Barnwell[146] compile the generally accepted values for various reactions and use these as default values in the QUAL 2e model. The QUAL II temperature correction coefficients are compiled in Table 24 along with the values compiled in a review by Gromiec et al.[147] The few default values used in the WQRRS model are also included. In the latest development of the QUAL 2e model, Barnwell[148] notes that physical reactions such as settling are assumed to require a temperature correction coefficient of $\theta = 1.024$ probably based on a somewhat tenuous analogy with the reaeration process that has been extensively studied. Biologically mediated reactions are assumed to be described with $\theta = 1.047$ based on studies of biochemical oxygen demand decay. Exceptions are that larger values have been derived in the laboratory for SOD, nitrification, and nutrient release rate coefficients and these are used as default values in the QUAL 2e model. Specific values will be discussed further in the sections that follow at the point where the constants are introduced and in Volume II of this series.

The choice and use of values of the temperature correction coefficient for an application should be guided by the fact that values of θ are slightly dependent upon temperature and the temperature range for which the constant is derived. Furthermore, it should be noted that variations in the reported values of θ for different ranges of temperatures[73] occur because of this small temperature dependence. Therefore, when stream temperature changes by more than 5°C and the range of temperature falls outside 15 to 25°C, model default values should be carefully considered.

Assuming that the activation energy and the gas constant do not vary, then $E/(R\ T\ T_0)$ will vary by about 10% over the range of 5 to 35°C. This translates into an expected variation in $\theta = 1.05 \pm 0.0053$ ($\pm$ 0.5%). However, Bowie et al.[73] reports much greater variation in experimental estimates of θ, especially for biochemical

TABLE 24
Generally Accepted Values of the Temperature Correction Coefficient, θ

Rate coefficient	Default values in versions of the QUAL II model		Review by Gromiec et al.[147]	WQRRS model[49]
	SEMCOG	QUAL2e		
CBOD decay	1.047	1.047	1.047—1.075	—
CBOD settling	—	1.024	1.00—1.075	—
Reaeration	1.0159	1.024	1.0241	1.022
SOD uptake	—	1.060	—	—
Organic nitrogen decay	—	1.047	—	—
Organic nitrogen settling	—	1.024	—	—
Ammonia decay	1.047	1.083	1.106	—
Benthic ammonia source	—	1.074	—	—
Nitrite decay	1.047	1.047	1.072	—
Organic phosphorus decay	—	1.047	—	—
Organic phosphorus settling	—	1.024	—	—
Benthic phosphorus source	—	1.074	—	—
Algae growth	1.047	1.047	—	—
Algae respiration	1.047	1.047	—	—
Algae settling	—	1.024	—	—
Coliform decay	1.047	1.047	—	1.04

Note: SEMCOG was a previous version of QUAL II that is replaced by QUAL2e.

reactions such as decay of carbonaceous biochemical oxygen demand (CBOD). Therefore, it is not clear that estimates of θ are precise and accurate. In the case of biochemical reactions where different mixtures of diverse species of bacteria are involved and the substrate (carbonaceous material in the case of CBOD decay) is a variable mixture of compounds that changes as the reaction proceeds, it may well be possible that the activation energy is not constant for all possible combinations of bacteria and substrate.

The effect of temperature on reaction constants is occasionally reported in a number of different and confusing forms. Experimentally determined relationships between temperature and pseudo-first-order rate constants, maximum growth constants, Michaelis constants, oxidation rates, or growth rates have been reported in terms of the temperature correction factor, $θ$,[6,145,149-151] as Q_{10} ($θ^{10}$, the ratio of a change in the rate constant caused by a 10°C change from the reference temperature),[73,140,152,153-155] in the form equal to the natural exponent of a constant [exp(constant)],[156,157] as a percentage change in the constant or coefficient per degree temperature change,[156,158,159] and as a plot of the logarithm of the rate coefficient or constant as a function of the inverse of temperature or as a function of temperature.[153,160-168] In addition, activation energies have been reported as well.[163]

Q_{10} is related to θ and reaction coefficients or constants as

$$Q_{10} = θ^{10} = \frac{K_r}{K_{(-10)}} = \frac{K_{10}}{K_r} \qquad (101)$$

where K_r is a reference rate constant at the reference temperature (usually 20°C), $K_{(-10)}$ is the rate constant at the reference temperature –10°C, and K_{10} is the rate constant at the reference temperature plus 10°C. In addition, Thomann et al.[145] gives the following comparison (see Table 25) of θ and Q_{10} from which activation energies, E, can be computed.

Reports of percent change in reaction rates or constants per degree temperature change

TABLE 25
Comparison of the Temperature
Correction Coefficient, θ, Q_{10}, and
Activation Energies
[E = −R(293.16)² ln θ]

θ	Q_{10}	E (kcal/mol)
1.01	1.10	1.70
1.03	1.34	5.05
1.05	1.63	8.33
1.07	1.97	11.55
1.09	2.37	14.72
1.10	2.59	16.28

From Thomann, R. V., O'Connor, D. J., and Di
Toro, D. M., Mathematical Modeling of Natural Systems, Technical Report, Environmental Engineering
and Science Program, Manhattan College, New York,
1976.

are perhaps the most confused and difficult to interpret. It is especially important that
reports of percent change in rate constants per degree temperature change be given with a
reference temperature and a range of applicability. The value of $(K_T/K_{T0}) \times 100\%$ is
valid only to describe the change in K_T for a 1° *increase* in temperature and is related to
θ as

$$\text{rate of increase } (\%/°C) = (θ − 1)100\% \tag{102}$$

It should be clearly noted that Q_{10} is not equal to 10 times (%/°C)/100%. Furthermore,
the rate of decrease in rate constants per degree is not equal to the rate of increase as
shown below:

$$\text{rate of decrease (in } \%/°C) = [(1 − θ)/θ]100\% \tag{103}$$

Values of θ or exp(constant) can be derived from a plot of ln(K) or $\ln(K_T/K_{T0})$ vs. T
or 1/T. In Figure 84, the data of Wild et al.[168] are plotted as $\ln(K_T/K_{T0})$ vs.
temperature. For limited ranges of temperature, θ is constant and is equal to exp(Δ)
where Δ is the slope of the line or

$$θ = \left[\frac{(K_T/K_{T_0})_1}{(K_T/K_{T_0})_2}\right]^{[1/(T_1 − T_2)]} = \left[\frac{K_{T_1}}{K_{T_2}}\right]^{[1/(T_1 − T_2)]} \tag{104}$$

where the subscripts 1 and 2 refer to any two pairs of data from the straight line shown
in Figure 84, i.e., $[(K_T/K_{T0})_1, T_1]$ and $[(K_T/K_{T0})_2, T_2]$. A similar plot of ln(K) vs.
temperature would be handled in the same manner. Equation 104 is also useful when
ln(K) is plotted vs. temperature and can be used when oxidation rates for concentrated
systems are plotted vs. temperature.

Concentrated systems are those with very high concentrations of substrate (energy
source such as ammonia for nitrifying bacteria) and bacteria. For example, such
conditions occur when ammonia and *Nitrosomonas* concentrations are on the order of 10
mg-N/l or greater (these conditions are frequently encountered in some batch
wastewater studies). In concentrated systems, the applicable zero-order kinetics are
expressed as

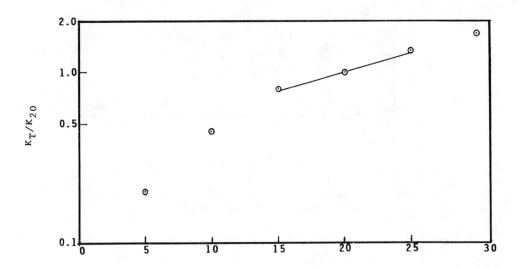

FIGURE 84. Graphical relationship between the rate coefficient for nitrification and water temperature. Data collected by Wilde et al.[169]

$$\frac{\partial N}{\partial t} = K'_n \qquad (105)$$

where N is substrate or cells grown and K'_n is a zero-order rate constant. Temperature corrections for concentrated systems cannot be determined from dilute systems because a linear relationship between oxidation rates and rate coefficient does not exist. Examples where this is the case included carbonaceous biochemical oxygen demand exertion, organic nitrogen mineralization in streams, and nitrite oxidation. Boon and Laudelout[160] seem to indicate that nitrite oxidation is a first-order process over much of the range of concentration of nitrite and *Nitrobacter* encountered.

If ln(K) is plotted versus 1/T where T is in degrees Kelvin or centigrade, then θ is exp(A) where A is the slope of the resulting line divided by $T_1 T_2$ or

$$\theta = \left[\frac{(K_T/K_{T_0})_1}{(K_T/K_{T_0})_2}\right]^{\{-1/[(1/T_1 - 1/T_2)T_1 T_2]\}} = \left[\frac{K_{T_1}}{K_{T_2}}\right]^{\{-1/[(1/T_1 - 1/T_2)T_1 T_2]\}} \qquad (106)$$

where the subscripts 1 and 2 refer to any two pairs of data from the straight line shown in Figure 85, i.e., $[(K_T/K_{T_0})_1, 1/T_1]$ and $[(K_T/K_{T_0})_2, 1/T_2]$.

When Equation 105 is valid it is also possible to derive θ as

$$\theta = \left[\frac{M_{T_1}}{M_{T_2}}\right]^{[1/(T_1 - T_2)]} \qquad (107)$$

where M is the substrate oxidized or bacteria cells grown and the subscripts 1 and 2 refer to any two pairs of data, i.e., (M_{T_1}, T_1) and (K_{T_2}, T_2).

For more precise estimates, it is useful to fit data for rate constants and temperature to the form[167]

$$\ln(K) = AT - B \qquad (108)$$

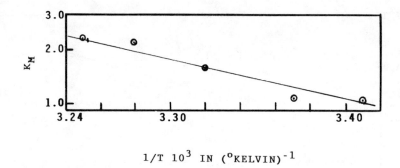

FIGURE 85. Relationship between Michaelis-Menten half-saturation constant for nitrification and the inverse of the absolute temperature.

where it can be shown that $\theta = \exp(A)$ and that B is a function of θ, the reference temperature, and the reference rate constant at the reference temperature.

Given that $\exp(A)$, Q_{10} and percent change per °C are related to θ, it should also be noted that it is important that experimental values of these parameters be associated with a reference temperature and a temperature range of applicability to be properly interpreted and compared.

In the cases where a maximum rate coefficient occurs over the range of temperatures being considered, some investigators have evidently assigned the rate coefficients a constant value for limited calibrations. One exception is the empirical temperature correction developed by Thornton and Lessum[140] that can be expressed as

$$K_A(T) = \frac{K_1 \exp(\gamma_1[T - T_1])}{1 + K_1[\{\exp(\gamma_1[T - T_1])\} - 1]} \qquad \text{where} \qquad T_1 \leq T \leq T_{opt} \qquad (109)$$

where $K_A(T)$ is a reaction rate multiplier that is multiplied times the optimum reaction rate to correct the rate to some temperature, T, within the range of applicability; K_1 is a reaction rate multiplier near the lower threshold temperature; γ_1 is a specific rate coefficient; T_1 is a lower threshold temperature; and T_{opt} is the optimum temperature at which the maximum rate occurs. For temperatures greater than optimum, the second half of the expression is written as

$$K_A(T) = \frac{K_4 \exp(\gamma_2[T_4 - T])}{1 + K_4[\{\exp(\gamma_2[T_4 - T])\} - 1]} \qquad \text{where} \qquad T_{opt} < T \leq T_4 \qquad (110)$$

where K_4 is a reaction rate multiplier near the upper threshold temperature, γ_2 is a specific rate coefficient, and T_4 is an upper threshold temperature. Thornton and Lessum[140] illustrate the use of these equations in Figures 83(c) and 86 in predicting growth and respiration rate multipliers. Mortality rate multipliers (not included here) were not predicted very well. In addition, Bowie et al.[73] note that Grenney and Kraszewski[73,169] used this method in the SSAM-IV model to correct sediment oxygen demand rates over the temperature range of 5 to 30°C. Furthermore, Grenney and Kraszewski developed a similar expression to correct the carbonaceous biochemical oxygen demand decay rate coefficient that is written as

$$\frac{K_T}{K_{20}} = \frac{0.1393 \exp[0.174(T - 20)]}{0.9 + 0.1 \exp[0.174(T - 20)]} \qquad (111)$$

where K_T and K_{20} are the CBOD decay rate coefficients at an arbitrarily selected temperature T and 20°C, respectively.

In addition to the temperature correction forms reviewed here, Bowie et al.[73] review other corrections for chemical reactions and other processes.

C. HEAT BALANCE

The important factors in the heat balance are the surface flux and other boundary conditions. The temperatures of all inflows are measured and these measurements are converted to heat using the physical properties of water that include heat capacity, C_p (heat per unit length per degree of temperature), and density, ρ (mass per unit volume), or

$$H = C_p\rho T \tag{112}$$

where T is water temperature.

In addition, the flux of heat to and from bottom sediments must be included for most small streams if hourly temperatures are to be accurately simulated. Models that average over steady-state periods of a day or more are rarely sensitive to the bottom flux, and large rivers involving large volumes of water will not be as sensitive to the bottom flux as compared with shallower rivers.

The surface heat flux expression is thought to be as complete as presently necessary to completely simulate changes in water temperature if the following form is employed:

$$H = Q_s + Q_{sr} + Q_{lr} + Q_{rl} + Q_{br} + Q_e + Q_c \tag{113}$$

where H is the net surface heat flux, Q_s is the short wave radiation striking the water surface, Q_{sr} is the short wave radiation reflected from the surface, Q_{lr} is the incoming long wave radiation from the atmosphere, Q_{rl} is the long wave radiation reflected at the water surface, Q_{br} is the long wave back radiation emitted by the water body, Q_e is the energy required for evaporation and carried to the atmosphere by water vapor, and Q_c is the energy convected to or from the water body as sensible heat. Typical values of the surface heat flux are given in Figure 87 from Parker and Krenkel.[28]

If terms are added for conduction of heat to and from the bed; for the addition of heat from waste sources, tributaries, and precipitation; and for the loss of heat as water leaves the model domain or a segment, then the complete source-sink term for the full heat balance or one-dimensional advective-dispersive equation can be written as

$$S = \frac{1}{C_p\rho} (Q_s + Q_{sr} + Q_{lr} + Q_{rl} + Q_{br} + Q_e + Q_c + Q_t + Q_{bc}) \tag{114}$$

where Q_t is the amount of heat entering or leaving at the boundaries through inflows or outflows, and Q_{bc} is the bed conduction term. S is substituted into Equations 35 or 39, and these equations are solved to predict one-dimensional distributions of temperature in streams. The same procedure is followed to solve lateral and vertical mixing problems when Equation 114 is substituted into the two- and three- dimensional heat balance equations.

In performing a heat balance, it is occasionally useful to be able to compute the measured change in heat in a water body from the average temperature of the water body. This is computed from[28]

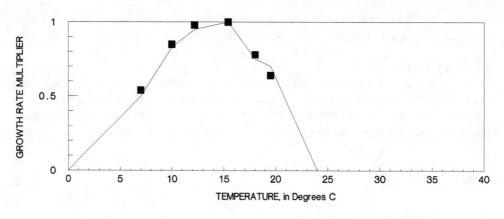

a

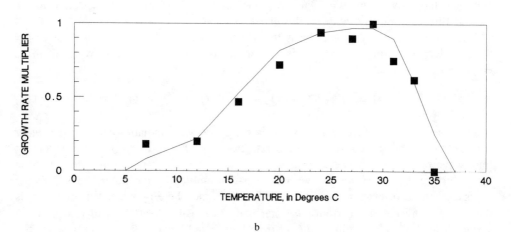

b

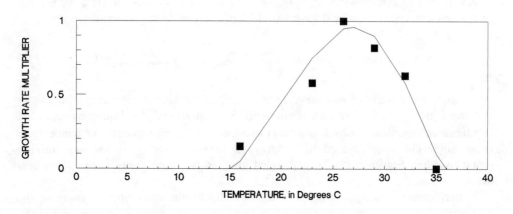

c

FIGURE 86.   Growth rate multiplier (GRM) curve for (a) brook trout, (b) emerald shiner, (c) smallmouth bass, and (d) channel catfish.[140]

CHANNEL CATFISH

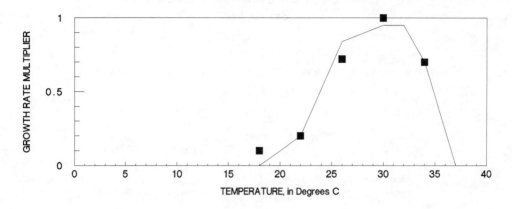

FIGURE 86d

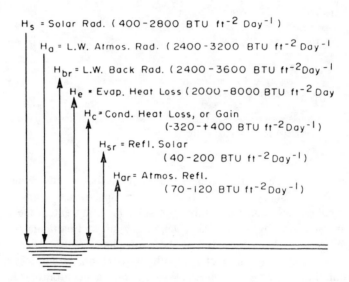

NET RATE AT WHICH HEAT CROSSES WATER SURFACE

FIGURE 87. Components of the surface heat flux along with typical values.[28]

$$\Delta Q = \frac{C_P}{A_{st}} [\rho_1 V_1 (T_1 - T_0) - \rho_2 V_2 (T_2 - T_0)] \tag{115}$$

where A_s is the average surface area over the period of time for which the heat change is computed, ρ_1 and ρ_2 are the densities of water at temperatures T_1 and T_2 measured at the beginning and end of the computation period, and T_0 is a reference temperature usually taken as 20°C.

One of the more difficult aspects of analyzing the heat balance involves interpretation of past studies reported in diverse units of measure. Bowie et al.[73] note, as is immediately obvious from the review of the literature, that temperature modeling has not been based on consistent units. The heat flux units are so difficult to apply that several models have suffered from simple conversion errors. This is a difficult problem that cannot fully be addressed here but Table 26 assists by giving the most commonly used conversion factors for heat flux and pressure.

TABLE 26
Commonly Used Units Conversions for Heat Transfer
Calculations[a]

Heat flux converted to:	BTU/ft²/d	W/m²	Langleys/d	kcal/m²/h
From:				
BTU/f²/d	1	0.131	0.271	0.113
W/m²	7.61	1	2.07	0.86
Langleys/d	3.69	0.483	1	0.42
kcal/m²/h	8.85	1.16	2.40	1
Pressure converted to:	**in Hg**	**kPa**	**mbar**	**mm Hg**
From:				
in Hg	1	3.3	33.0	25.4
kPa	0.303	1	10	7.69
mbar	0.03[b]	0.1	1	0.769
mm Hg	0.039	0.13	1.3	1

Note: 1 W = 1 j/s.
   1 Langley = 1 cal/m².
   °C = (5/9) (°F − 32).

[a]   Adapted from Reference 73.
[b]   Example: 1 mbar = 0.03 in Hg.

1. Net Solar Radiation

Short wave radiation originates with the sun and can be directly related to the solar constant (amount of energy radiating across empty space that seems to be constant enough for simulating water temperature). As solar radiation enters the atmosphere, it is absorbed and scattered by ozone, water vapor, dry air, and particulates. The decrease in short-wave radiation from the edge of the atmosphere to the water surface should therefore be related to the length of the path of radiation, cloud cover, and the air quality. Fluctuation of the solar constant and ozone concentrations have not been observed to be important in so far as this writer is aware.

While short-wave radiation can be measured with a Pyrheliometer,[28] it is usually better to rely on predictive methods to avoid complication of data collection surveys. Exceptions occur if similar measurements are required for phytoplankton modeling, but even then it should usually be more convenient to predict short-wave radiation and check the measurements if the predictions of temperature or phytoplankton prove to be sensitive to estimates of short-wave radiation. It has even proven possible to correct for shading effects of canyon walls and trees, and this can be done easier than attempting to make observations at one or two locations that may only have a limited chance of representing the typical condition for short-wave radiation reaching the water surface.

The reflected, short-wave radiation can be measured with some difficulty, but it turns out that estimates that assume that reflected, short-wave radiation is a constant percentage of the short-wave radiation reaching the water surface are more than adequate for modeling. If it is assumed that a constant percentage is reflected, net short wave radiation penetrating the water surface in Btu/ft²-h is written as[4]

$$Q_{sn} = Q_o a_t (1 - R_s)(1 - 0.65 C_L^2) \qquad (116)$$

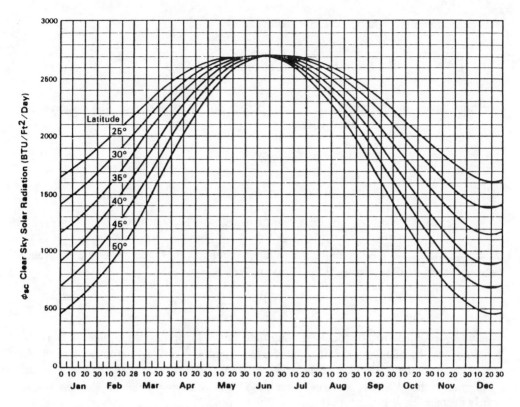

FIGURE 88. The relationship between latitude, time of year, and clear sky solar radiation according to Harmon, Weiss, and Wilson.[171] (From Bowie et al.[73])

where Q_o is the radiation flux reaching the edge of the earth's atmosphere in Btu/ft²-h, a_t is an atmospheric transmission term, R_s is albedo or a coefficient of reflection of solar radiation by the water surface, and C_L is cloudiness expressed as a fraction of the sky covered. The radiation flux reaching the atmosphere (Q_o) can be expressed in terms of the solar constant, Q_{sc}, the normalized radius of the earth, r, the latitude of the modeling site in degrees, ϕ, the declination angle of the sun in degrees, δ, the beginning and ending angles of the sun for the time period over which solar radiation is computed, α_b and α_e, and a correction factor for diurnal exposure, Γ, as follows

$$Q_o = \frac{Q_{sc}}{r^2}\left\{ \sin\frac{\pi\phi}{180}\sin(\delta)(\alpha_e - \alpha_b) + \frac{12}{\pi}\cos\frac{\pi\phi}{180}\cos(\delta)\left[\sin\frac{\pi\alpha_e}{12} - \sin\frac{\pi\alpha_b}{12}\right]\right\}\Gamma \quad (117)$$

where the solar constant Q_{sc} has a value of 438.0 Btu/ft²-h. The solar constant is also expressed as 2.00 cal/cm²-min which is an update of previous estimates (1.94 cal/cm²-min) used in a number of tables and empirical formulas published prior to the late 1960s.[170] Figure 88 shows the effect of latitude on Q_o at different times during the year.

The relative earth-sun distance is approximately

$$r = 1.0 + 0.017 \cos\left[\frac{2\pi}{365}(186 - \text{Day})\right] \quad (118)$$

where Day is the number of days since December 31 (i.e., Day = 1 for January 1 and Day = 32 for February 1). Wunderlich[170] and Smith[49] give convenient tables for this purpose. Correction for leap years is normally not necessary.[170]

The declination angle of the sun is expressed as

$$\delta = \frac{23.45}{180} \pi \cos\left[\frac{2\pi}{365}(173 - \text{Day})\right] \tag{119}$$

The hour angles for the beginning, α_b, and end, α_e, of the period over which short-wave radiation is to be calculated are written as

$$\alpha_b = t_{sb} - \Delta t_s + E_T - 12 \tag{120}$$

and

$$\alpha_e = t_{se} - \Delta t_s + E_T - 12 \tag{121}$$

where t_{sb} and t_{se} are the beginning and ending standard times, and Δt_s is the difference between standard time and local civil time written as

$$\Delta t_s = (\epsilon_a/15)(L_{sm} - L_{lm}) \tag{122}$$

where ϵ_a is equal to -1 for west longitude and $+1$ for east longitude; L_{sm} is the longitude of the standard meridian in degrees; and L_{lm} is the longitude of the local meridian in degrees. E_T is an expression for time from a solar ephemeris that is the difference in hours between "true solar time" and time computed on the basis of a yearly average. E_T is given for each day of the year, Day, as

$$E_T = 0.000121 - 0.12319 \sin\left[\frac{2\pi}{365}(\text{Day} - 1) - 0.0714\right] =$$

$$0.16549 \sin\left[\frac{4\pi}{365}(\text{Day} - 1) - 0.3088\right] \tag{123}$$

The correction factor for diurnal exposure is a discontinuous step function written as

$$\Gamma = 1 \quad \text{when} \quad t_{sr} \leq t_{sb} \quad \text{or} \quad t_{se} \leq t_{ss} \tag{124}$$

and

$$\Gamma = 0 \quad \text{when} \quad t_{ss} \leq t_{sb} \quad \text{or} \quad t_{se} \leq t_{sr} \tag{125}$$

where t_{sr} and t_{ss} are the standard times of sunrise and sunset, respectively, and can be written as

$$t_{sr} = 12 - \frac{12}{\pi} \arccos\left[\tan\left(\frac{\pi\phi}{180}\right)\tan(\delta)\right] + \Delta t_s \tag{126}$$

and

$$t_{ss} = 24 - t_{sr} + 2\Delta t_s \tag{127}$$

For definition of unfamiliar terms used in this section the reader should refer to Wunderlich.[170] Hour angles, standard time, time from a solar ephemeris, local civil time, and longitude of the standard and local meridian are explained in some detail.

TABLE 27
Total Dust Depletion Coefficient d

Season	Washington, D.C.		Madison, WI		Lincoln, NE	
	$\theta_{am} = 1$	$\theta_{am} = 2$	$\theta_{am} = 1$	$\theta_{am} = 2$	$\theta_{am} = 1$	$\theta_{am} = 2$
Winter	—	0.13	—	0.08	—	0.06
Spring	0.09	0.13	0.06	0.10	0.05	0.08
Summer	0.08	0.10	0.05	0.07	0.03	0.04
Fall	0.06	0.11	0.07	0.08	0.04	0.06

Note: θ_{am} is the optical air mass.

From Wunderlich,[170] originally from Bolsenga,[172] based on data from Kimball.[173]

This concludes a rather extensive, but highly useful development of the method to compute solar radiation at the edge of the atmosphere. Next, the effect of scattering and absorption, cloudiness, and reflection can be expressed.

Water Resources Engineers[4] have developed an empirical expression for the atmospheric scattering and absorption term, a_t, required in Equation 116. The expression is somewhat related to Beer's Law[170] and can be written as

$$a_t = \frac{a'' + 0.5(1 - a' - d)}{1 - 0.5R_s(1 - a' + d)} \tag{128}$$

where d is the dust attenuation coefficient based on the assumption that dust does not absorb significant amounts of solar radiation. d is specified for each individual application from the range 0 to 0.13 depending on the dusty, hazy nature of the atmosphere. For perfectly clear conditions, 0 would be an appropriate specification. For very poor air quality causing a significant reduction in light transmittance, specification of a value of 0.13 would be appropriate. Usually, a typical value on the order of 0.07 is chosen and only checked if temperature seems to be sensitive to this value. Wunderlich[170] gives typical values for three American cities in Table 27.

The mean atmospheric transmission coefficient, a'', in Equation 128 is written as

$$a'' = \exp\{-[0.465 + 0.0408P_{wc}][0.179 + 0.421\exp(-0.721\Theta_{am})]\Theta_{am}\} \tag{129}$$

where P_{wc} is the average daily precipitable water content in the atmosphere, expressed as

$$P_{wc} = 0.00614\exp(0.0489T_d) \tag{130}$$

In Equation 130, T_d is the dew point of the atmosphere in degrees F that can be calculated from

$$T_d = \frac{\ln\left(\dfrac{e_a + 0.0837}{0.1001}\right)}{0.03} \tag{131}$$

in which e_a is the water vapor pressure of the air. The optical air mass, Θ_{am}, in Equation 129 can be expressed as

TABLE 28
Empirical Coefficients A and B Required to Estimate the
Reflection Coefficient, R_s[174]

Cloudiness C_L	Clear 1.0		Scattered 0.1—0.5		Broken 0.6—0.9		Overcast 1.0	
Coefficients	A	B	A	B	A	B	A	B
	1.18	−0.77	2.20	−0.97	0.95	−0.75	0.35	−0.45

$$\Theta_{am} = \frac{\exp(-Z/2531)}{\sin(\alpha) + 0.15\left(\dfrac{180\alpha}{\pi} + 3.885\right) - 1.253} \tag{132}$$

where Z is the elevation of the study site, and α is the sun altitude expressed in radians as

$$\alpha = \arcsin\left[\sin\frac{\pi\phi}{180}\sin(\delta) + \cos\frac{\pi\phi}{180}\cos(\delta)\cos\frac{\pi t}{12}\right] \tag{133}$$

The absorptive effect of clouds on solar radiation in Equation 116 is expressed as[4]

$$C_s = 1.0 - 0.65C_L^2 \tag{134}$$

where C_L is the decimal fraction of the sky covered that is reported by the U.S. National Weather Service at most major and minor American airports. Typically, this information is reported as sky cover in tenths every 3 h. These and other meteorological data for the U.S. are available from National Climatic Center, Federal Building, Asheville, North Carolina, 28801, telephone (704)259-0682.

Equation 134 has been derived for determining the effect of cloud cover on daily values of solar radiation, and it is not clear that shorter-term calculations over periods of hours are precise. However, typical simulations of water temperature do not seem to suffer because Equation 134 has not been closely investigated to determine if it is adequate to make hourly simulations.

The reflection coefficient, R_s, used in Equation 116, can be expressed in an approximate manner as a function of the solar altitude, α in degrees, using Anderson's[174] empirical formula that is given as

$$R_s = A\alpha^B = \frac{Q_{sr}}{Q_s} \tag{135}$$

where A and B are empirical coefficients related to cloudiness, C_L, in Table 28. As an alternative for simplified modeling, R_s is assumed to be 0.94 on average[73,174,175] and a_t is assumed to be 1.

2. Net Long-Wave Radiation

Long-wave radiation can be measured, but it is more conveniently computed. In developing a computational formula, Anderson[174] notes that long-wave radiation varies directly with atmospheric moisture. Humidity and air temperature have a major influence on long-wave radiation while ozone, carbon dioxide, and possibly other materials have a minor influence.[4] Anderson[174] also reports that atmospheric radiation may also be affected by the height of clouds. The amount of long-wave radiation that is reflected at the water surface has been determined to be 0.03 (i.e., 3%) of the incoming long-wave radiation.

A useful approach for computing long-wave radiation involves Swinbank's formula that is based in part on the Stefan-Boltzmann Law. The atmosphere is a selectively radiating body that can only be treated approximately in applying the Stefan-Boltzmann Law.[170] Bowie et al.[73] indicate that the Swinbank formula is probably accurate to ±5%. For net, long-wave radiation, $(Q_{ln} = Q_{lr} - Q_{rl}$ in Equation 113) expressed in Btu/ft^2-h, the Swinbank formula is written as

$$Q_{ln} = (2.89 \times 10^{-6})\sigma(T_a + 460)^6(1.0 + 0.17C_L^2)(1 - R_L) \tag{136}$$

where σ is the Stefan-Boltzmann constant that has a value of 1.73×10^{-9} Btu/ft^2/h/°Rankine[4] (°Rankine is the absolute temperature scale consistent with the Fahrenheit scale, i.e., temperatures in °F can be converted to °R by adding 460); T_a is the air temperature measured at 2 m (6 ft) above the water surface; and R_L is the reflectivity of the water surface that has been found to be 0.03. Use of air temperatures measured at the nearest airport in Equation 136 has proven to be practical except in mountainous areas where meteorological data may need to be measured on site. Equation 136 can also be converted to metric units by converting the Stefan-Boltzmann constant, using the conversion to degrees Kelvin [T_a(°K) + 273] and the correct conversion factor in place of 2.89×10^{-6}.

Bowie et al.[73] note that recently Hatfield et al.[73,176] found that Brunt's formula[73,177] gives more accurate results over the latitudes 26° 13' N to 47° 45' N, and between elevations of −30 to 3342 m (−98 to 10,965 ft). This includes most locations (except some high altitudes in the Rocky Mountains) in the continental U.S. Morgan et al.[178] compared the formulas by Swinbank, Brunt, and others using clear sky data collected at Davis, CA and concluded that Brunt's formula best matched the data collected. Brunt's equation for net, long-wave atmospheric radiation in Btu/ft^2/d is written as

$$Q_{ln} = (2.05 \times 10^{-8})(T_a + 460)^4(1.0 + 0.17C_L^2)[1 + 0.149(e_2)^{1/2}](1 - R_L) \tag{137}$$

where e_2 is the air vapor pressure in mm Hg measured at 2 m (6 ft) above the water surface or computed from dew point temperature measurements (collected by the National Weather Service). See Wunderlich[170] or Linsley et al.[179] for the appropriate equation.

There are other clear sky formulations for atmospheric, long-wave radiation reviewed and compared by Wunderlich, but these equations are not adapted to handle the effect of clouds well enough to warrant mention here. The reader is referred to Wunderlich[170] for further information.

3. Back Radiation from the Waterbody

Like the atmosphere, a water body also emits long-wave radiation and this can be a significant part of the heat loss from the river as indicated in Figure 89. Parker and Krenkel[28] indicate that radiation is the second most important component in dissipating elevated, surface-water temperatures.

Long-wave radiative heat loss is best described by the Stefan-Boltzmann, Fourth-Power, Radiation Law for a blackbody because the emissivity of water is well understood. The resulting equation is written as

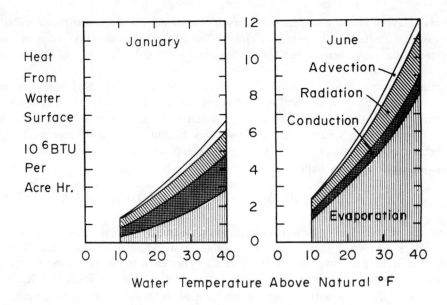

Water Temperature Above Natural °F

FIGURE 89. Typical heat dissipation from water surface by evaporation, radiation, conduction, and advection during January and June. Note that °C = (5/9) (°F − 32). (From Parker and Krenkel.[28])

$$Q_{br} = \epsilon\sigma(T_s + 460)^4 \qquad (138)$$

where ϵ is the emissivity of the water surface, and T_s is the temperature of the water surface in °F. The emissivity is usually taken as a constant equal to 0.97 based on the measurement of Anderson[174] where it was found that $\epsilon = 0.971 \pm 0.005$. Wunderlich,[170] however, notes that ϵ may vary somewhat due to local obstructions and surface contamination (i.e., oily films have an effect) and recommends a value of 0.96. Equation 138 is easily solved except that T_s is the variable of interest. To aid in the solution, Equation 138 is simplified to a linear form that has proven quite useful. Bowie et al.[73] note that a linearized form deviates from Equation 138 by less than 2.1% over the range of 0° to 30°C when $\epsilon = 0.97$. The linear form investigated by Bowie et al.[73] is

$$Q_{br} = 73.6 + 1.17T_s \qquad (139)$$

A more accurate approach is utilized in the QUAL 2e model[4] in that the linear form is fitted to Equation 138 for each 5°F temperature range from 35 to 135°F. These linear coefficients are used in place of the coefficients shown in Equation 139 (i.e., 73.6 and 1.17) for each 5°F temperature range.

4. Evaporation

Evaporation is another significant heat flux process governing the exchange of heat across the surface. Typical heat transfer calculations from Parker and Krenkel[28] (see Figure 89) indicate that the evaporative heat flux is usually the most important factor in dissipating elevated surface-water temperature. Brown and Barnwell[4] note that the evaporation of 1 lb of water carries away its latent heat of vaporization which is approximately 1050 Btu at 60°F. Unfortunately for the purpose of prediction however, evaporation is not quite as well understood as some of the other heat transfer processes.

Evaporation is caused by water vaporizing into the atmosphere. The process is driven by a vapor gradient between the water surface and the dry air above the water. When the air becomes saturated with water vapor, evaporation ceases. If the air becomes supersaturated with respect to water vapor because of decreases in air or water temperature, then condensation at the water surface can reverse the effect of evaporation and transmit heat to the water body.

Any liquid exposed to a gas or vacuum that contains none of the vapor phase of the liquid begins to volatilize. This volatilization is caused by heat in the liquid that causes molecular motion. This motion overcomes the molecular attraction in the liquid phase and causes the water molecules to escape to the gas or vacuum above. If there are few water molecules in the air above, there is a greater tendency for the water molecules to stay in the vapor phase. Molecular motion and collision in the air will also lead to the reintroduction of some water molecules back into the water body. The escape and reintroduction of molecules will occur simultaneously but at different rates. If the air becomes laden with water molecules, then the reintroduction of water molecules into the water body will balance the escape of molecules. When this occurs, the volatilized species is said to have achieved its saturation vapor pressure. Because transfer of molecules across the water surface is governed by the intensity of molecular motion, which is in turn influenced by temperature, the saturation vapor pressure must be different for different water surface temperatures (assuming that the very thin layer of air next the water surface stays at the temperature of the water at the surface). Furthermore when pressures are constant, no other processes are known to be affecting volatilization, so temperature must be the only parameter that influences saturation vapor pressures. As a result, a relationship can be written to relate air temperature in °C and saturation vapor pressure, e_o in mbar, as follows

$$e_o = \exp\left[2.3026\left(\frac{aT_s}{T_s + b} + c\right)\right] \tag{140}$$

Wunderlich[170] notes that this is the Magnus-Tetens formula and that it is very similar to the Goff-Gratch formula. The coefficients in this formula are a = 7.5, b = 237.3, and c = 0.7858 when air is over water and $T_s > 0°C$. Different values are used in this formula to describe saturation vapor pressures for sublimation.

Condensation occurs when air temperature drops and causes the saturation vapor pressure to be lower than the vapor pressure existing before the change in temperature. If the water vapor condenses at the water surface, then the heat of condensation returns to the water body. More often, however, the condensation occurs on minute particles in the air causing the fog that is often observed. The fact that fog occurs frequently indicates that vapor transfer is not a process that takes place instantaneously. Furthermore, the occurrence of low-lying fogs in river valleys and near other water bodies, indicates that water vapor is not quickly mixed throughout the atmosphere. In fact, it is rare not to find a gradient of water vapor decreasing away from the water surface. This indicates the importance of mixing and transport in the atmosphere that has been observed to control evaporation and leads to what is known as gas film limited mass transfer that will be discussed more in the next section on volatilization. In the water body itself, there are no gradients of water and thus mass transfer between the air and water can only be inhibited by transport processes in air. In the section on volatilization, it will be possible to formulate a conceptual model of the transport of materials like water in the two thin films on either side of the water surface, and this interest in the transport limitations in the vapor phase and liquid phase will become clearer.

The transport processes in the gas above the water are important in evaporation because water vapor quickly reaches an equilibrium in the layer at the interface and impedes further transport across the interface until the water vapor in the lower layer of the air is mixed into the air above. Dalton's Law indicates that the water vapor flux across the lowest layer or any layer of air is governed by the difference in vapor pressure or the gradient of water vapor. Typically, these gradients lead to transport of the vapor phase by turbulent mixing (although viscous mixing can be important as well). Knowledge of this turbulent mixing process in the air allows the formulation of equations to describe the water flux away from the surface. The close linkage between the water vaporized and the heat required for vaporization allows calculation of the heat flux due to evaporation.

Molecules in the liquid phase require a certain amount of energy to overcome the weak bonds that bind molecules together in a liquid. This energy is known as the heat of vaporization and must be furnished by heat added to or present in the liquid phase. Thus each time a given amount of water vaporizes, the heat of vaporization for that amount of mass must be subtracted from the total heat of the water body or, on a localized basis, the heat of the water near the surface. Therefore, if the amount of net evaporation is known, the effect of evaporation on the heat balance can be quantified. The heat of vaporization, L in Btu/lb, is related to water surface temperature in °F as

$$L = 1084 - 0.5T_s \qquad (141)$$

The evaporative heat flux can be expressed as

$$Q_e = \gamma LE + Q_v \qquad (142)$$

where γ is the specific weight of the water in lb/ft,^{3} and E is the evaporation rate in ft/h. Q_v is the sensible heat carried with water vapor when evaporation occurs and the evaporated water is at a higher temperature than the river water.[28] The sensible heat is the subject of some controversy regarding whether this type of energy is part of the conducted heat flux or part of the evaporative flux. Parker and Krenkel indicate that it is part of the evaporative process and note that it can be computed as

$$Q_v = \rho_w C_P E(T_e - T_s) \qquad (143)$$

where ρ_w is the density of water vapor in g/cm,^{3} C_P is the specific heat in cal/g; E is the volume of evaporated water in g/cm^2/d; T_e is the temperature of evaporated water in °C; and T_s is the temperature of the water surface in °C. Wunderlich[170] derives a similar expression assuming that sensible heat transfer is similar to heat transfer from a flat plate. Wunderlich expresses sensible heat transfer as

$$Q_v = \alpha_v(T_s - T_a) \qquad (144)$$

where α_v is a sensible heat transfer coefficient derived in greater detail by Wunderlich. Brown and Barnwell[4] found, however, that for practical applications Q_v is negligible. Furthermore, the controversy about how to treat sensible heat may be most important when evaporation into still air is important.

The evaporation rate expression is derived by assuming that turbulent mixing controls the transport of momentum, heat, and water vapor in the atmospheric boundary layer and by writing the standard eddy viscosity and diffusivity expressions for the turbulent fluxes as given in Chapter 1. If it is assumed that the eddy-viscosity and eddy-diffusivity for water vapor are approximately equal in the shallow lowest layers of the atmosphere and the logarithmic velocity profile for aerodynamically rough surfaces can be used to describe the wind profile near the surface for adiabatic conditions (unstratified), then the evaporation rate can be expressed as

$$E = \phi u_2(e_1 - e_2) \tag{145}$$

where u_2 is wind speed measured at a small distance above the water surface (usually 2 m or 6 ft), e_1 and e_2 are vapor pressures or concentrations of water vapor at two different levels, and ϕ is an evaporation coefficient related to surface roughness. The evaporation coefficient, ϕ, is expressed as

$$\phi = \frac{\rho_a}{\rho_w} \frac{\kappa^2 \left(1 - \dfrac{u_1}{u_2} \right)}{\ln \left(\dfrac{z_2}{z_1} \right)^2} \tag{146}$$

in which ρ_a and ρ_w are the densities of moist air and water vapor, respectively; κ is a constant equal to 0.41 known as the von Karman's universal turbulence constant determined from the derivation of the logarithmic velocity profile; and u_1 and u_2 are wind speeds at elevations above the water surface, z_1 and z_2. Equation 145 is attributable to Thornthwaite and Holzman and Pasquil. Deacon and Swinbank derived a very similar form by introducing an expression for the surface roughness coefficient into the eddy-diffusivity relationships defined in Chapter 1.[170]

The derivation of Equation 145 sketched here indicates that evaporation formulas are approximate and quite dependent on the location that wind speeds were originally measured to determine the evaporation coefficient. The use of the logarithmic velocity profile to derive Equation 145 ignores the effect of stratification in the atmosphere. Therefore, the wind speed measurement must be made as near the surface as possible to minimize the effect of stability (stratification). Whether the influence of stratification can be minimized depends on how high the wind speed measurements are made above the surface and the degree of stratification. The degree of stratification will vary continuously with at least a minor effect down to the 2-m (6-ft) level where most wind speed measurements are collected. Unstable stratification will cause the velocity at 2 m (6 ft) to decrease and thus cause the evaporation rate to be underpredicted. Stable stratification will cause the velocity at 2 m (6 ft) to increase and thus cause the evaporation rate to be overpredicted. Unstable stratification occurs more frequently than stable and neutral stratification combined, so there may be an overall tendency for the evaporation coefficients to be underpredicted.

A related data collection problem is that wind speed measurements are almost never collected at the stream of interest but are taken up on the bank in a location protected from vandalism (inside a wastewater treatment plant or similar compound).[9] Even more often, wind speed measurements from the nearest airport are used. Little work has been undertaken, it seems, to relate remote wind speed measurements to the 2 m (6 ft) height above the stream water surface. Wunderlich[170] notes that limited study has been undertaken to develop wind speed functions for lakes that are based on data collection sites located on the bank. On smaller rivers, however, bank roughness is relatively more important, and it is not clear if this uncertainty is more important compared to lake data collection problems.

The result of expressing the evaporation coefficient, ϕ (see Equation 146), as a function of wind speed at two levels (u_1 and u_2), indicates that the result must be dependent on surface roughness. Because the development of boundary layers on water bodies of limited extent cannot be independent of the roughness characteristics of the banks and surrounding obstructions, then each evaporation coefficient must be related to the unique topography of each river (and for important and dominant wind directions). This indicates that the extensive calibration of wind speed functions (evaporation coefficients) for lakes are not necessarily very useful for river modeling because of the very different fetch and roughness characteristics of rivers vs. lakes.

Related to the effect of surface roughness on water-vapor transport in the air is the expectation that turbulence, especially that in smaller rivers, may have some effect on evaporation.[28,180] Equation 145 does not seem to support such a hypothesis. The derivation of Equation 145 assumes that surface roughness is very small, and the surface roughness on a river would be extremely small compared to the roughness presented by nearby banks and trees. Therefore, any water turbulence resulting in a rougher surface would seem to be unimportant. Clearly, increased turbulence in the water cannot have any effect on the water transport in the water phase. As a result, it is not clear how such expectations could be supported by the understanding of the processes.

Finally, these formulations do not take into account the evaporation that will occur when the horizontal wind speed is actually zero or recorded as zero. Vertical convection that arises when the temperature of the water and air are different also causes turbulent mixing not accounted for in the derivation of Equation 145.

Eddy-viscosity formulations only represent turbulent mixing due to horizontal shear of wind over the water surface. A related experimental problem is that wind speed anemometers have bearing friction that prevents the standard instruments from recording velocities below the starting speed. At least for older instruments the starting speeds have been as high as 0.7 m/s (2.3 ft/s or 1.6 mi/h).[170]

Regarding evaporation in still air, Wunderlich[170] derives a plausible, but evidently untested, formula for the evaporation rate, E, of the form

$$E = \alpha_e(e_o - e) \tag{147}$$

where α_e is a free convection evaporation coefficient derived from the eddy-diffusivity coefficient for heat transfer. Eddy-diffusivity equations, however, are not applicable for convective flows of this type that are very definitely not a boundary-layer type flow.[181] Horizontal air movements at ground level set up classical type boundary layer flows, but vertical convective currents are very different in the way that turbulent mixing occurs.

Applications of Equation 145 expressing the evaporation rate typically take the form of

$$E = (a + bu_2)(e_o - e_2) \tag{148}$$

where it is assumed that the vapor pressure at the lowest level near the water surface can be represented by the saturation vapor pressure, e_o, and a and b are site specific empirical coefficients that must be determined by calibration. The water vapor pressure is not frequently measured by the National Weather Service but can be calculated from barometric pressure, P_a, in in of Hg; wet bulb temperature, T_{wb} in °F; and dry bulb temperature, T_a in °F, as[4]

$$e = 0.1001\exp(0.03T_{wb}) - 0.0837 - 0.000367P_a(T_a - T_{wb})\left(1 + \frac{T_{wb} + 32}{1571}\right) \tag{149}$$

where the first two terms on the right-hand side are an approximation of the saturation vapor pressure in inches of Hg at the wet bulb temperature. See Equation 150 below. Wunderlich notes that saturation vapor pressure can be adequately computed with Equation 140, but in practice a slightly simpler form is used[4] that is written as

$$e_o = 0.1001\exp(0.03T_s) - 0.0837 \qquad (150)$$

For further computation ease, this expression is linearized over 5°F increments for the range of 35 to 135°F and written as

$$e_o = \alpha_1 + \beta_1 T_s \qquad (151)$$

where α_1 and β_1 are empirical coefficients determined for each 5°F increment of the total range of temperatures. See Bowie et al.[73] for example values of α_1 and β_1 for Equation 151 expressed in metric units.

In a limited review of saturation vapor pressure equations, Bowie et al.[73] indicate that the formula of Thackston[182] is the most convenient and is accurate (standard error of prediction reported as 0.00335). This equation is written as

$$e_o = \exp\left(17.62 - \frac{9501}{T_s + 460}\right) \qquad (152)$$

where saturation pressure, e_o, is in units of in of Hg and the water surface temperature, T_s, has units of °F.

Wind speed coefficients that have been determined for field conditions are not, in general, applicable to other sites for the reasons discussed above. It is useful, however, to compile the existing coefficients for guidance in choosing initial values for calibration or in choosing approximate values for screening level calculations. Furthermore, at some sites conditions may occur so that the heat balance may not be sensitive to evaporation. In such an event, an estimate of the wind speed coefficients from the literature may be appropriate.

Wind speed functions have been compiled by two investigators. Bowie et al.[73] report in Table 29, the functions compiled from the literature by Ryans and Harleman[175] for lake and reservoir studies. In addition, Wunderlich[170] compiles other functions not contained in Table 29. This additional work is given in Table 30.

More importantly, Bowie et al.[73] report the only rigorous studies of wind speed functions for open-channel flows. In a study of the San Diego Aqueduct using an energy balance method for estimation, the following form was determined as most appropriate for the evaporation rate:

$$E = (3.01 + 1.13u_2)(e_o - e_2) \qquad (153)$$

where the wind speed was measured in m/s and the vapor pressures are in terms of kPa. The Aqueduct was in an open arid region of the U.S. (southern California) and measurements were made on-site near the water surface. In a study of the Chattahoochee River upstream of Atlanta, GA, the effects of tree-lined banks, some channel meanders, and other effects led to the calibration of a dynamic temperature model based on the full heat balance that required that the wind speed function in Equation 153 be reduced by 30%. This same wind speed function, $0.70(3.01 + 1.13u_2)$, was later applied to reach downstream of Atlanta where the characteristics of the banks and channel alignment were somewhat similar. Temperature regimes were significantly different, however. In

TABLE 29
Evaporation Formula Developed for Lakes and Reservoirs[175]

Investigator	Evaporation rate expression in original form[a]	Units for E, u, and e	Observation levels	Time scale	Water body	Formula at sea-level, measurement height as specified and units of (BTU/ft²/d, mi/hr, mm Hg)	Remarks
Marciano & Harbeck[174]	$E = 6.25 \ 10^{-4} \ u_8$ $(e_0 - e_8)$	cm/3 h, kn, Hg, mbar	8 m-wind 8 m-e_a	3 h Daily	Lake Hefner	$12.4W_8 \ (e_s - e_8)$ $17.2W_2 \ (e_s - e_2)$	Good agreement with Lake Mead, Lake Eucumbene, and Russian Lakes data. Essentially the same as the Lake Hefner Formula.
Kohler[183]	$E = .0034u_4 \ (e_0 - e_2)$	in/day, mi/d, in Hg	4 m-wind 2 m-e_a	Daily	Lake Hefner OK, 2587 acres	$15.9W_2 \ (e_x - e_2)$	
Zaykov[184]	$E = [.15 + .108u_2]$ $(e_0 - e_2)$	mm/d, m/s, mbar	2 m-wind 2 m-e_a		Ponds and small reservoirs	$(43 + 14W_2) \ (e_s - e_8)$	Based on Russian experience. Recommended by Shulyakovskiy.
Meyer[185]	$E = 10 \ (1 + .1u_8)$ $(e_0 - e_8)$	in/month, mi/ hr, in Hg	25 ft-wind 25 ft-e_a	Monthly	Small lakes and reservoirs	$(73 + 7.3W_3) \ (e_s - e_8)$ $(80 + 10W_2) \ (e_s - e_2)$	e_a is obtained daily from mean morning and evening measurements of T_a, and relative humidity. Increase constants by 10% if average of maximum and minimum used.
Morton[186]	$E = (300 + 50u_8)$ $(e_0 - e_a)/p$	in/month, mi/ hr, in Hg	8 m-wind 2 m-e_a	Monthly	Class A pan	$(73.5 + 12.2W_8) \ (e_s - e_2)$ $(73.5 + 14.7W_2) \ (e_s - e_2)$	Data from meteorological stations. Measurement heights assumed.
Rohwer[187]	$E = .771[1.65 - .0186B] \times [.44 + .118u]$ $(e_0 - e_a)$	in/d, mi/hr, in Hg	0.5 — 1 ft-wind 1 in-e_a	Daily	Pans, 85-ft-diameter tank, 1300-acre reservoir	$(67 + 10W_2) \ (e_s - e_2)$	Extensive pan measurements using several types of pans. Correlated with tank reservoir data.

(where B = atmos. press.)

[a] For each formula, the units are for evaporation rate, wind speed, and vapor pressure.

TABLE 30
Evaporation Formula Compiled by Wunderlich[170]

Investigator	Evaporation rate expression $E = f(u, e_o, e, etc.)$	Units for E, u, and e	Time scale	Type of water body
Penman[188]	$0.35(0.5 + 0.01u_2)(e_o - e_2)$	mm/d, mi/d, @ 2 m, mm Hg	—	Lake, met. data collected on land
Meyer[185]	$0.36(1 + 0.1u_{7.6})(e_o - e_{7.6})$	in/month, mi/hr @ 7.6 m, in Hg	daily	Small lakes and reservoirs, pan evaporation
Harbeck[189]	$0.078u_2(e_o - e_2)$	in/d, mi/hr @ 2 m, in Hg	daily	Lake Mead, NV
Turner[190]	$0.00030u_2(e_o - e_2)$	ft/d, mi/hr @ 2 m, in Hg	—	Lake Michie, NC
Fry[191]	$0.0001291u_2(e_o - e_2)$	cm/d, km/d @ 2 m, mbar	—	—
Easterbrook[191]	$0.000302u_2(C_o - C_2)$ $0.00000194u_2(C_o - C_2)$ (C is relative humidity, unitless)	gm/cm²/s, ft/s	—	Lake Hefner mid-lake, Lake Hefner combined data

the upper reach, the stream temperatures are depressed below natural conditions by the release of cold water from an upstream reservoir. In the downstream reach, temperatures were elevated above those entering the reach by cooling water discharges from two power plants near the head of the reach. That second model was a dynamic model but of the equilibrium temperature type. The predictions agreed with measurements equally well in the second study. Later McCutcheon[2] compared four independent sets of data that were collected at approximately steady-state conditions over the same reach of the lower Chattahoochee with the predictions of the full steady state heat balance by the QUAL II model (after errors in the heat balance were corrected) using the same wind speed function. Meteorological data for the first lower Chattahoochee study were collected at a local sewage treatment plant. During the later steady-state studies, meteorological data were collected at the Atlanta airport some miles away. It was found that the calibration wind speed function matched the original upper Chattahoochee function to within 7% but the root mean square errors were almost the same. These studies indicate that it is possible to extrapolate wind speed coefficients from similar reaches on the same river. If the largest influence on wind speed coefficients is roughness in the vicinity of the stream and channel alignment, it may be possible to develop better guidance for temperature modeling in rivers.

In a study of the West Fork Trinity River in Fort Worth, TX,[9] this writer calibrated a steady flow, equilibrium temperature model and found that the optimum wind speed functions for three reaches were about 40 to 50% of wind speed function in Equation 153 and varied on the order of 10% from subreach to subreach over a total reach length of 21.6 km (13.4 mi). This stream meandered much more than the Chattahoochee, but in places did not seem to have as much tree cover. Meteorological data were collected at two sites near the stream but not at the optimum 2 m (6 ft) height over the water surface. As a result, studies of the three channel flows (San Diego Aqueduct, Chattahoochee River, and West Fork Trinity River) do not seem to follow a consistent pattern regarding roughness condition in the vicinity of the channel. The general form of the wind speed function (a + bu₂) does seem appropriate but this has not been fully investigated.

5. Conduction

Conduction is a gradient-driven process that occurs in any solid, liquid or gas, or any combination of the phases like that found at the water surface. Conduction tends to become more important than other heat transfer processes when water bodies and the adjacent air mass are at a quiescent state.

Sensible heat is transferred between water and air when the temperatures are different. The rate of transfer is known from Fourier's classical heat transfer studies to be the product of a heat transfer coefficient times the gradient of heat, or in more practical terms, the gradient of temperature.

Conductive heat transfer cannot be measured in the field[28] and must therefore be computed from other types of measurements. The most convenient computation method makes use of what is known as the Bowen ratio that results from a number of observations that conductive and convective heat transfer are related. In the case of heat conduction across the water surface, it has been found that the heat flux due to conduction can be related to the heat flux due to evaporation as follows:

$$\text{Bowen ratio} = \frac{Q_c}{Q_e} = C_B\left(\frac{T_s - T_a}{e_o - e}\right)\frac{P_a}{29.92} \tag{154}$$

where C_B is a coefficient that is approximately 0.01, and P_a is the local barometric pressure. If Equations 142 (ignoring sensible heat flux) and 148 are substituted into Equation 154, the resulting expression for the conductive heat flux in Btu/ft^2/h is

$$Q_c = \gamma L(a + bu_2)\left(0.01\frac{P_a}{29.92}\right)(T_s - T_a) \tag{155}$$

The reliance on the Bowen ratio to express the conductive flux compounds the effect of uncertainty in the wind speed function.

For practical purposes, the ratio $P_a/29.92$ is taken to be 1. In lower elevations, this is quite useful to simplify calculations without loss of accuracy. In higher elevations, however, this can be a source of large error that will result in calibration errors in the wind speed coefficients (a and b) if the simulation is sensitive to calculations of conductive heat transfer. Higher elevations referred to here are not limited to remote mountainous areas. McCutcheon[2] found that values of $P_a/29.92$ decreased to 0.83 in a study on the Great Plains of the U.S. In the study of the Arkansas River in Colorado east of the Rocky Mountains, the elevation was 1340 to 1460 m (4400 to 4800 ft) above mean sea level (U.S. National Geodetic Vertical Datum of 1929). Discrepancies in high elevation areas will be greater on average and some discrepancy may occur down to elevations of 300 to 600 m (1000 to 2000 ft). In addition, $P_a/29.92$ is used in the calculation of dissolved gas saturation in water (i.e., dissolved oxygen), and these calculations may be even more sensitive to the ratio (see McCutcheon's[2] study of the Arkansas River).

6. Bed Conduction

Bed conduction is important when predicting the diel temperature of a stream that is less than 3 m (10 ft) deep. In such a case, dynamic prediction may be useful if maximum or minimum temperatures violate criteria designed to protect fisheries and stream habitats. Furthermore, Jobson[192-195] has found it necessary when accurately calibrating a dynamic model to include this effect. The bed conduction flux can be expressed in terms of gross thermal properties such as heat capacity of the sediments and thermal diffusivity. As a result, the temperature of the bed sediments need not be measured. Therefore, bed conduction can be easily simulated based on general knowledge of the type of sediments involved without requiring additional field data collection. The reader is referred to Jobson for details of the computation method.

7. Other Heat Sources

The final component of a heat balance that may be of importance is additions or subtractions of heat that occur as water moves into and out of a reach of a river (in and out of the model domain) as flow or enters as precipitation. Parker and Krenkel[28] note that the flux due to these sources can be written as

$$Q_t = \frac{C_P \rho}{A_s} [V_{si}(T_{si} - T_o) + V_{gi}(T_{gi} - T_o) + V_{so}(T_{so} - T_o)$$

$$+ V_{go}(T_{go} - T_o) + V_p(T_p - T_o)] \tag{156}$$

where A_s is the surface area of the river reach receiving the heat inflow or from which the heat outflow is leaving, V is volume of water, and T is temperature of the water. The subscripts refer to surface inflows (si), surface outflows (so), groundwater inflows (gi), ground water outflows (go), and precipitation (p). T_o is a reference temperature usually taken as 20°C. Some of these terms are unmeasurable, such as V_{gi} and V_{go} and, as a result, it is virtually impossible to measure evaporation by a water balance method.

D. EQUILIBRIUM TEMPERATURE AND OTHER APPROXIMATE METHODS

The methods given in the section on the heat balance are very useful for understanding the process of temperature modeling. These methods are, however, tedious to implement for simpler conditions where the heat balance may not be very dynamic. In addition, the form of the equations is very complex and more difficult to implement than other water quality process formulations seem to be. As a result, it is useful to consider alternative approximations that are simplified but still accurate for most modeling studies. One approximation that seems well suited for this purpose is the equilibrium temperature model. There are different ways of estimating equilibrium temperature that lead to further simplifications. Finally, there is at least one other entirely empirical method used in the past worth mentioning — the harmonic analysis.

1. Equilibrium Temperature

The equilibrium temperature of a stream is that temperature at which the net heat flux becomes zero and the stream has neither the tendency to warm nor cool. If the water temperature is less than the equilibrium temperature, the water will become warmer until the meteorological conditions change. If the water temperature is warmer than the equilibrium temperature, the water will cool down until meteorological conditions change to reverse this. Equilibrium temperatures can, therefore, only be measured under certain conditions that are difficult to determine. This condition is that the net heat flux is zero. Faye et al.[195] show that equilibrium temperatures can be estimated by measuring temperature vs. time at two river sites and plotting the two curves on the same graph. The downstream curve is lagged by the time of travel between the two stations. When the curves for the two sites cross, that is a measurement of the equilibrium temperature in the reach between the two sites. This technique is illustrated in Figure 90.

In itself, the definition of the equilibrium temperature does not lead to an approximation of the full heat balance. The full equation can be simply rewritten in terms of the equilibrium temperature and the same tedious solution would be required that is necessary to solve the original heat balance equation. Where the rewriting in terms of the equilibrium temperature does prove useful is that it makes it obvious how the full equations can be linearized and then easily solved.

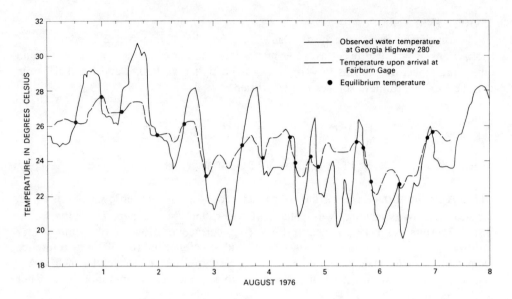

FIGURE 90. Observed temperature of the Chattahoochee River at Georgia Highway 280 and the observed temperature of the same water upon arrival at the Fairburn Gage Downstream. The mean time of travel between the stations was 15 ± 2 h. The average equilibrium temperature over the period is 24.8 ± 1.4°C.[195]

The linearization results in the following simple form for the heat flux, H, that can be written as

$$H = K(T_s - T_e) \tag{157}$$

where K is defined as a surface heat exchange coefficient, T_s is water temperature (surface temperature for stratified water bodies), and T_e is the equilibrium temperature. The heat exchange coefficient can be derived as

$$K = \frac{-W}{AC_p}\left[4\sigma\epsilon(T + 273.16)^3 + L(a + bu_2)\left(\frac{\partial e_o}{\partial T} + \gamma\right)\right] \tag{158}$$

where W is the width of the stream, A_s is the area of the cross section, γ is the psychrometric constant, and $\delta e_o/\delta T$ is the slope of the saturation vapor pressure curve that must be empirically approximated. T is temperature in °C that has been taken as water temperature, T_s,[196] or the average of water temperature and equilibrium temperature.[111] The other terms are as defined previously in the section on the heat balance. The slope of the saturation vapor pressure curve can be approximated as

$$\frac{\partial e_o}{\partial T} = \frac{1.1532 \times 10^{11}\exp[-4271.1/(T + 273.16)]}{(T + 273.16)^2} \tag{159}$$

The equilibrium temperature can be computed on a daily basis by solving the following equation iteratively

$$\text{Heat in} = \text{Heat out} \tag{160}$$

or

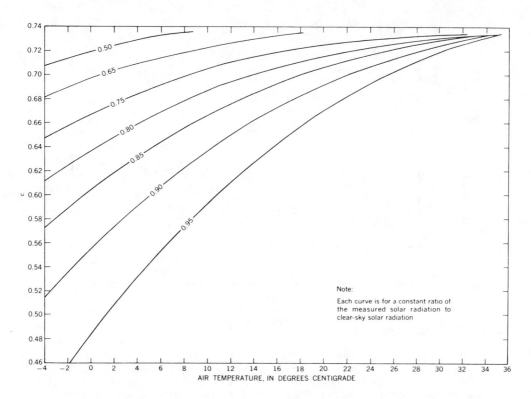

FIGURE 91. Family of curves defining the value of the empirical constant, c_c, in Brunt's equation for atmospheric infrared radiation.[197]

$$Q_s + Q_{sr} + Q_{lr} + Q_{rl} = \sigma\epsilon(T_e + 273.16)^4$$

$$+ (a + bu_2)(595.9 - 0.545T_e)\left[\frac{e_o - e}{e_o} + 0.06(T_e - T_a)\right] \qquad (161)$$

where the emissivity is assumed to be 0.97. Solar radiation, Q_s, is taken from the nearest U.S. National Weather Service measuring site (these stations are widely dispersed in the U.S.; for example, Montgomery, AL, is the nearest station to Atlanta, GA). In most cases, however, Q_s can be taken from the plot of clear sky radiation vs. time of the year and latitude of the site in Figure 88 and a correction for cloudiness may be included for more precise estimates. The reflected solar radiation, Q_{sr}, is typically assumed to be 6% of Q_s.

Long wave radiation from the atmosphere is typically computed from Koberg's approximation[197] in place of the less approximate Equation 136. This form is useful for estimating a daily average long wave radiation from daily average air temperature, saturation vapor pressure, and cloudiness. Koberg's equation is as follows:

$$Q_{lr} = \sigma(T_a + 273.16)^4(c_c + 0.0263[e_o]^{1/2}) \qquad (162)$$

where c_c is a coefficient that is a function of cloudiness and temperature as shown in Figure 91. Hourly values of air temperature, T_a, and relative humidity, R_H, are used to compute saturation vapor pressure, e_o, and vapor pressure, e. Vapor pressure is computed from $e = e_o R_H$. Equations 140, 150 or 152 can be used to compute vapor pressure (e_o). In addition, the following equation is useful for computing hourly values of e_o in millibars from hourly measurements of air temperature (T_a) in °C. That equation is written as

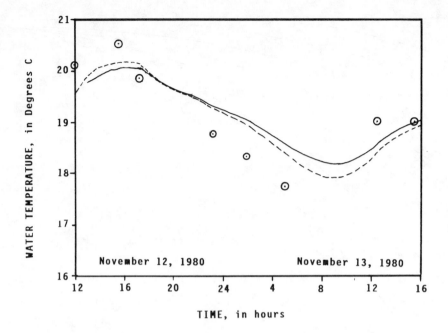

FIGURE 92. Water temperature measurements from a Lagrangian experiment on the West Fork Trinity River in Fort Worth, TX. Water temperature was measured as the peak concentration of a dye cloud passed the Beach Street Bridge and sites downstream. The solid curve is predicted using an equilibrium temperature model and a wind speed function that was chosen to be 25% of the function defined by Equation 153. The dashed line illustrates the sensitivity to the choice of wind speed coefficients. (34% of the function defined in Equation 153 is employed). 34% minimizes the root-mean-square error between measurements and predictions. 25% is the optimum wind speed function over a longer reach of this river. Note that the initial temperature at the beginning of the experiment was selected as 19.6 vs. 20.1°C measured. This was done to optimize the agreement between predictions and all measurements.

$$e_o = \exp\left[54.721 - \frac{6788.6}{T_a + 273.16} - 5.0016\ln(T_a + 273.16) \right] \qquad (163)$$

Hourly values of saturation vapor pressure (e_o) are averaged to obtain the daily average in Equation 162. Once Q_{lr} is computed, the reflected, long-wave radiation, Q_{rl}, is assumed to be 3% of the estimated value, and the heat into the water can be computed. It then becomes a trial-and-error procedure to solve for equilibrium temperature (T_e) on both sides of Equation 161.

To aid in a more efficient application of the equilibrium temperature model, there is a need to evaluate the various predictive equations for saturation vapor pressure (e_o) to determine which is most useful for various modeling needs. At present, it is not clear which forms are redundant and which are the most practical for various applications.

Figure 92 illustrates the precision of the method outlined above. The data shown were obtained from a Lagrangian experiment on the West Fork Trinity River[9] by measuring the water temperature of a dyed parcel of water as it moved downstream. The flow was steady. Measurements of solar radiation, air temperature, relative humidity, and wind speed were used to make the predictions shown. The computations, based on the use of an approximation for the saturation vapor pressure curve and the computation of daily long wave radiation, only allows the model to capture about 2 of the 2.75°C diel temperature variation at the site. The method does, however, do well in simulating

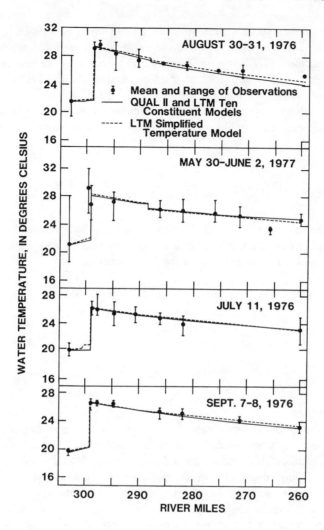

FIGURE 93. Observed and predicted water temperatures in the Chattahoochee River, GA, for four periods of steady flow and relatively steady thermal loading. The solid curve is predicted using the QUAL II model that employs a solution of the full heat balance equation. The LTM hybrid model is a Lagrangian solution of a precise linearization (equilibrium temperature approximation) of the heat balance. The simplified model is an equilibrium temperature model that assumes that the equilibrium temperature is equal to the air temperature. (Courtesy of D. H. Schoellhamer, U.S. Geological Survey.)

the daily average temperature. In a more extensive test on the Chattahoochee River during steady flow conditions, it was shown[111] that the model captures even more of the diel variation in temperature in a river reach that is on the order of 1 to 2 m (3 to 7 ft) deep compared to the West Fork Trinity River that is much shallower in many places. In addition, Schoellhamer[198] shows that the model works well for steady flow and constant heat loading. Schoellhamer was able to reproduce the results from a model of the full heat balance (QUAL II) almost exactly such that the results of the two computations could not be distinguished. This is shown in Figure 93. Finally, Faye et al.[195] and others[111] show that this method also works well when the flow is highly and moderately unsteady.

Perhaps the most useful aspect of the equilibrium temperature equation is the similarity in form to the first-order kinetic expressions used for other water quality parameters. Equation 157 is very similar to the expression for surface reaeration where K is analogous to the reaeration coefficient, and the equilibrium temperature is analogous to the saturation concentration of dissolved oxygen.

There are additional approximations of equilibrium temperature and the surface heat exchange coefficient attributable to Brady et al.[73] Equilibrium temperature in °F is expressed as

$$T_e = \frac{Q_{sn}}{23 + (a + bu_2)(\beta + 0.255)} + T_d \tag{164}$$

where Q_{sn} is the net short-wave radiation in Btu/ft^2/d, β is a proportionality factor that is a function of temperature with units of mm of Hg/°F, and T_d is the dew point temperature of air in °F. The expression for β is written as

$$\beta = 0.255 - 0.0085T_* + 0.000204T_* \tag{165}$$

where

$$T_* = \frac{1}{2}(T_s + T_d) \tag{166}$$

2. Air Temperature

There is one additional simplification of the equilibrium temperature method that is useful under some circumstances. This is an approximation that assumes that the equilibrium temperature is equal to the air temperature. When applicable, this method is quite easy to apply, especially in a descriptive study where the value of the heat exchange coefficient, K, can be determined by calibration and the temperature predictions are daily averages.

The equation where air temperature is used to approximate equilibrium temperature is written as[198]

$$\frac{\partial T_s}{\partial t} = \frac{-W}{A\rho C_p} K(T_s - T_a) \tag{167}$$

where temperatures are in °C. K can probably be expressed in a form similar to Equation 158, but the uncertainty in the approximation of the $T_e = T_a$, usually leads to K being treated as a calibration parameter. Application of Equation 167 is probably limited to steady-state simulations or average calculations over 1 or more d. In Figure 93, Schoellhamer shows that the method can be used with ease for steady-state calculations in a moderately shallow (2 m, or 6 ft depth) river like the Chattahoochee River. McCutcheon[196] found that dynamic simulations using time steps of hours rather than days was not possible in a pool-and-riffle stream (depth typically less than 1 m or 3 ft).

3. Excess Temperature

An approximate model, similar to the equilibrium temperature model, has also been applied in the past. This is the excess temperature model that assumes that elevated (or depressed) stream temperatures due to thermal discharges decrease exponentially back to the natural temperature. This method is applied by assuming that temperatures measured upstream of the discharge are the natural temperatures. Therefore, the effect of several overlapping heat sources may be difficult to simulate. In addition, it may be difficult to define a natural temperature without an equilibrium temperature analysis. As a result, the distinction between equilibrium temperature models and excess temperature models may begin to fade as advanced planning or crude design modeling is undertaken.

Krenkel and Novotny[149] apply the excess temperature approach by writing a heat balance equation (see Equation 35, written for excess temperature, $T - T_n$), in the following fashion:

$$\frac{\partial(T - T_n)}{\partial t} = D_x \frac{\partial^2(T - T_n)}{\partial x^2} + U \frac{\partial(T - T_n)}{\partial x} + K(T - T_n) \tag{168}$$

where D_x is the longitudinal dispersion coefficient, U is average velocity, and K is the heat exchange coefficient. Under steady-state conditions where dispersion is negligible, the excess heat is dissipated according to

$$\frac{(T - T_n)}{(T - T_n)_i} = \exp(-Kt) \tag{169}$$

where $(T - T_n)_i$ is the initial temperature difference due to the thermal discharge, and t is the travel time from the point of discharge. Krenkel and Novotny[149] tested Equation 169 and found that it works well in a laboratory flow as illustrated in Figure 94. In addition, they tested the full dynamic equation and found that it provided accurate predictions in at least one field study. This field study was the inverse problem of a cold water discharge from Wolf Creek Dam on the Cumberland River, KY. Figure 95 shows the agreement obtained between predictions based on the excess temperature model and measurements downstream of Wolf Creek Dam.

Bauer and Yotsukura[199] have also incorporated Equation 169 in a stream tube model to estimate lateral and longitudinal temperature variability in streams. To account for variability in depth, they write Equation 169 as

$$\frac{(T - T_n)}{(T - T_n)_i} = \exp\left(-\int_{t_o}^{t} K(t) \frac{W}{A} d\tau\right) \tag{170}$$

where t_o is the initial time of the discharge. The heat exchange coefficient, K, is defined as a function of temperature, T in °C, as follows:

$$K(T) = \frac{(4\epsilon\sigma/\rho)(T + 273.16) + LT(a + bu_2)(\partial e_o/\partial T) + \beta\gamma P_a(a + bu_2)}{C_p 30.48} \tag{171}$$

where L = 584.9 at 20°C, u_2 is in mi/h, e_o is in mbars, γ is a constant in Bowen's ratio equal to 0.00061 degree,$^{-1}$ P_a = 1013 mbars, ϵ = 0.97, σ = 1.17 × 10^{-7}/24 cal/cm^2/°K/h, C_p = 1 cal-g °C, and ρ = 1 g/cm.^{3}

4. Harmonic Analysis

The harmonic analysis is an empirical method of extrapolating daily water temperatures at a site from 1 or more years of measurements. This analysis is primarily of historical interest because it has been superseded by more deterministic methods based on the heat balance. In addition, any prediction is usually quite approximate.

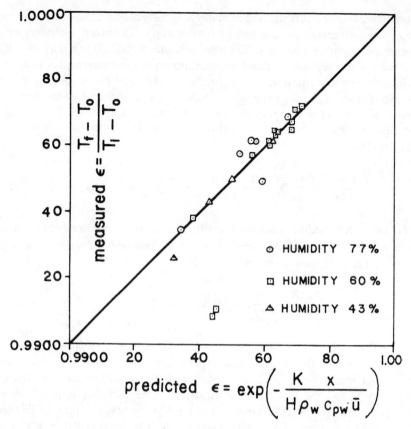

FIGURE 94. Cooling experiments in a laboratory flume verifying the excess temperature approach. (From Novotny and Krenkel.[200])

Computed and Predicted Water Temperatures

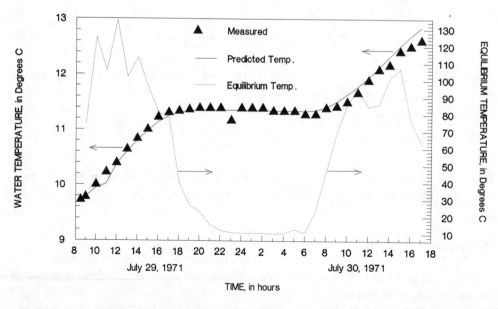

FIGURE 95. Comparison of computed and predicted river temperatures, Cumberland River below Wolf Creek Dam, KY, July 29 to 30, 1971. (From Novotny and Krenkel.[200])

This method is of some interest when interpreting geochemical monitoring data. In such an analysis, the harmonic analysis is consistent with the crude interpretations of other parameters such as average metals concentration at a site.[201]

Velz[6] notes that the harmonic analysis for temperature and geochemical data at a site has the form:

$$T(t) = A_1 + B_1 \sin(b_1 t + c_1) \tag{172}$$

where $T(t)$ is the mean daily temperature or other parameter of interest at day, t; A_1 is the harmonic mean; B_1 is the amplitude; $b_1 = 2\pi/365$; and c_1 is the phase angle in radians. The empirical constants (A_1, B_1, b_1, and c_1) are determined by curving fitting a time series of data like the daily temperature measurements shown in Figure 96. Obviously, such an approach is limited to describing parameters that are measured (extrapolation over a well-defined range) and have a harmonic response like an annual series of daily temperature measurements. Such estimates are simple to perform but tend to be imprecise. Steele et al.[201] show that the standard error for daily temperature at Maybell, CO, on the Yampa River is 2.3°C.

IV. GAS TRANSFER

A. PREDICTION OF THE FLUX OF OXYGEN, AMMONIA, CARBON DIOXIDE, AND VOLATILE ORGANIC CHEMICALS

Knowledge of gas transfer between water in a stream and the atmosphere has proven to be very important in predicting the mass balance for a number of important chemicals. Perhaps most important is the effect that reaeration has on the dissolved oxygen balance. In many cases where the effect of conventional organic pollution has been analyzed, oxygen transfer from the atmosphere is the only important component of the stream waste assimilative capacity that offsets the consumption of dissolved oxygen by biodegradation. Conventional organic wastes are an energy source for heterotrophic and nitrifying bacteria that consume dissolved oxygen in breaking down the wastes. Dissolved oxygen is important to the maintenance and growth of aquatic animals and is very frequently used as the primary regulatory standard to determine what level of protection should be afforded aquatic wildlife. In addition, the occurrence of dissolved oxygen determines whether the stream and its associated benthic sediments are an oxidizing or reducing environment. The existence of oxidizing or reducing conditions affect various geochemical cycles of such nutrients as nitrogen and phosphorus and affect the fate and transport of some toxic organic chemicals and metals. As a result, the prediction or the measurement of reaeration is both of primary and secondary importance in determining the health of a stream, as gaged by the health of the aquatic biota. In determining the fate and transport of conventional and toxic pollutants, gas transfer may be of secondary importance. In addition, the transfer of other gases such as ammonia and carbon dioxide can be important as well.

Occasionally, when the pH is high (8 to 9) volatilization of ammonia can be important. This primarily affects predictions of the nitrogen balance in the stream and almost always involves the loss of ammonia from the stream because the atmospheric concentration is usually low.

The transfer of carbon dioxide is also quite important. The exchange of carbon dioxide is an integral component of the carbonate balance in a stream. As the major buffering system in most water bodies,[73] the carbonate balance usually influences the pH of a stream. pH in turn influences chemical and biochemical reactions. Ammonia volatilization and nitrification are sensitive to pH changes over an extensive range of

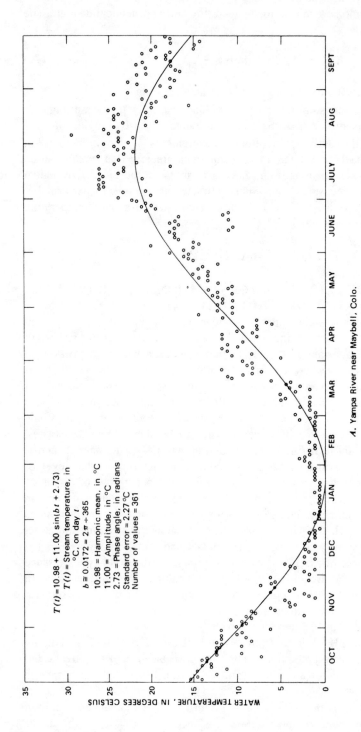

$T(t) = 10.98 + 11.00 \sin(bt + 2.73)$

$T(t)$ = Stream temperature, in
　　　°C, on day t

$b \cong 0.0172 = 2\pi \div 365$

10.98 = Harmonic mean, in °C
11.00 = Amplitude, in °C
2.73 = Phase angle, in radians
Standard error = 2.27°C
Number of values = 361

A. Yampa River near Maybell, Colo.

FIGURE 96. Use of the harmonic temperature analysis to reproduce seasonal temperature patterns during the 1963 water year at Maybell, CO.[201]

pH. Heterotrophic bacteria are influenced to some extent in degrading organic matter, and specialized bacteria acclimated to degrade a single compound can be quite sensitive to pH changes. Dissolution and precipitation of metals, hydrolysis of organic chemicals, and ionization reactions are other examples of processes that may be influenced by carbon dioxide exchange when it controls pH. The major source of carbon dioxide for algae growth is gas transfer from the atmosphere.[73]

For some toxic chemicals, volatilization can be very important in determining the fate of the contaminant. Many organic chemicals volatilize so quickly that any prediction must accurately characterize gas exchange, and in quite a few cases, it may be the only important process to simulate. In other cases, the persistence of an organic contaminant may be very directly related to limited volatilization.

Except for evaporation (volatilization of water vapor) and reaeration (exchange of dissolved oxygen), gas transfer has not been extensively studied. There have been relatively few field studies of volatilization of ammonia, carbon dioxide, and organic chemicals. There has been, however, sufficient study to develop methods to relate volatilization rate coefficients of various gases that make it possible to relate volatilization to water vapor and oxygen transfer, of which much is understood. These relationships are of the form

$$k_g = Ck_2 \tag{173}$$

where k_g is a volatilization rate coefficient of some compound of interest, C is a constant, and k_2 is the reaeration coefficient. Equation 173 can be used to relate volatilization coefficients to measurements or predictions of the reaeration coefficient if gas is limited by transport through the upper most layer of water. Similar methods[202] are available to relate gas transfer to water evaporation if transfer is limited by transport through the lowest layer of the atmosphere in contact with the water body.

B. TWO-FILM THEORY

The two-film theory is a conceptual or phenomenological theory first published by Lewis and Whittman to describe gas transfer between liquids and gases. Lewis and Whittman theorize that the interface consists of a liquid film and gas film as shown in Figure 97. It is assumed that the bulk liquid below the liquid film is well mixed and has a constant concentration. Likewise the bulk of the gas above the gas film is assumed to be well mixed. Given the likelihood of stream flows and the air above the water being at a turbulent state and relatively well mixed, the assumption that compounds are evenly mixed in the air and water not immediately in contact with the interface seems to be more than adequate to describe expected conditions. Linear concentration gradients are assumed to extend across the two films where diffusion is assumed to control transport.

Henry's Law and expressions for the flux of gas through the liquid and gas films are the basis for the two-film theory of gas exchange. Henry's Law indicates at equilibrium there is a partitioning of a dilute gas between the bulk gas solution and the liquid. At the interface, the gas concentration (C_i) and the partial pressure (P_{ci}) of the gas (concentration of the gas in the atmosphere) are proportional, and the constant of proportionality is the Henry's Law constant, H. Henry's Law is written as

$$P_{ci} = HC_i \tag{174}$$

where the saturation vapor pressure, P_{ci}, is usually expressed in atmospheres or torr (there are approximately 760 torr per atmosphere), and C_i is expressed as moles per cubic meter or moles per cubic feet.

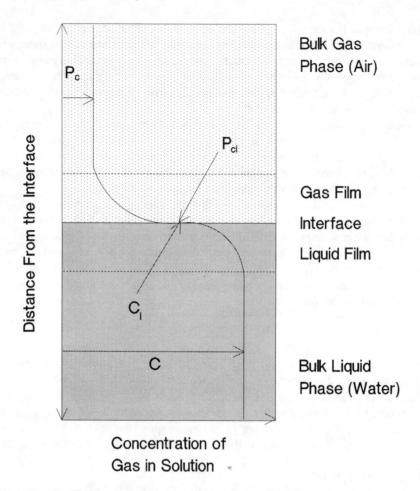

FIGURE 97.   Two-film theory of gas exchange between air and water.

Mackay and Shiu[203] note that Henry's Law constant can be expressed in a number of ways. The dimensionless form is H/R_gT where R_g is the ideal gas constant in standard atmospheres cubic meter per gram mole °Kelvin (or in English units: standard atmospheres cubic feet per pound mole °Rankine), T is absolute temperature in °K (°R); and H is Henry's Law constant in standard atmospheres cubic meter per gram mole (standard atmospheres cubic foot per pound mole). In some industrial applications, the constant is approximately the ratio of gas and liquid fractions. In addition, gas solubility has been expressed as Bunsen, Ostwald, or absorption coefficients that are expressed in terms of mass of dissolved gas per unit mass of water. These expressions are also alternative forms of the Henry's Law constant.

Mackay and Shiu[203] refer to sources of measurements of the Henry's Law constant and review measurement techniques. In addition, Mills et al.[202] list values for hydrocarbons and other selected toxic chemicals. For compounds with limited solubility, Mills et al. note that the Henry's Law constant can be computed from

$$H \text{ (atm-m}^3/\text{mo)} = \frac{P_sM_w}{760S_w} \qquad (175)$$

where P_s is the saturation vapor pressure of the pure compound in torr, M_w is the molecular weight, 760 is the standard atmospheric pressure in torr, and S_w is the

TABLE 31
Typical Values of Pollutant Volatilization Rates in Surface Waters

K_H(atm $\cdot$ m³/mol)	K_H(dimensionless)	k_v(cm/h)[a]	k_v(1/d)[b]	
10^0	41.6	20.	4.8	Liquid-film limited
10^{-1}	4.2	20.	4.8	
10^{-2}	4.2×10^{-1}	19.7	4.7	
10^{-3}	4.2×10^{-2}	17.3	4.2	
10^{-4}	4.2×10^{-3}	7.7	1.8	
10^{-5}	4.2×10^{-4}	1.2	0.3	
10^{-6}	4.2×10^{-5}	0.1	0.02	Gas-film limited
10^{-7}	4.2×10^{-6}	0.01	0.002	

[a]   Using $k_G = 3000$ cm/h.
        $k_L = 20$ cm/h.
[b]   For water depth = 1 m.

From Mills, W. B., Porcella, D. B., Ungs, M. J., Gherini, S. A., Summers, K. V., Mok, L., Rupp, G. L., and Bowie, G. L., Water Quality Assessment, Part 1 (Revised 1985), EPA/600/6-85/002a, U.S. Environmental Protection Agency, Athens, GA, 1985.

solubility in water in parts per million. Mills et al. explain how to determine if the gas of interest is slightly soluble and note that other methods should be employed if highly soluble gases are encountered.

In the absence of accumulation at the interface or elsewhere, the flux of mass through the water film and air film must be equivalent. The equivalence is expressed as

$$\frac{\partial C}{\partial z} = \frac{k_G}{R_g T} (P_c - P_{ce}) = k_L(C - C_e) \qquad (176)$$

where P_c is the partial pressure of gas in the atmosphere; P_{ce} is the partial pressure in equilibrium with the bulk concentration of gas dissolved in the water, C; C_e is the dissolved concentration in equilibrium with the partial pressure, P_c; k_L is the bulk liquid film mass transfer coefficient; and k_G is the bulk gas film mass transfer coefficient. When combined with Henry's Law (Equation 174), the overall mass transport through the two films can be expressed as

$$\frac{1}{K_r} = \frac{1}{k_L} + \frac{R_g T}{Hk_G} = r_T = r_L + r_G \qquad (177)$$

where K_r is the overall mass transfer coefficient. The mass transfer coefficients have units of length per time (meters per day, centimeters per hour, and feet per day are typically used). If the two-film theory is expressed in terms of mass transfer resistance, the total resistance ($r_T = 1/K_r$) is the sum of the resistance due to gas transfer ($r_G = R_g T/Hk_G$) and liquid transfer ($r_L = 1/k_L$).

The relative importance of the terms, r_L vs. r_G determines whether or not mass transfer is considered to be limited by transport through the liquid film or transport through the gas film. When the resistance to mass transfer in the liquid film is much greater than the resistance in the gas film ($k_L \ll Hk_G/R_g T$), transfer is considered to be liquid-film limited. When resistance to mass transfer is limited by the gas film ($Hk_G/R_g T \ll k_L$), exchange is considered to be gas-film limited. When the resistance to mass transport in the gas film and liquid film are comparable, gas exchange is neither gas-film nor liquid-film limited. Table 31 illustrates that gas exchange is typically liquid-film limited when the Henry's Law constant is greater than 10^{-3} and gas-film limited when Henry's Law constant is less than 10^{-5}. These typical ranges are based on the assumption that the gas transfer rate is typically 3000 cm/h (4.1 ft/d) and the liquid gas transfer rate is typically 20 cm/h (0.027 ft/d). The exchange of many gases, such as relatively insoluble gases like oxygen, carbon dioxide, nitrogen, and hydrogen sulfide[204] are liquid-film limited. Gas-film limitations and the combined gas-film and liquid-film limitations do seem to be encountered too frequently. The most significant gas-film limitation arises with the evaporation of water, but other gases such as the vapor phases of phenol and propionic acid are gas-film limited as well. Methanol encounters significant resistance in both the liquid and gas films.[204]

The overall mass transfer rate coefficient given in Equation 177 can be determined from estimates or measurements of the Henry's Law constant and estimates of the liquid-film and gas-film transfer coefficients, k_L and k_G. The liquid-film and gas-film transfer coefficients can be determined from estimates based on evaporation rates of water and reaeration.

The preceding section on the heat balance reviewed a number of evaporation formulas (see Equations 145, 148, and 153). In addition, Mills et al.[202] indicate that there are even cruder formulas of the form:

$$E = C_E u \tag{178}$$

where E is the evaporation or water vapor transfer rate in centimeters per hour, u is wind speed in meters per second, and C_E is a constant that Mills et al. determined to be 700. Liss conducted measurements in an experimental basin and determined that C_E was 1000. Mills et al. note that Rathbun and Tai develop other formulations as well.

There are a number of conceptual formulations relating the mass transfer of different gases in the films. First, the two-film theory and the penetration theory[205] indicate that mass transfer coefficients are proportional to $(D_m)^n$ where D_m is molecular diffusivity. Therefore,

$$\frac{(K_r)_1}{(K_r)_2} = \frac{(k_G)_1}{(k_G)_2} = \frac{(k_L)_1}{(k_L)_2} = \left[\frac{(D_m)_1}{(D_m)_2}\right]^n \tag{179}$$

where $(K_r)_1$, $(k_G)_1$, $(k_L)_1$, and $(D_m)_1$ are the mass transfer coefficients and molecular diffusivity for the same compound, and the subscript 2 refers to a different compound. For example, the gas transfer coefficient for a compound, $(k_G)_1$, can be related to the evaporation rate, E, using the ratio of the molecular diffusivity of the compound, $(D_m)_1$, to the diffusivity of water vapor in air, $(D_m)_{wv}$ as

$$\frac{(k_G)_1}{E} = \left[\frac{(D_m)_1}{(D_m)_{wv}}\right]^n \tag{180}$$

The exponent n varies significantly. There are at least three conceptual models that can be used to predict n.[204] The film theory of both Nernst and Whittman indicates that n = 1. The penetration theory developed by Higbie and the surface-renewal theory by Danckwerts indicate that n = 0.5. Boundary-layer theory indicates that n = 2/3.

Each of these conceptual models is predicated on the behavior of the interface between the air and water. The film theory assumes that a stagnant laminar film resides at the interface. The penetration theory assumes that turbulent eddies continually penetrate the film, and gas diffuses out of or into the eddies. The surface renewal theory extends the penetration theory to quantify surface film renewal. The boundary layer theory relates mass transfer to momentum transfer in a boundary layer of moving fluid. As a result, the field or laboratory conditions of interest must exactly match the conceptual assumptions behind various estimates of the exponent, n, if n is to be reliably predicted. This indicates that experiments in quiescent vessels cannot be used to predict n for natural conditions where boundary layers may be more prevalent. Because boundary layers are prevalent at stream surfaces, Thibodeaux[204] recommends that generally n should be chosen as 2/3 if field or laboratory experiments are not possible to define n for the unique geometry at a site. In the final analysis, it must be realized that n = 2/3 is only a crude approximation. Volatilization in various areas of a stream and at different times will resemble all of the proposed theoretical models: stagnant films near banks and during quiescent periods, boundary layers away from banks, and intense mixing and surface renewal in rapids and at weirs and dams.

Equation 179 is applicable to both gas-film transfer and liquid-film transfer. Application to estimate liquid-film resistance from the characteristics of another compound requires that the liquid molecular diffusivities be measured or estimated.

There are a number of methods to estimate molecular diffusivity. Thibodeaux[204] notes that Graham observed that molecular diffusivity is proportional to the square root of the inverse of the density or molecular weight of the compound of interest. As a result, Equation 179 can be simplified to

$$\frac{(K_r)_1}{(K_r)_2} = \frac{(k_G)_1}{(k_G)_2} = \frac{(k_L)_1}{(k_L)_2} = \left[\frac{(D_m)_1}{(D_m)_2}\right]^n = \left[\frac{(M_w)_2}{(M_w)_1}\right]^{n/2} \tag{181}$$

For simple screening analyses, Mills et al.[202] recommend that Equation 180 be applied with n = 1/2. Thibodeaux[204] indicates that Graham's law is a reasonable approximation for diffusion in gases, but he notes that the more complex Chapman-Enskog formula may also be necessary for more precise estimates.

There are, in addition, more complex methods available to estimate molecular diffusivity in liquids. The approximate relationship based on the Stokes-Einstein equation is frequently used for dilute solutions of nondissociating solutes. The expression is written as[204]

$$D_m = (7.4 \times 10^{-8}) \frac{(\Psi_{H_2O} Mw_{H_2O})^{1/2} T}{\mu_{H_2O} V_A^{0.6}} \tag{182}$$

where D_m is in centimeters squared per second, Ψ_{H2O} is the association parameter for water (= 2.6), Mw_{H2O} is the molecular weight of water, T is the absolute temperature in °Kelvin, μ_{H2O} is the viscosity of the solution in centipoises, and V_A is the molar volume of the solute A in cubic centimeters per mole.

In addition to the relationships between mass transfer coefficients and molecular diffusivity (Equation 180) and molecular weight (Equation 181), mass transfer coefficients are also related to the inverse of the molecule diameter.[206,207] This relationship is based on Einstein's law of diffusion but has not proven extremely practical.[206] Information on molecular diameters is limited, and questions arise regarding what volume is appropriate to use in computing average diameters. Longest dimensions have been used to represent long chain molecules, but it is not clear what dimensions are appropriate to use when relating dissimilar molecules.

Earlier it was noted that the overall mass transfer coefficient could be estimated if the gas-film and liquid-film mass transfer coefficients were known. As an example, it was shown that the gas-film mass transfer coefficient could be estimated from the evaporation rate. This works well because the resistance to vaporization of water is exclusively gas-film limited. The liquid-film coefficient can be estimated in the same way using oxygen. Oxygen is almost completely limited by liquid-film resistance and reaeration rate coefficients have been measured extensively. As a result the liquid film mass transfer coefficient can be estimated as

$$k_L = k_a \left[\frac{32}{M_w} \right]^{n/2} \tag{183}$$

where k_a is the mass transfer coefficient for oxygen, 32 is the molecular weight of oxygen molecules, and M_w is the molecular weight of the compound of interest. Mills et al.[202] recommend that n be taken as 1/2 for crude screening estimates, but Thibodeaux[204] indicates that n = 2/3 is probably more realistic. The gas-film mass transfer coefficient is written explicitly as

$$k_G = E \left[\frac{18}{M_w} \right]^{n/2} = 700u \left[\frac{18}{M_w} \right]^{n/2} \tag{184}$$

where E is the evaporation rate that Mills et al.[202] indicate can be approximately represented as 700 u, and u is wind speed. Mills et al. recommend that n = 1/2 in this screening level calculation.

Because values of n are difficult to estimate, mass transfer coefficients are related to the mass transfer coefficients for other compounds, reaeration coefficients, or evaporation rate coefficients by experimentation. Equations 182 and 183 indicate, as Rathbun et al.,[208] Tsivoglou and Wallace,[207] and Bales and Holley[209] have observed, that the ratio of volatilization coefficients for different compounds are equal to a constant (at a constant temperature for methylchloride).

The existing experimental knowledge primarily focuses on the relationship of reaeration coefficients to volatilization coefficients of other compounds. Some of the data available are summarized in Table 32. It is believed there are a number of other studies not included, and it is not clear that the data presented are fully representative and accurate. Some experiment results such as those relating carbon dioxide exchange to reaeration are diverse, and the discrepancies do not seem to have been fully investigated.

C. GAS FLUX

Once the gas transfer coefficient can be estimated, it is possible to easily incorporate volatilization into standard models. Typically, the gas flux is computed as

$$\frac{\partial C}{\partial t} = \frac{-K_r A}{V} \left(C - \frac{P}{H} \right) = -K(C - C_e) \tag{185}$$

where A is the water surface area, V is the volume of the stream segment of interest, C is the concentration of dissolved gas in the water, P is the partial pressure of the gas in the atmosphere, H is Henry's Law constant, and C_e is the equilibrium concentration of the gas in water. For gases that are typically present in the atmosphere such as oxygen, C_e is the saturation concentration. When a gas is not present in the atmosphere in any significant quantities (like many toxic chemicals), $C_e = 0$ and P = 0.

The volatilization coefficient, K (reaeration coefficient for dissolved oxygen, K_2 or K_a), is usually treated as a lumped parameter because the mass transfer coefficient, K_r in length per day, is difficult to measure or estimate, and the area of the water surface is difficult to estimate. When the water surface is quiescent, the water surface area of a

TABLE 32
Ratios of Reaeration Coefficients to Volatilization Coefficients

Compound	Ratio	Ref.	Comment
Ethylene	1.15 ± 0.02	Rathbun et al.[208]	Experimental determinations.
Propane	1.39 ± 0.03		Found no effect of oil films or surfactants.
Ethylene	1.13 ± 0.11	Bales and Holley[209]	Determined in laboratory
Propane	1.36 ± 0.13		flume.
Ethylene	1.14 ± 0.09	Rainwater and Holley[210,211]	Determined in a mixing tank.
Propane	1.36 ± 0.09		
Methyl chloride	1.41	Wilcock[212,213]	At 20°C.
Krypton	1.20 ± 0.07	Tsivoglou and Wallace[207]	
1,1,1-tri-Chloroethane	1.68 ± 0.08	Rathbun and Tai[206]	
1,2-di-Chloroethane	1.61 ± 0.04		
Carbon dioxide	1.04 ± 0.02	Ljubisavljevic[214]	
Benzene, chloroform, methylene chloride, and toluene	1.53	Rathbun and Tai[202]	
Vinyl chloride	0.43 ± 0.07	Hill et al.[215]	

stream reach divided by stream reach volume is the inverse of the depth, D. The depth is the average depth of a well-mixed stream or the depth of the well mixed upper layer if the water body is stratified. Unfortunately, the water surface is only rarely flat. Typically, the water surface is very irregular and surface area cannot be reasonably estimated. As a result, it is difficult to estimate gas exchange in turbulent streams. Generally, the volatilization coefficient or reaeration coefficient, K, must be estimated from empirical equations that are not directly related to the mass transfer coefficient, K_r, because the surface area cannot be precisely predicted .

D. REAERATION COEFFICIENT

Equation 185 is used to simulate the exchange of oxygen with the atmosphere. Generally, dissolved oxygen concentrations in streams are less than the saturation concentration that results when dissolved oxygen is at equilibrium with the partial pressure of oxygen in the atmosphere. On occasion, there is a flux of oxygen from the stream into the atmosphere when the water becomes supersaturated because of photosynthesis or increasing water temperature. In any event, Equation 185 is fully valid whether oxygen is transported into the stream or away from the stream. The direction of the oxygen (or gas) flux has no effect on the manner of estimating the reaeration coefficient, K_2 (or K).

Calculation of the flux of oxygen depends on measurements or estimates of the reaeration coefficient, K_2, and the saturation concentration of dissolved oxygen in water (or the Henry's Law constant). This section will discuss methods of estimating the reaeration coefficient, and the next section will discuss methods of estimating the saturation concentration of dissolved oxygen for streams.

For precise studies where model results are sensitive to reaeration calculations, volatilization measurements using standard tracers are necessary. These methods are discussed in a following section. Occasionally, the reaeration coefficient can be computed from semi-empirical or empirical equations when the modeling results are not very sensitive to reaeration or when screening level calculations are being performed. There are a number of formulas that have been published to relate K_2 to physical and chemical characteristics of streams that will be reviewed here. The sheer number of formulas that has been proposed is perhaps the best indication that reaeration is still only poorly understood.

Reaeration in rivers and streams has been the focus of much of the past reaeration research. As a result, as many as 34 or more different reaeration formulas for predicting K_2 have been developed. These 34 formulations are listed in Table 33. The limited research on reaeration in lakes and estuaries does not seem to provide much additional understanding of the basic process except that the experience in these areas indicates that wide, open streams may be affected by wind.

In interpreting the past work, it is important to note that prior to the widespread use of calculators and digital computers, reaeration coefficients were sometimes reported in terms of base 10 instead of base e as required by Equation 185. Calculation in terms of base 10 facilitated the use of the slide rule. The conversion is a simple one but has been a source of computational error. The conversion is

$$K_2 = \ln(10)k_2 = 2.303k_2 \tag{186}$$

where K_2 is the base e coefficient, and k_2 is the base 10 coefficient.

The units of K_2 and parameters used in predictive equations are also a source of misinterpretation worthy of note. Generally, values of K_2 are reported in terms of 1/d. However, units of 1/h have been used occasionally.[141] When listing K_2 formulas like those in Table 33, it is usually important that measurements or estimates of velocity, depth, and other physical-chemical parameters be specified in the correct units. Only a few of the available predictive equations are dimensionally homogenous. Most K_2 equations are empirical or semi-empirical and thus contain constants and coefficients that are consistent with only certain sets of units that were used to derive the coefficients. To facilitate the proper interpretation, the K_2 equations in Table 33 are written in terms of units of 1/d, and the units of all equation parameters are specified. Finally, K_2 values are generally reported for application to conditions where stream temperatures are assumed to be 20°C. This convention is used in Table 33 as well except where conversion may introduce ambiguity or it is not clear how the conversion should be performed.

In applying reaeration equations, temperature corrections are necessary if temperature changes are significant (1 to 2°C). In addition, corrections are needed when comparing equations and results. In most cases, the correction is assumed to be

$$K_2 = (K_2)_{20}(1.0241)^{(T-20)} \tag{187}$$

where T is water temperature in °C, and $(K_2)_{20}$ is the reaeration coefficient at 20°C. A number of studies have estimated the temperature correction parameter as shown in Table 34. However, the results of the American Society of Civil Engineers Committee on Sanitary Engineering Research[249] has generally been accepted as the most useful estimate.

It is clear that Equation 187 is applicable to empirical and semi-empirical K_2 equations, but corrections to equations based on conceptual models (i.e., the O'Conner-Dobbins and Dobbins equations in Table 33) are not as straightforward. The O'Conner-Dobbins equation has been used extensively and temperature corrections have been made in two ways. Typically, Equation 187 is employed, but on occasion, the O'Conner-Dobbins equation has been applied by specifying the molecular diffusivity at the temperature of interest and recomputing the constant 12.9 for the temperature of interest.[142] These procedures differ because the correction for molecular diffusivity, D_m, seems to be the following:

$$D_m = (D_m)_{20}(1.1)^{(T-20)} \tag{188}$$

TABLE 33
Formulas to Predict Reaeration Coefficients for Rivers and Streams

Citation	K_2 base e at 20°C (l/d)	Units	Applicability
	Derived from Conceptual Models		
O'Conner and Dobbins[216]	$\dfrac{12.8\ U^{1/2}}{D^{1.5}}$ $\dfrac{3.90\ U^{1/2}}{D^{1.5}}$	U: ft/s D: ft U: m/s D: m	Conceptual model based on the film penetration theory for moderately deep to deep rivers. 1 ft $\leq$ D $\leq$ 30 ft (0.3 m $\leq$ D $\leq$ 9.1 m), 0.5 ft/s $\leq$ U $\leq$ 1.6 ft/s (0.15 m/s $\leq$ U $\leq$ 0.49 m/s), 0.005/d $\leq$ K_2 $\leq$ 122/d. O'Conner and Dobbins developed a second formula but O'Conner[217] noted that the difference between the two formulas was insignificant and recommended the use of this form.
Dobbins[218]	$\dfrac{C_1[1 + F^2(US)^{0.375}]}{(0.9 + F)^{1.5}D}\ \coth\left[\dfrac{4.10(US)^{0.125}}{(0.9 + F)^{0.5}}\right]$ (coth [ ] is the hyperbolic cotangent) For $C_1 = 117$ For $C_1 = 6.24$	U: ft/s D: ft S: ft/ft U: m/s D: m S: m/m	Based on film penetration model combined with data from natural streams and the flume data of Krenkel and Orlob.[219]
	Semi-Empirical Models		
Krenkel and Orlob[219,220]	$\dfrac{C_2(US)^{0.408}}{D^{0.660}}$ Where $C_2 = 234$ Where $C_2 = 174$ or	U: ft/s S: ft/ft D: ft U: m/s D: m S: m/m	Energy dissipation model calibrated by multiple correlation analysis using 1-ft (0.3-m) wide flume data; 0.08 ft $\leq$ D $\leq$ 0.2 ft (0.02 m $\leq$ D $\leq$ 0.06 m).

TABLE 33 (continued)
Formulas to Predict Reaeration Coefficients for Rivers and Streams

Citation	K_2 base e at 20°C (1/d)	Units	Applicability
	$\dfrac{8.4\,(D_x)^{1.321}}{D^{2.32}}$ or $\dfrac{0.0024(D_x)^{1.321}}{D^{2.32}}$ or $\dfrac{26(D_y)^{1.237}}{D^{2.087}}$	D_x: ft²/s D: ft D_x: m²/min D: m D_y: m²/s D: m	Based on correlation with longitudinal and vertical dispersion and calibration with data from 1-ft (0.3-m)-wide flume with deoxygenated water. Other similar forms were also reported. The flume D_x was less than that typically encountered in streams.
Thackston and Krenkel[321]	$\dfrac{C_3(1 + F^{1/2}u_*)}{D}$ Where $C_3 = 24.9$ Where $C_3 = 24.9$ or A_4Q^B A_4 and $B =$ constants	u, ft/s D: ft u, m/s D: m	Calibrated with measurements of deoxygenated water in a 2-ft (0.61-m)-wide flume; 0.05 ft $\leq D \leq$ 0.23 ft (0.015 m $\leq D \leq$ 0.091 m). Derived from the original equation given above.
Cadwaller and McDonnell[222]	$\dfrac{C_4(US)^{1/2}}{D}$ Where $C_4 = 336$ Where $C_4 = 185$	U: ft/s D: ft S: ft/ft U: m/s D: m S: m/m	Form determined by multivariate analysis of reaeration data, including the data of Churchill et al.,[223] Owens et al.,[224] and the Water Pollution Research Laboratory channel data.

Reference	Equation	Units	Comments
Parkhurst and Pomerory[225]	$\dfrac{C_5(1 + 0.17F^2)(US)^{3/8}}{D}$ Where $C_5 = 48.8$ Where $C_5 = 23.0$	U: ft/s D: ft S: ft/ft U: m/s D: m S: m/m	Developed from data collected in 12 sewers and in natural streams.
Tsivoglou and Wallace[207]	(4700)US or 0.054(Δh/Δt) at 25°T (15,300)US or 0.18(Δh/Δt) at 25°T	U: ft/s S: ft/ft Δh: ft Δt: d U: m/s S: m/m Δh: m Δt: d	Energy dissipation model calibrated from radioactive tracer measurements in five rivers.
Tsivoglou and Neal[226]	* 0.11(Δh/Δt) or 9500 US For 1 ≤ Q ≤ 10 ft³/s 0.36(Δh/Δt) or (31,200)US For 0.028 ≤ Q ≤ 0.28 m³/s	Δh: ft Δt: d U: ft/s S: ft/ft U: m/s S: m/m Δh: m Δt d	Calibrated with data collected on 24 streams using the radioactive tracer technique.
	*0.054(Δh/Δt) For 25 ≤ Q ≤ 3000 ft³/s (0.70 ≤ Q ≤ 84 m³/s)	Δh: ft Δt: d	See Tsivoglou and Wallace above for equation written in terms of (U)(S) and in SI units.
Grant[227]	0.09(Δh/Δt) at 25°C	Δh: ft Δt: d	Calibrated with radioactive tracer data from 10 small streams in Wisconsin. 2.1/d ≤ K_2 ≤ 55/d; 0.00023 ≤ S ≤ 0.0132; 0.3 ft³/s ≤ Q ≤ 37 ft³/s; (0.0084 m³/s ≤ Q ≤ 1.0 m³/s).
Grant[228]	0.06(Δh/Δt) at 25°C 5200 US at 25°C	Δh: ft Δt:d U: ft/s S: ft/ft	Calibrated with radioactive tracer data from Rock River, WI, and Illinois 0.1/d ≤ K_2 ≤ 0.8/d; 0.000038 ≤ S ≤ 0.0006;

TABLE 33 (continued)
Formulas to Predict Reaeration Coefficients for Rivers and Streams

Citation	K_2 base e at 20°C (l/d)	Units	Applicability
	$0.20(\Delta h/\Delta t)$ at 25°C	Δh: m Δt: d U: m/s S: m/m	0.25 ft/s $\leq$ U $\leq$ 1.6 ft/s (0.076 m/s $\leq$ U $\leq$ 0.49 m/s); 260 ft³/s $\leq$ Q $\leq$ 1030 ft³/s (7.3 m³/s $\leq$ Q $\leq$ 28.9 m³/s). Calibration by statistical analysis of radioactive tracer data collected for rivers in 7 states.
	$(17,000)US$ at 25°C	Δh: ft Δt: d U: ft/s S: ft/ft	
Shindala and Truax[229]	$0.08(\Delta h/\Delta t)$ at 25 °C	Δh: m Δt: d U: m/s S: m/m	
	$6900\ US$ at 25°C		
	$0.26(\Delta h/\Delta t)$ at 25°C		
	$(22,700)US$ at 25°C		
Neal[230]	$C_e(\Delta h/\Delta t)$ or $(86,400)C_eUS$ $C_e = 0.039$ to $0.35/ft$	Δh: ft Δt: d S: ft/ft U: ft/s	Calibration of the Tsivoglou–Wallace energy dissipation equation for swamp streams indicates that the escape coefficient, C_e, decreases with increasing Reynolds number as the work of Tsivoglou and Neal[226] and Grant[228,229] seems to indicate.
	$C_e(\Delta h/\Delta t)$ or $(86,400)C_eUS$ $C_e = 0.128$ to $1.15/m$	Δh: m Δt: d U: m/s S: m/m	1.6 ft $\leq$ D $\leq$ 5.8 ft (0.49 m $\leq$ D $\leq$ 1.8 m); 0.00017 $\leq$ S $\leq$ 0.00072; 0.059 ft/s $\leq$ U $\leq$ 0.66 ft/s (0.018 m/s $\leq$ U $\leq$ 0.20 m/s); 4 ft³/s $\leq$ Q $\leq$ 220 ft²/s (0.11 m³/s $\leq$ Q $\leq$ 6.2 m³/s).
McCutcheon and Jennings[231]	$$\dfrac{-\ln\left[1 - 2\left(\dfrac{D_m I\,24}{\pi(30.48D)^2}\right)^{1/2}\right]}{I}$$ $$D_m = 1.42(1.1)^{T-20}$$ $$[I = 0.0016 + 0.0005\,D]\quad D \leq 2.26\ \text{ft}$$	D: ft T: °C	Originally, derived by Hirsch[232] to replace the Velz[6] iterative method. Expressions for the mix interval, I, are derived from the extensive experience in applying the iterative method. The underlying concept is similar to the surface-renewal theory.

Churchill et al.[223]	$[I = 0.0097 \ln(D) - 0.0052]$ $\quad D > 2.26$ ft $\dfrac{0.035U^{2.695}}{D^{3.085}S^{0.823}}$ $\dfrac{0.746U^{2.695}}{D^{3.085}S^{0.823}}$	U: ft/s D: ft U: m/s D: m	Based on dimensional analysis. Derived from data collected in rivers below Tennessee Valley Authority dams.
Bennett and Rathbun[233]	$\dfrac{106U^{0.413} S^{0.273}}{D^{1.408}}$ $\dfrac{32.8U^{0.413} S^{0.273}}{D^{1.408}}$	U: ft/s D: ft S: ft/ft U: m/s D: m S: m/m	Based on reanalysis of data from Churchill et al.[223] and Owens et al.[224] This equation with a slope parameter predicts slightly better than velocity-depth formulations.
Lau[234]	$2515\left(\dfrac{u_*}{U}\right)^3 \dfrac{U}{D}$	u$_*$: ft/s U: ft/s D: ft or u$_*$: m/s U: m/s D: m	Developed by dimensional analysis from the data of Churchill et al.[223] for natural streams, and flume data of Krenkel[235] (1 ft or 0.3 m wide) and Thackston and Krenkel[221] (2 ft or 0.61 m wide).
Ice and Brown[236]	$\dfrac{36W^{2/3}S^{1/2}U^{7/6}G^{1/2}}{Q}$	W: ft S: ft/ft U: ft/s Q: ft³/s	Based on data collected on several small Oregon streams.
Streeter and Phelps[237]	$\dfrac{C_{SP}^2 U^n}{D^2}$ Rough irregular channels: $C_{SP} = 1.15^{2.3}$ Smooth regular channels: $C_{SP} = [0.39(10^{1.16S}) + 17]$ C_{SP} varied from 0.23 to 131 for the Ohio River	U: ft/s D: ft S: ft/mi	Developed from data collected on the Ohio River using Fick's Law of diffusion and the penetration theory. n is a coefficient related to a stage-velocity relationship that had values of 0.57 to 5.40 for the Ohio River. n varied from one reach to the next. D is depth above low water stage.

Empirical Formulas

Churchill et al.[223]	$\dfrac{11.6U^{0.969}}{D^{1.673}}$ $\dfrac{5.01U^{0.969}}{D^{1.673}}$ or $\dfrac{6.91U}{D^{1.67}}$	U: ft/s D: ft U: m/s D: m	See Churchill et al. above. This form is almost as good and is recommended by Churchill et al. 2 ft (0.61 m) ≤ D ≤ 11 ft (3.35 m); 1.8 ft/s ≤ U ≤ 5 ft/s (0.55 m/s ≤ U ≤ 1.5 m/s).

TABLE 33 (continued)
Formulas to Predict Reaeration Coefficients for Rivers and Streams

Citation	K_2 base e at 20°C (l/d)	Units	Applicability
Langbein and Durum[238]	$\dfrac{7.6U}{D^{1.33}}$ $\dfrac{5.1U}{D^{1.33}}$	U: ft/s D: ft U: m/s D: m	Regression of field data Churchill et al., data from O'Conner and Dobbins,[216] laboratory data from Streeter et al.,[239] and flume data (1 ft or 0.3 m wide from Krenkel and Orlob.[219]
Owens et al.[224]	$\dfrac{21.7U^{0.67}}{D^{1.85}}$ $\dfrac{5.32U^{0.67}}{D^{1.85}}$	U: ft/s D: ft U: m/s D: m	Developed from oxygen recovery data collected on 6 English streams following deoxygenation with sodium sulfide by Gameson et al.[240] and Owens et al.[224] and collected below TVA dams by Churchill et al.[223] 0.1 ft/s ≤ U ≤ 5 ft/s (0.03 m/s ≤ U ≤ 1.5 m/s).
	$\dfrac{23.3U^{0.73}}{D^{1.75}}$ $\dfrac{6.92U^{0.73}}{D^{1.75}}$	U: ft/s D: ft U: m/s D: m	This second formula was developed for 0.1 ft/s ≤ U ≤ 1.8 ft/s (0.03 m/s ≤ U ≤ 0.55 m/s); 0.4 ft ≤ D ≤ 1.5 ft (0.12 m ≤ D ≤ 0.46 m) from a restricted data set at the Water Pollution Research Laboratory.
Isaacs and Gaudy[241]	$\dfrac{C_6 U}{D^{1.5}}$ Where $C_6 = 8.62$ Where $C_6 = 4.75$	U: ft/s D: ft U: m/s D: m	Form determined by dimensional analysis. $C_6 = 7.03$ (3.88 in SI units) for data collected in a 1-ft (0.3-m)-wide annular flume with a 4-ft (1.2-m) inside diameter. Analysis of Krenkel's 1-ft (0.3-m)-wide flume data[219,235] indicates $C_6 = 5.62$ (3.10 in SI units). Analysis of 2-ft (6.1-m)-wide flume data of Thackston and Krenkel[221] indicates $C_6 = 6.75$ (3.73 in SI units). $C_6 = 8.62$ (4.75 in SI units) was obtained for selected data from Churchill et al.[223] (29 of 30 field experiments). Using all of the Churchill et al. data, $C_6 = 9.26$ (5.11 in SI units).

Reference	Equation	Units	Description
Isaacs and Maag[242]	$\dfrac{2.980\alpha_1\alpha_2 U}{D^{1.5}}$ $\dfrac{1.645\alpha_1\alpha_2 U}{D^{1.5}}$	U: ft/s D: ft U: m/s D: m	α_1 is a nondimensional variable that changes with channel geometry, and α_2 is a nondimensional variable that is a measure of surface velocity.
Negulescue and Rojanski[243]	$10.9\left(\dfrac{U}{D}\right)^{0.85}$	U: ft/s D: ft or U: m/s D: m	Developed from a recirculating flume with depths less than 0.5 ft (0.15 m).
Padden and Gloyna[244]	$\dfrac{6.9U^{0.703}}{D^{1.054}}$ $\dfrac{4.52U^{0.703}}{D^{1.054}}$	U: ft/s D: ft U: m/s D: m	Regression analysis of data collected in a research flume with high K_2: $9.8 \le K_2 \le 28.8/d$; 0.1 ft/s $\le U \le 0.46$ ft/s (0.03 m/s $\le U \le 0.14$ m/s); 0.1 feet $\le D \le 0.62$ ft (0.03 m $\le D \le 0.19$ m).
Bansal[245]	$\dfrac{4.67U^{0.6}}{D^{1.4}}$ $\dfrac{1.80U^{0.6}}{D^{1.4}}$	U: ft/s D: ft U: m/s D: m	Based on reanalysis of extensive data for numerous rivers.
Bennett and Rathbun[233]	$\dfrac{20.2U^{0.607}}{D^{1.689}}$ $\dfrac{5.58U^{0.607}}{D^{1.689}}$	U: ft/s D: ft U: m/s D: m	Based on reanalysis of data from Churchill et al.[223] and Owens et al.[224] This empirical equation is almost as good a predictor as the semiempirical equation given above with the stream slope.
Long[246]	$\dfrac{4.021U^{0.273}}{D^{0.894}}$ $\dfrac{1.923U^{0.273}}{D^{0.894}}$	U: ft/s D: ft U: m/s D: m	Developed from radioactive tracer measurements in Texas streams. Known as the Texas equation.
Foree[247]	$0.30 + 0.19\,S^{1/2}$ at 25°C $0.116 + 2147\,S^{1/2}$ at 20°C	S: ft/mi S: m/m	Developed from radioactive tracer measurements in small Kentucky streams. 0 ft/mi $\le S \le 42$ ft/mi (0.0080 m/m).
Foree[248]	$0.888(0.63 + 0.4s^{1.15})q^{0.25}$ At 25°C for $0.05 \le q \le 1$ $0.888(0.63 + 0.4S^{1.115})$ At 25°C for $q \ge 1$ $0.42(0.63 + 0.4S^{11.5})$ At 25°C for $0.05 \ge q$	S: ft/mi q: ft³/s/mi²	Developed from the reanalysis of Foree's 1976[247] data to include watershed unit discharge.

TABLE 33 (continued)
Formulas to Predict Reaeration Coefficients for Rivers and Streams

Citation	K_2 base e at 20°C (l/d)	Units	Applicability
Wilcock[213]	$$\dfrac{12 \times 10^4 (D_m U)^{1/2}}{D^{3/2}}$$	D_m: m²/s U: m/s D: m	Modification of the O'Conner-Dobbins equation to agree with methyl chloride tracer measurements in a number of New Zealand rivers.

Note: U = average stream velocity; D = average depth of stream flow; F = $U/(gD)^{1/2}$ = Froude number. g = gravitational constant. S = slope of water surface or bed. D_x = longitudinal dispersion coefficient. D_y = averaged vertical eddy-diffusivity. u_* = $(gDS)^{1/2}$ = shear velocity. Q = stream discharge. Δh = change in water surface elevation in a reach (between two points). Δt = time of travel in the reach over which change in elevation is measured. D_m = molecular diffusion coefficient for oxygen in water. T = water temperature. I = mix interval. W = stream width. q = stream discharge per unit area of the watershed.

[a] Adapted from References 73, 141, and 147.

TABLE 34
Reported Values of the Temperature Correction Coefficient for Reaeration[a]

Ref.	Coefficient	Comment
250	1.0159	Proposed as a preliminary estimate using stream data and later revised.
239	1.047	Estimate based on questionable data from channel lacking temperature control. It is not clear if $\theta = 1.016$ was intended. Phelps[251] interprets the results to indicate $\theta = 1.047$ and O'Conner and Dobbins[216] interpret the results to indicate $\theta = 1.016$ and derive a similar value from the Stokes-Einstein equation.
252	1.022 ± 0.005	Based on experiments with stirred water in beakers. A dependence on stirring rate was noted. At 12.8 rpm, $\theta = 1.0202$. At 38 rpm, $\theta = 1.0238$. At 86 rpm, $\theta = 1.0173$, and at 115 rpm, $\theta = 1.0206$. These investigators indicated that a linear equation better represents the effect of temperature changes on K_2.
253	1.018, 1.015, 1.008, and 1.018	Determined from channel experiments like those of Streeter et al.[216] Also found that a linear expression better fitted their data.
249	1.0241	Determined by mixing in flask with large and small impeller at 5, 10, 15, 20, 25, and 30°C using triplicate experiments. Distilled water was mixed in an 18-l flask. Large impeller experiments gave $\theta = 1.0226$ and a vortex was present that was not present in the small impeller experiments. Data follows the Arrhenius formula.
254	1.013 (10—20°C) 1.016 (20—30°C)	Values related to degree of mixing in a stirred vessel. Experimental values seem to be in agreement with values derived from effect on molecular diffusion, film thickness, and mixing rate. Data follow the Arrhenius formula.
207	1.022 ± 0.004	Determined by Tsivoglou[255] using *Standard Methods*[256] saturation data. Chlorides had no effect.
149	1.029	Recommended value of unknown origin.

[a] Adapted from References 73 and 249.

where the constant 1.1 was determined by Velz in 1939[6] and may be the subject of some controversy. Rich[142] indicates that a more accepted form is

$$D_m = (D_m)_{20}(1.037)^{(T-20)} \tag{189}$$

where $(D_m)_{20} = 2.037 \times 10^{-5}$ cm^2/s. Chao et al.[136] indicate that other forms are also employed. In any case, however, Equations 188 and 189 produce different results from that of Equation 187 when temperature corrections are estimated.

One of the original forms of the O'Conner-Dobbins equation is as follows:

$$K_2 = \frac{(D_m U)^{1/2}}{D^{3/2}} \tag{190}$$

where U is the average stream velocity, and D is average depth. From Equation 190 it is clear that the application of Equations 187 and 189 are inconsistent.

The generally accepted use of 1.0241 as the temperature correction coefficient seems to derive from the 1961 findings of the American Society of Civil Engineers Committee on Sanitary Engineering Research.[249] However, it is not clear how valid it is to assume that the temperature correction parameter is constant for all conditions encountered in streams. The studies of Metzger[254] indicate that the coefficient changes somewhat with degree of mixing. However, unlike Chao et al.,[136] Metzger indicates that θ increases with decreased mixing. Furthermore, Thackston and Krenkel[257] note that Dobbins[218] postulates a reliance on hydraulic conditions, and they found that θ varied with Froude number. Therefore, it is not clear that the effects of mixing and other hydraulic

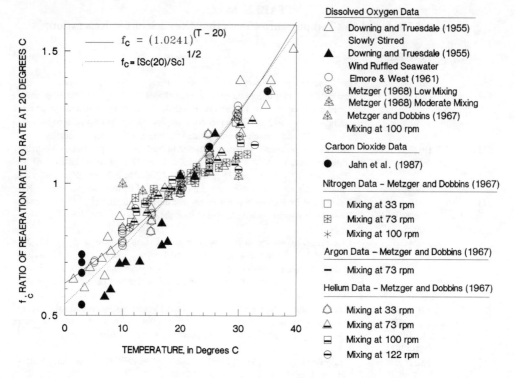

FIGURE 98. Comparison of experimental data with $f_c = \theta^{T-20}$; $f_c = [S_c(20)/S_c]^{1/2}$. Downing and Truesdale,[252] and Elmore and West[258] collected the oxygen transfer data. Metzger and Dobbins[261] investigated transfer of oxygen, nitrogen, helium, and argon. Jahne et al.[262] investigated carbon dioxide transfer. (Original figure from Daniil and Gulliver.[137])

processes are fully incorporated into present methods of correcting for changes in temperature.

The experimental procedure used to determine dissolved oxygen for the widely accepted study by the Committee on Sanitary Engineering Research has also been questioned. The methods used to establish $\theta = 1.0241$ are the same as those used by Elmore and West[258] to determine oxygen saturation values. Montgomery[259] and Landine[260] note that the results using these experimental methods are the subject of significant questions regarding experimental accuracy. The use of uncapped samples fixed with iodine is subject to volatilization losses if not analyzed quickly. Therefore, it is not fully clear that experimental methods are beyond question. These matters seem to require some additional investigation.

In the final analysis, however, the use of a temperature correction factor of $(1.0241)^{T-20}$ seems adequate. Except for Metzger's estimates, the crude estimates of Streeter et al., and the unexplained recommendation of Krenkel and Novotny, the studies that are readily available indicated that $\theta = 1.02 \pm 0.01$. Furthermore, Daniil and Gulliver[137] present the most accurate results of stirred reactor experiments involving argon, nitrogen, helium, oxygen, and carbon dioxide that illustrate that the value 1.0241 is adequate for a number of gases that are liquid-film limited. These results are presented in Figure 98 where the scatter in the data seems to indicate that significant improvements in this area may be impeded by the uncertainty in existing measurement techniques.

In a previous section on the derivation of the form Equation 187, it was shown that the temperature correction coefficient varies somewhat with water temperature.

Schneiter and Grenney (see Bowie et al.[73]) have taken this into account by developing an approach that allows the temperature correction coefficient to vary with temperature. This approach does not seem to have been fully evaluated, however.

Other alternative approaches to specifying temperature corrections include the recent observation that gas mass transfer coefficients are related to the Schmidt number (v/D_m where v is kinematic viscosity) and the linear relationships proposed by Downing and Truesdale,[252] Truesdale and Van Dyke,[253] and Elmore and West[258] some time ago. The linear relationship between water temperature and reaeration coefficients is as follows:

$$\frac{(K_2)_T}{(K_2)_{20}} = 1 + \theta_L(T - T_r) \tag{191}$$

where θ_L is a constant dependent on the reference temperature, T_r. Experiments by the Committee on Sanitary Engineering Research[249] and Metzger[254] seem to discount the validity of the linear relationship in Equation 191.

Daniil and Gulliver[137] have recently investigated the dependence of the liquid-film coefficient on the Schmidt number (Sc). They note that the observation of Jahne et al.[262] that the mass transfer coefficient is related to $Sc^{-1/2}$ for a wavy surface agrees very well with Equation 187 over the temperature range of 10 to 40°C. Discrepancies over the range 1 to 10°C are small (≤ 11 percent). The effect of temperature on molecular diffusivity, viscosity, and density is expressed as

$$\frac{(k_2)_T}{(k_2)_{20}} = \left(\frac{Sc_{20}}{Sc_T}\right)^{1/2} = \left(\frac{T + 273}{293}\right)^{1/2}\left(\frac{v_{20}}{v_T}\right)\left(\frac{\rho_{20}}{\rho_T}\right)^{1/2} \tag{192}$$

where $(K_2)_T$, Sc_T, v_T, and ρ_T are reaeration coefficient, Schmidt number, kinematic viscosity, and density, respectively, at some arbitrary temperature, T, and $(K_2)_{20}$, Sc_{20}, v_{20}, and ρ_{20} are the reaeration coefficient, Schmidt number, kinematic viscosity, and density, respectively, at 20°C. Equation 192 is similar to forms derived by Metzger assuming that reaeration can be described by the film theory.

Table 33 lists the readily available equations to predict K_2. There are three classes of equations in the listing. First, there are equations derived from conceptual models such as the film theory, surface renewal theory, and penetration theory. Second, there are semi-empirical formulas based on dimensional analysis and similarity to theoretical forms that require evaluation with data to select empirical coefficients. Third, there are empirical formulas derived by statistical analysis.

Experience indicates no individual or class of equations can be applied under all circumstances encountered in streams. Equations based on conceptual models do not encompass all conditions because no single mass transfer theory can presently describe all conditions encountered. (Existing mass transfer theories seem to be overly simplistic, and there seems to be a need to develop a more comprehensive conceptual approach to mass transfer at the air-water interface.) Semi-empirical and empirical models are limited to describing conditions that are very similar to the data base used to calibrate these equations. (Available data from tracer measurements are now extensive but should be evaluated to determine if the data are representative of all categories of streams, including the larger, deep rivers.) Unfortunately, Table 33 does not adequately describe the limits of the data base used to derive each equation and it is not possible to fully explore the limitations of each mass transfer equation, used to derive reaeration equations. As a result, the predictive limitations of reaeration equations are not well understood. Nevertheless, there is some experience to guide the use of reaeration equations. The experience that is useful to guide in application of reaeration equations is compiled in Table 35 from Bowie et al.[73] In addition, Covar[270] describes one approach to predicting K_2 that will be reviewed later.

TABLE 35
Summary of Reviews and Evaluations of Stream Reaeration Coefficients

Bennett and Rathbun[233]

Thirteen equations were evaluated.

The standard error of the estimate was used as a measure of the difference between predicted values and data.

The equation which provided the best fit to their original data set was that of Krenkel.[235]

The equations which best fit the entire range of data were: O'Connor and Dobbins,[216] Dobbins,[213] Thackston and Krenkel.[221]

Of the 13 equations the Churchill et al.[223] formula provided the best fit to natural stream data.

The Bennett and Rathbun formula, developed from the data evaluated during their review, provided a smaller standard error for natural streams than the other 13 equations.

There was a significant difference between predictions from equations derived from flume data and equations derived from natural stream data.

The expected root-mean-square error from different measurement techniques is 15% using the radioactive tracer technique; 65% using the dissolved oxygen mass balance, and 115% using the disturbed equilibrium method.

Lau[234]

Both conceptual and empirical models were reviewed.

Conclusions reported were similar to those of Bennett and Rathbun.

It was found that no completely satisfactory method exists to predict reaeration.

Wilson and MacLeod[263]

Nearly 400 data points were used in the analysis.

16 equations were reviewed.

The standard error of estimate and graphical results were both used in error analysis.

It was concluded that equations which use only depth and velocity are not accurate over the entire range of data investigated.

The methods of Dobbins[218] and Parkhurst and Pomeroy[225] gave the best fits to the data investigated.

Rathbun[141]

19 equations were reviewed.

Equation predictions were compared against radioactive tracer measurements on five rivers (Chattahoochee, Jackson, Flint, South, Patuxent).

The best equations in terms of the smallest standard error estimates were Tsivoglou-Wallace[207] (0.0528), Parkhurst-Pomeroy[225] (0.0818), Padden-Gloyna[244] (0.0712), and Owens et al.[224] (0.0964).

No one formula was best for all five rivers.

Rathbun and Grant[264]

Compared the radioactive and modified tracer techniques for Black Earth Creek and Madison Effluent Channel in Wisconsin.

Differences in Black Earth Creek were -9 to 4% in one reach and 16 to 32% on another reach attributable to increased wind during the latter part of the test.

Unsteady flow during the Madison Effluent Channel tests led to differences of as much as 25 to 58% in one case and -5 to 3% in another.

Shindala and Truax[229]

Reaeration measurements for streams in Mississippi, Wisconsin, Texas, Georgia, North Carolina, Kentucky, and New York were made using the radioactive tracer technique.

The energy dissipation model resulted in the best correlation for reaeration coefficient prediction for small streams.

The following escape coefficients (defined as the coefficients times the $\frac{\Delta h}{t}$ in energy dissipation models for reaeration coefficients) were recommended:

$$0.0802/\text{ft for Q} < 10 \text{ ft}^3/\text{s}$$
$$0.0597/\text{ft for } 10 \leq Q \leq 280 \text{ ft}^3/\text{s}$$

TABLE 35 (continued)
Summary of Reviews and Evaluations of Stream Reaeration Coefficients

NCASI Bulletin[265]

Six reaeration formulas were compared against measurements made using radioactive tracer techniques and hydrocarbon tracer techniques for a reach of the Ouachita River, Arkansas.
The hydrocarbon tracer technique produced reaeration rates higher than both the radioactive tracer and empirical formulas.
The O'Connor-Dobbins[216] equation was chosen as the best empirical equation.

Kwasnik and Feng[266]

13 reaeration formulas were reviewed and compared against values measured using the modified tracer technique for 2 streams in Massachusetts.
The equations of Tsivoglou-Wallace[207] and Bennett-Rathbun[233] gave the closest predictions to the field values.
The study indicates that results using the modified tracer technique are reproducible.

Grant and Skavroneck[267]

4 modified tracer methods and 20 predictive equations were compared against the radioactive tracer methods for 3 small streams in Wisconsin.
Compared to the radioactive tracer method the errors in the modified tracer techniques were the following:
  11% for the propane-area method
  18% for the propane-peak method
  21% for the ethylene-peak method
  26% for the ethylene-area method
Compared to the radioactive tracer method, the equations with the smallest errors were the following:
  18% for Tsivoglou-Neal[226]
  21% for Negulescu-Rojanski[243]
  23% for Padden-Gloyna[244]
  29% for Thackston-Krenkel[221]
  32% for Bansal[245]

House and Skavroneck[268]

Reaeration coefficients were determined on 2 creeks in Wisconsin using the propane-area modified tracer technique and compared against 20 predictive formulas.
The top five predictive formulas were as follows:
  Tsivoglou-Neal,[226] 34% mean error
  Foree,[248] 35% mean error
  Cadwallader and McDonnell,[222] 45% mean error
  Isaacs-Gaudy,[241] 45% mean error
  Langbein-Durum,[238] 49% mean error

Zison et al.[269]

13 reaeration formulas were reviewed, but none were compared against historical data.
Covar's method[270] was discussed which shows how stream reaeration can be simulated by using three formulas (O'Connor-Dobbins,[216] Churchill et al.,[223] and Owens et al.,[224] each applicable in a different depth and velocity regime.

Yotsukura et al.[271]

Developed a steady injection method to avoid uncertainty in dispersion corrections.
Determined reproducibility to be 4%.
Found negligible effect of wind where stream banks are high.

Ohio Environmental Protection Agency[272]

18 reaeration coefficient equations were compared against data collected in 28 Ohio streams.

TABLE 35 (continued)
Summary of Reviews and Evaluations of Stream Reaeration Coefficients

Ohio Environmental Protection Agency[272]

The streams were divided into four groups based on slope and velocity. The best predictive equations for each group are shown below:

Group	Slope (ft/mi)	Flow (ft³/s)	Preferred equation
1	<3	All data	Negelescu-Rojandki[243] Krenkel-Orlob[220]
2	3—10	≤30	Parkhurst-Pomeroy[225]
3	3—10	>30	Thackston-Krenkel[221]
4	>10	All data	Parkhurst-Pomeroy[225] Tsivoglou-Neal[226]

From Bowie, G. L., Mills, W. B., Porcella, D. B., Campbell, C. L., Pagenkopf, J. R., Rupp, G. L., Johnson, K. M., Chan, P. W. H., and Gherini, S. A., Rates, Constants, and Kinetics Formulations in Surface Water Quality Modeling, 2nd ed., EPA/600/3-85/040, U.S. Environmental Protection Agency, Athens, GA, 1985.

The review of Bennett and Rathbun[233] seems to be the best review of the theories behind some of the most important equations. In addition, they illustrate the divergence obtained from predicting the reaeration coefficient with various equations. In Figure 99, the 13 equations that they evaluated show almost an order of magnitude difference in prediction of K_2 for all depths assuming an average velocity of 0.3 m/s (1 ft/s) and a slope of 0.0001. Bowie et al.[73] note, however, that the smallest divergence occurs for typical stream depths (0.3 to 3 m or 1 to 10 ft).

A similar conclusion can be reached from the evaluation by Wilson and MacLeod.[73,263] The spread of the approximately 400 measurements available about the best predictive equations approach an order of magnitude for K_2. They also concluded that relations based on velocity and depth did not predict well over the full range of conditions they encountered with the data base available.[73]

Later Rathbun[141] evaluated nineteen equations using tracer measurements obtained from five rivers in the southern U.S. None of the equations evaluated consistently gave the best predictions. Ranked best was the Tsivoglou and Wallace[207] equation developed from the data set used by Rathbun for the evaluation. All of the equations that reasonably matched the data were of the energy dissipation-type (K_2 proportional to the product of the slope and velocity such as the Tsivoglou-Wallace and Parkhurst-Pomeroy equations) or developed from field (Owens et al.) or flume (Padden-Gloyna) data biased by measurements collected in shallow turbulent flows. Following Rathbun's evaluation, Shindala and Truax[73,229] found that the energy dissipation type equations are best for small shallow streams. Kwasnik and Feng,[266] Grant and Skavroneck,[267] House and Skavroneck,[268] the Ohio Environmental Protection Agency,[272] Bauer et al.,[274] and Goddard[275] reached similar conclusions about the applicability of the energy dissipation equations to small streams.

Deeper, slower moving streams do not seem to have been studied as frequently, but there are indications that the O'Conner-Dobbins equation may be the most appropriate. This equation seemed to best fit the diverse data base compiled by Bennett and Rathbun,[233] and this equation best fit the tracer measurements from the slow moving deep Ouachita River in Arkansas.[281]

Deeper, fast-moving rivers have not been studied very extensively either. However, there is some indication that the Churchill et al. equation calibrated to the few data available may be the best available. Covar[270] observed that there seems to be three K_2 regimes in streams in which certain equations better predict the reaeration coefficient.

Covar[270] notes that the data of O'Connor-Dobbins, Churchill et al., and Owens et al.

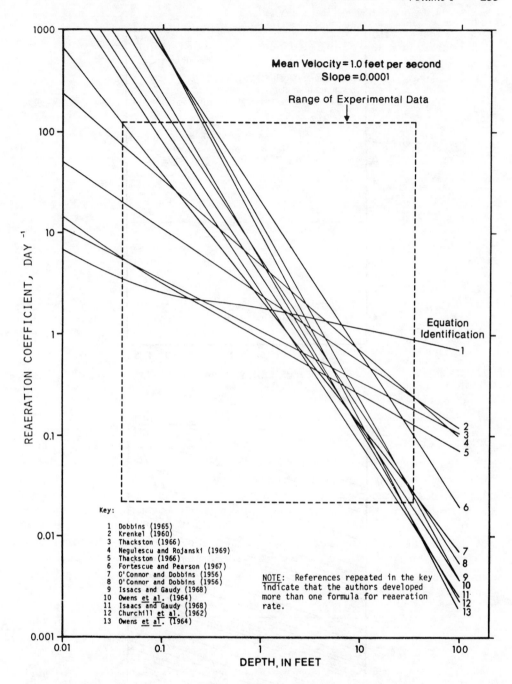

FIGURE 99. Predicted reaeration coefficients as a function of depth from 13 predictive equations.[73,233] Note that the equation numbers are associated with References 218, 235, 89, 243, 273, 216, 216, 241, 224, 241, 223, and 224, respectively.

define three separate predictive regimes as shown in Figure 100. The O'Conner-Dobbins equation seems adequate to predict reaeration coefficients for deeper streams with slower velocities while the Churchill et al. equation better predicts faster moving deep streams. Covar[270] noted that the boundary between the O'Conner-Dobbins and Churchill et al. equations followed naturally from the series of points (straight line in Figure 100) where the predictions were equivalent. This line is also a reasonable boundary between the two data sets.

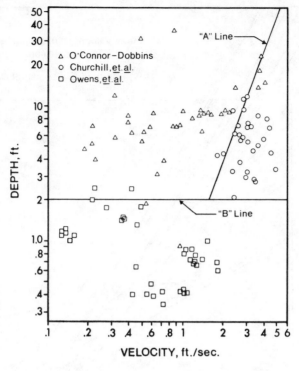

a

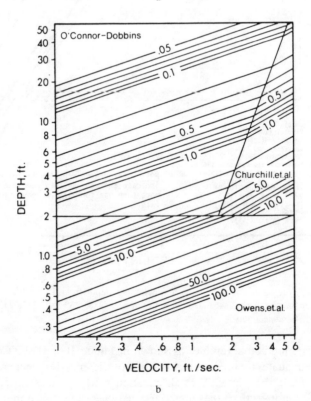

b

FIGURE 100. Covar's method[270] for predicting the reaeration coefficient in streams. (a) Data used to define three regimes. (b) Reaeration coefficient in per day vs. stream velocity and depth.[73]

Covar's separation of shallow flows from deeper flows is an arbitrary one that seems even more reasonable in light of more recent studies that show that the O'Conner-Dobbins and Churchill et al. equations are not fully applicable (see Table 35). At the time, the Owens et al. equation seemed to be the best shallow flow predictor of the reaeration coefficient. Since that time, a number of studies indicate that the Tsivoglou-Wallace equation may be the best in this regime, but this has not been fully investigated.

As a result, Covar's method seems to be the clearest distillation of the extensive study and evaluation that has continued intensely for at least 30 years. The definition of three predictive regimes overcomes the troubling problem that no single equation does well in predicting all conditions encountered in existing data sets. In addition, it can be seen that Covar's method is consistent with the rule of thumb that suggests the use of the O'Conner-Dobbins equation when there is no other information available to guide the selection of a reaeration formula. The O'Conner-Dobbins regime defined by Covar seems to cover a majority of stream conditions expected to occur. Thackston and others have suggested that there are also other regimes that can be defined for other combinations of equations, but alternative approaches have not been fully developed and evaluated.

While Covar's method seems to be the best available guidance, there continues to be a pressing need to develop comprehensive predictive methods. It is anticipated that the next surge in reaeration research may investigate more fully the mass transfer theory behind existing formulations and develop more consistency between the separate conceptual models that are used to derive the conceptual and semi-empirical equations such as the O'Conner-Dobbins equation, energy dissipation type equations, and others. Covar's three regimes probably define three mass transport regimes that should be investigated more fully. The next phase should also focus on the re-evaluation of the extensive data base that has been accumulated (see Table 36). Bowie et al.[73] indicate that much of the flume data may not be representative of stream reaeration. Thus, this data should be segregated from the data base of existing K_2 measurements, as should the equations based on these data until the basic process is better understood. Evaluation of the extensive historical data base is presently being undertaken by the National Council for Air and Stream Improvement at Tufts University, and these results are eagerly awaited.

There is also a need to investigate minimum values of the reaeration or liquid-film coefficients. Bowie et al.[73] reviewed reaeration in lakes and concluded that a minimum K_2 value in streams may be 0.01 to 0.05/d. However, refinements of the tracer methods seem to continually reduce the perception of what should be a minimum value for reaeration. Terry et al.[278] note that the equation based on the Velz method,[231] which is a function of depth, may best predict minimum values in deeper, slowly moving backwaters and sluggish streams. This approach has not yet been contrasted with the work of Neal[230] in shallower swampy streams and marsh areas. There are indications that most other equations underpredict the minimum reaeration coefficient because K_2 seems less dependent upon, or perhaps independent of, very low velocities.

There are also several factors that are not included in existing formulas. The effect of this lack of full appreciation of what is not known about reaeration probably contributes to the order of magnitude prediction errors expected from most K_2 equations. The factors that are known to be occasionally important include hydraulic structures (dam reaeration), wind, suspended solids, and surfactants.

The effects of small dams, weirs, and rock barriers on reaeration are not explicitly included in typical stream water quality models. These effects can be significant, however. Bowie et al.[73] note that typical weirs and dams can increase dissolved oxygen

TABLE 36
Sources of Stream Reaeration Data

Source	Contents
Owens et al.[224]	Reaeration coefficients using disturbed equilibrium technique for six rivers in England (Ivel, Lark, Derwent, Black Beck, Saint Sunday's Beck, Yewdale Beck), and associated hydraulic data.
O'Connor and Dobbins[216]	Reaeration data for Clarion River, Brandywine Creek, Illinois River, Ohio River, and Tennessee River.
Churchill et al.[223]	Reaeration data using dissolved oxygen balance downstream from deep impoundments for Clinch River, Holston River, French Broad River, Watauga River, and Hiwassee River.
Tsivoglou and Wallace[207]	Hydraulic properties and radioactive tracer measured reaeration coefficients for Flint, South, Patuxent, Jackson, and Chattahoochee Rivers.
Bennett and Rathbun[233]	Summaries of data from Churchill et al.,[223] Owens et al.,[224] Gameson et al., [276] O'Connor and Dobbins[216] Tsivoglou et al.,[255,277] Negulescu and Rojanski,[243] Thackston,[89] and Krenkel[235].
Foree[247]	Radioactive tracer measurements and reaeration hydraulic characteristics for small streams in Kentucky, and reaeration measurements for small dams in Kentucky.
Grant[227]	Reaeration measurements and hydraulic characteristics for 10 small streams in Wisconsin.
Grant [228]	Reaeration measurements and hydraulic characteristics for Rock River, Wisconsin.
Zison et al.[269]	Summary of reaeration coefficients and hydraulic characteristics for rivers throughout the U.S.
Kwasnik and Feng[266]	Reaeration data using the modified tracer technique on selected streams in Massachusetts.
Grant and Skavroneck[267]	Reaeration data from three small streams in Wisconsin.
House and Skavroneck[268]	Reaeration data for two small streams in Wisconsin.
Shindala and Truax[229]	Radioactive tracer measurements of reaeration rates and escape coefficients, plus hydraulic data, for rivers in Mississippi, Wisconsin, Texas, Georgia, North Carolina, Kentucky, and New York.
Terry et al.[278]	Hydrocarbon tracer measurements of k_2 and hydraulic data for Spring Creek, Osage Creek, and Illinois River, Arkansas. Bennett-Rathbun[233] best fit all three streams. Eight equations were tested.
Bauer et al.[274]	Hydrocarbon tracer measurements of k_2 and hydraulic data for the Yampa River, Colorado best matched the Tsivoglou-Neal and the Thackston and Krenkel energy dissipation type equations. Lau's equation was extremely error prone. 19 equations were tested.
Goddard[275]	Hydrocarbon tracer measurements of k_2 and hydraulic data from the Arkansas River in Colorado were used to test 19 equations. The best fitting equations were those by Dobbins, Padden and Gloyna, Langbein and Durum, and Parkhurst and Pomeroy.
Hren[279]	Radioactive tracer measurements for the North Fork Licking River, Ohio.
Rathbun et al.[280]	Hydrocarbon tracer measuremants of West Hobolochitts Creek, Mississippi.
NCASI[281]	Hydrocarbon tracer measurements for Ouachita River, Arkansas, and Dugdemona River, Louisiana.
Parkhurst and Pomeroy[225]	Reaeration coefficients were determined by a deoxygenation method in 12 sewers in the Los Angeles County Sanitation District.
Ice and Brown[236]	Reaeration coefficients were determined using sodium sulfite to deoxygenate the water in small streams in Oregon.
Ohio Environmental Protection Agency[272]	Reaeration coefficients were determined for 28 different streams in Ohio using predominantly the modified tracer technique, and in one case, the radioactive tracer technique.
Long[246]	Reaeration coefficients, hydraulic data, and time of travel data collected on 18 streams in Texas.

From Bowie, G. L., Mills, W. B., Porcella, D. B., Campbell, C. L., Pagenkopf, J. R., Rupp, G. L., Johnson, K. M., Chan, P. W. H., and Gherini, S. A., Rates, Constants, and Kinetics Formulations in Surface Water Quality Modeling, 2nd ed., EPA/600/3-85/040, U.S. Environmental Protection Agency, Athens, GA, 1985.

TABLE 37
Compilation of Equations that Predict the Effects of Small Dams on Stream Reaeration

Ref.	Predictive equation	Units	Source
Gameson[282]	$r = 1 + 0.5abh$	h, in meters	Field survey
Gameson et al.[276]	$r = 1 + 0.11ab(1 + 0.046T)h$	h, in feet	Model
Jarvis[283]	$r_{15} = 1.05\ h^{0.434}$	h, in meters	Model
Holler[284]	$r_{20} = 1 + 0.91h$	h, in meters	Model
Holler[284]	$r_{20} = 1 + 0.21h$	h, in meters	Prototype
Department of the Environment[285]	$r = 1 + 0.69h(1 - 0.11h)(1 + 0.046T)$	h, in meters	Model
Department of the Environment[285]	$r = 1 + 0.38abh(1 - 0.11h)(1 + 0.046T)$	h, in meters	Model
Nakasome[286]	$\log_e(r_{20}) = 0.0675h^{1.28}\ q^{0.62}\ d^{0.439}$	d, h, in meters q, in m²/hr	Model
Foree[247]	$r = \exp(0.1bh)$	h, in feet	Field survey

Note: $r_T = \dfrac{C_s - C_u}{C_s - C_d}$

C_s = dissolved oxygen saturation
C_u, C_d = concentration of dissolved oxygen upstream and downstream of dam, respectively
a = measure of water quality (0.65 for grossly polluted; 1.8 for clear)
b = function of weir type
h = water level difference
d = tailwater depth below weir
q = specific discharge
T = water temperature, °C

From Bowie, G. L., Mills, W. B., Porcella, D. B., Campbell, C. L., Pagenkopf, J. R., Rupp, G. L., Johnson, K. M., Chan, P. W. H., and Gherini, S. A., Rates, Constants, and Kinetics Formulations in Surface Water Quality Modeling, 2nd ed., EPA/600/3-85/040, U.S. Environmental Protection Agency, Athens, GA, 1985.

by 1 to 3 mg/L. Larger structures on rivers can also cause significant increases in dissolved oxygen. McCutcheon[2] notes that the Willamette Falls and fish ladder increase the dissolved oxygen by 0.35 mg/L in the Willamette River (Oregon).

In some models it is possible to define a very short reach at the dam or weir and increase K_2 until the predicted increase in dissolved oxygen matches that observed (i.e., see McCutcheon[2]) or that predicted with an empirical equation (see Equation 193). In some cases, this is difficult to do. The QUAL 2e model automatically smooths the transition from one K_2 in a reach to the different K_2 in the next downstream reach. The K_2 in the first and last elements in adjoining reaches are averaged. Therefore, specification of K_2 in a short reach will be automatically diffused into the adjoining reaches.

There are methods to predict reaeration at small dams and weirs but not larger structures. Bowie et al.[73] review the methods available in Table 37 and note the important reviews of Avery and Novak[73,287] and Butts and Evans.[73,288] Butts and Evans reviewed the predictive methods and collected additional reaeration data for 54 small dams located in Illinois. As a result of their study, nine classes of structures have been identified, and it has been determined that the change in dissolved oxygen concentration over the dam could be predicted using the following empirical equation:

$$r = \frac{C_s - C_u}{C_s - C_d} = 1 + 0.38abh(1 - 0.11h)(1 + 0.046T) \qquad (193)$$

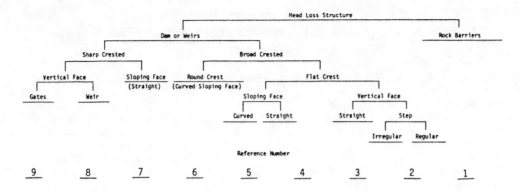

FIGURE 101.   Classification of heat loss structures by dam type and definition of the aeration coefficient, b.[73]

where C_s is the saturation concentration of dissolved oxygen, C_u is the upstream concentration, C_d is the downstream concentration, a is a water quality factor varying between 0.65 for grossly polluted streams and 1.8 for clean streams, h is the static head loss in meters, T is the water temperature in °C, and b is a weir-dam aeration coefficient defined in Figure 101 for nine classes of structures.

The effect of wind on reaeration is a subject of controversy (see various contributions in Bruseart and Jirka[289]). Since oxygen transfer is liquid-film controlled, any effect must be indirect. In fact, it is suspected that the effect is to increase the surface area over which the exchange occurs by causing waves and to increase surface renewal by increasing turbulence.

There is very definite evidence from wind tunnel-flume studies that the influence of wind can be significant. This is illustrated in Figure 102. However, the wind tunnel studies seem to be very atypical of stream conditions. First, there is no evidence that stream reaeration is reproduced exactly in flumes. Second, it seems that sheltering effects and other influences are not taken into account in wind tunnel studies.

A number of reaeration equations have been developed from flume data. However, field applications seem to indicate that these formulas are not fully applicable because conditions in streams are not fully taken into account or because other influences on reaeration are present in flumes that are not important in streams. If the reaeration process in flumes is not fully understood, then it seems likely that the effect of wind on reaeration in flumes cannot be fully understood.

Jirka and Bruseart[290] note that smaller streams with higher slopes than average show little influence of wind, whereas moderate size rivers with lower slopes and increased exposure do show some influence of wind on reaeration. They seem to have developed

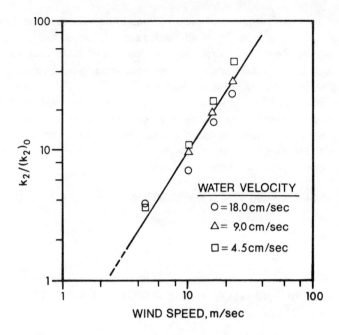

FIGURE 102. Ratio of reaeration coefficient under windy conditions to reaeration coefficient without wind, as a function of wind speed (based on laboratory studies).[73]

some indication of how important wind may be in large, open rivers where exposure approaches that obtained in lakes and estuaries.

Finally, surfactants and suspended solids seem to affect molecular diffusivity. These effects are not understood well enough to be considered in water quality models. For a review of these effects, see Zison et al.,[269] Tsivoglou and Wallace,[207] and Bowie et al.[73]

E. GAS SATURATION: PREDICTION

One important component of predicting the flux of gases that are present in the atmosphere (partial pressure in Equation 185 is non-zero) involves estimation of the equilibrium or saturation concentration of the dissolved gas in water. If gases are present in negligible quantities in the atmosphere, then the equilibrium concentration is zero and the gas volatilizes until a zero concentration in the water is reached or approached.

In general, specification of the saturation concentration at a given temperature involves determining the Henry's Law constant for a particular gas or compound at the temperature of interest. For dissolved oxygen, Henry's Law constants have been available since 1889.[260] For other volatile compounds, constants are available in various chemical engineering and chemistry handbooks. Mills et al.[202] list values for important toxic chemicals.

A less organized approach has been pursued to estimate saturation concentrations of dissolved oxygen. Saturation concentrations have been measured for various temperatures and tabulated. This is consistent with the general approach to determining equilibrium except Henry's Law constants have not always been reported.

Originally, the measurements of oxygen saturation from Whipple and Whipple were tabulated in *Standard Methods for the Examination of Water and Wastewater*.[291] Methods were developed to correct for differences in atmospheric pressure, and additional data were compiled to describe the effect of salinity on saturation concentrations. In 1961 Elmore and Hayes[292] made new measurements of saturation concentrations, and these were soon adopted as a modeling standard. Subsequently, significant concerns about those experimental methods[259] have been expressed. More recently, however, Benson and Krause[293] made what seem to be definitive measurements that have been adopted for *Standard Methods*.[294]

To facilitate modeling, it is more useful to have tabulations converted to continuous equations relating saturation concentrations of dissolved oxygen to water temperature, atmospheric pressure, and some measure of dissolved solids. Originally, this was done by Elmore and Hayes[292] for their questionable data who found that saturation concentrations were related to temperature as follows:

$$C_s = 14.652 - (0.41022T) + (0.007991T^2) - (7.7774 \times 10^{-5}T^3)$$

$$\text{for T in } °C \tag{194}$$

or

$$C_s = 24.89 - (0.4259T) + (0.003734T^2) - (1.328 \times 10^{-5}T^3)$$

$$\text{for T in } °F \tag{195}$$

The effects of atmospheric pressure and dissolved solids were not originally included.

As shown in Table 38, many older fresh water quality models use the out-of-date formulations given in Equations 194 and 195. More recent models include a correction for barometric pressure changes or elevation. Models applied in estuaries also include corrections for salinity or chloride concentrations.

The measurements of Elmore and Hayes[292] that were used to develop Equations 194 and 195 are suspect because some samples were evidently allowed to sit uncapped after they were fixed with iodine. Montgomery[259] and Landine[260] note that the resulting saturation measurements have been found to be significantly different from more recent measurements. The largest discrepancies occur at the highest temperatures[260] where the greater rates of volatilizations are expected to occur.

There seem to be two formulations that are more accurate than the approximations used in the past. Benson and Krause[293] developed a saturation equation from the recent measurements adopted by *Standard Methods*.[294] This equation is used in the QUAL 2e model.[4] Weiss[307] developed a similar form that has been evaluated and incorporated into U.S. Geological Survey models[302] and methods.[309]

Standard Methods[294] recommends that the relationship between temperature, chlorinity, and saturation concentrations be expressed according to Benson and Krause[293] as

$$\ln(C_s) = -139.34411 + \left(\frac{1.575701 \times 10^5}{T}\right) - \left(\frac{6.642308 \times 10^7}{T^2}\right) + \left(\frac{1.243800 \times 10^{10}}{T^3}\right)$$
$$- \left(\frac{8.621949 \times 10^{11}}{T^4}\right) - (\text{Chl})\left[(0.031929) - \left(\frac{19.428}{T}\right) + \left(\frac{3.8673 \times 10^3}{T^2}\right)\right] \tag{196}$$

where C_s is the dissolved oxygen saturation concentration computed at 1 atm of pressure in milligrams per liter, T is water temperature in °Kelvin (°C + 273.150), and Chl is chlorinity in parts per thousand. Equation 196 is valid over the range of 0.0 to 40.0°C and 0.0 to 28.0 ppt chlorinity.

Standard Methods[294] recommends that the effect of changing atmospheric pressure due to weather conditions and elevation changes be expressed as

TABLE 38
Methods Used by Selected Models to Predict Dissolved Oxygen Saturation

Eq. no	Model name (or description)	Model ref.	Equation for dissolved oxygen saturation C_s (mg/l)
1[a]	Limnological Model for Eutrophic Lakes and Impoundments	Baca and Arnett[295]	$C_s = 14.652 - (0.41022\,T) + (0.007991\,T^2) - (7.7774 \times 10^{-5}\,T^3)$ $T = °C$
1[a]	EXPLORE-1	Battelle[296]	Same as above
1[a]	Level III-Receive	Medina[297]	Same as above
1[a]	Water Quality Model for Large Lakes: Part 2: Lake Erie	Di Toro and Connolly[298]	Same as above
2	WRECEV	Johnson and Duke[299]	$C_s = 14.62 - 0.3898\,T + 0.006969\,T^2 - 5.897 \times 10^{-5}\,T^3$ $T = °C$
3	QUAL-II	Roesner et al.[300]	$C_s = 24.89 - 0.4259\,T + 0.003734\,T^2 - 1.328 \times 10^{-5}\,T^3$ $T = °F$
4	CE-QUAL-R1	U.A. Army COE[301]	$C_s = 14.6 \exp\left[-(0.027767 - 0.00027\,T + 0.000002\,T^2)\,T\right]$ $T = °C$
5[b]	One-Dimensional Steady-State Stream Water Quality Model[c]	Bauer et al.[302]	$C_s = (14.652 - .41022\,T + 0.007910\,T^2 - 7.7774 \times 10^{-5}\,T^3)\,(BP/29.92)$ $T = °C$ $BP = $ Barometric pressure (in Hg)
5[b]	HSPF (Release 7.0)	Imhoff et al.[303]	Same as above
6	DOSAG and DOSAG3	Duke and Masch[304]	$[14.62 - (0.3898\,T) + (0.006969\,T^2) - (5.897 \times 10^{-5}\,T^3)]$ $[1.0 - (6.97 \times 10^{-6}\,E)]^{5.167}$ $T = °C$ $E = $ Elevation, ft
7	Pearl Harbor Version of Dynamic Estuary Model (DEM)	Genet et al.[305]	$C_s = 14.5532 - 0.38217\,T + 0.0054258\,T^2 - CL(1.665 \times 10^{-4} - 5.866 \times 10^{-6}\,T + 9.796 \times 10^{-8}\,T^2)$ $T = °C$ $CL = $ Chloride concentration (ppm)
8[d]	RECEIV-II	Raytheon Co.[306]	$C_s = 1.4277 \exp[-173.492 + 24963.39/T + 143.3483 \ln(T/100.) - 0.218492\,T + S(-0.033096 + 0.00014259\,T - 0.00000017\,T^2)]$ $T = °K = °C + 273.15$ $S = $ Salinity (ppt)

TABLE 38 (continued)
Methods Used by Selected Models to Predict Dissolved Oxygen Saturation

[a] See also Equation 194 from the text.
[b] See also Equation 195 from the text.
[c] This model was updated in 1981 to use the Weiss[307] equation given as Equation 8. Equation 5 is no longer used.
[d] Equation 8 was developed by Weiss.[307] See Equation 203 from the text.

From Bowie, G. L., Mills, W. B., Porcella, D. B., Campbell, C. L., Pagenkopf, J. R., Rupp, G. L., Johnson, K. M., Chan, P. W. H., and Gherini, S. A., Rates, Constants, and Kinetics Formulations in Surface Water Quality Modeling, 2nd ed., EPA/600/3-85/040, U.S. Environmental Protection Agency, Athens, GA, 1985.

$$C'_s = C_s P_a \left[\frac{\left(1 - \frac{e}{P_a}\right)(1 - XP)}{(1 - P_e)(1 - X)} \right] \qquad (197)$$

where C'_s is the saturation concentration at nonstandard pressure in milligrams per liter, C_s is the saturation concentration at 1 atm of pressure in milligrams per liter (computed from Equation 196), P_a is atmospheric (or barometric) pressure in atmospheres, e is the partial pressure of water vapor in atmospheres, and X is a function of water temperature. Equation 197 is valid over the range of atmospheric pressures of 0.000 to 2.000 atm. The partial pressure of water vapor can be estimated as[73]

$$e = \exp\left[11.8571 - \left(\frac{3840.70}{T + 273.15}\right) - \left(\frac{216961}{(T + 273.15)^2}\right) \right] \qquad (198)$$

The partial pressure of water vapor can also be expressed in terms of wet bulb and air temperature and atmospheric pressure as shown in Equation 149. In addition, the vapor pressure can be computed from measurements of relative humidity, $R_H = e/e_o$, and calculations of the saturation vapor pressure, e_o (see Equations 140, 150, 151, and 152). The function X is expressed as

$$X = 0.000975 - (1.426 \times 10^{-5} \, T) + (6.436 \times 10^{-8} \, T^2) \qquad (199)$$

where T is in °C in both Equations 198 and 199.

At least two approximate forms of Equation 197 have been employed. *Standard Methods*[308] formerly recommended that Equation 197 with X = 0 be used to correct for barometric pressure differences. Because water vapor pressure in the resulting equation is normally small compared to atmospheric pressure, another approximation is $C'_s = C_s P$ where P is in atmospheres. This approach is employed in the modified Streeter-Phelps model devised by the U.S. Geological Survey[302] (see Table 38). McCutcheon[2] found that $C'_s = C_s P$ works well enough when elevations are on the order of 1400 m (4600 ft). McCutcheon[2] further notes that computational errors of as much as 0.5 mg/l occur at these elevations when the effects of barometric pressure are ignored. More severe computational errors are expected at higher elevations. Thus, most standard stream models intended to be used at high elevations should contain a correction for changes in atmospheric pressure. In light of this experience, it is not clear that the guidance from Bowie et al.[73] is correct. Bowie et al. indicate that pressure corrections can be ignored if the stream is at elevations less that 1200 m (4000 ft). Perhaps the best indication of the effect of decreasing barometric pressure on oxygen saturation is that of Bowie et al.[73] that states that saturation concentration of dissolved oxygen decreases approximately 7% for each 610 m (2000 ft) change in elevation. For streams with headwaters near saturation, a 7% error in saturation predictions can translate into a 7% error in the dilution capacity of the stream receiving point and nonpoint sources low in dissolved oxygen.

Chlorinity in Equation 196 can be related to measurements of salinity or specific conductance. This is important because specific conductance and salinity are more frequently measured and reported for streams than chlorinity. Salinity, S, in parts per thousand is related to chlorinity in milligrams per liter as

$$S = 0.03 + 0.001805 \, (\text{Chl}) \qquad (200)$$

or where chlorinity is in parts per thousand

$$S = 1.80655 \text{ Chl} \tag{201}$$

Equation 201 is from *Standard Methods.*[294] Also *Standard Methods*[294] notes that chlorinity is approximately equal to the chloride concentration in seawater in milligrams per gram of solution. The U.S. Geological Survey[309] recommends that the measurements of specific conductivity typically made during stream studies be related to salinity in parts per thousand according to

$$S = 5.572 \times 10^{-4} S_C + 2.02 \times 10^{-9} (S_C)^2 \tag{202}$$

where S_C is specific conductivity in micromhos per centimeter.

As an alternative to Equation 196, the Weiss[307] equation can also be used to compute equilibrium concentrations of dissolved oxygen in streams. The Weiss equation is written as

$$C_s = 1.4277 \exp\left[-173.492 + \frac{24963.39}{T} + 143.3483\ln\left(\frac{T}{100}\right) - 0.218492T \right.$$
$$\left. + S(-0.033096 + 0.00014259T - 0.00000017T^2) \right] \tag{203}$$

where T is in °Kelvin, and salinity, S, is in parts per thousand.

Bowie et al.[73] and Landine[260] have evaluated relative agreement between different tabulations and computational methods. Bowie et al. compared the formulations in Table 38, the Benson-Krause equation (Equation 196 which is the same as the tabulated values in *Standard Methods*[295]), and the old tabulations in *Standard Methods*[291] from Whipple and Whipple. This evaluation is shown in Table 39. Over the range encountered in streams (0 to 30°C), the Weiss equation (Equation 203) shows the closest agreement with the Benson-Krause equation (Equation 196) despite the fact that Mortimer[310] notes that there is a slight error in the Weiss equation. The Weiss equation predicts 0.01 to 0.03 mg/l less that the accepted standard. The former standard from the 1911 measurements of Whipple and Whipple are as much as 0.1 mg/l higher (or slightly more so) than the new standard. Equations 2 and 4 in Table 39 may be related to the Whipple and Whipple data because those computations are similarly as much 0.11 mg/l higher than those from the Benson-Krause equation. Equation 4 from Table 38 is also about 0.1 mg/l higher as well. Equations 1 and 3 (same as Equation 194), and Equation 2 (same as Equation 195) from Table 38 show the greatest discrepancies (i.e., 0.11 to 0.13 mg/l maximum difference). All three equations are based on the Elmore and Hayes polynominal which has been discounted as inaccurate.

Landine[260] also compares four methods that include the Whipple and Whipple data from *Standard Methods,*[311] the equation from Elmore and Hayes (Equation 193), an equation developed by Montgomery et al., and computations based on tabulated Henry's Law constants. The equation from Montgomery et al. is written as

$$C_s = \frac{468}{(31.6 + T)} \tag{204}$$

where T is water temperature in °C. Henry's Law constants were taken from the *Chemical Engineers Handbook*[144] and water vapor pressures were taken from *Mathematical and Physical Tables* by Clarke Irwin Co. Ltd. of Toronto.[312] The

TABLE 39
Comparison of Dissolved Oxygen Saturation Values from Ten Equations at 0.0 mg/1 Salinity and 1 atm Pressure

Temperature (°C)	1[a]	2	3[b]	4	5[c]	6	7	8[d]	APHA[291]	APHA[294,de]
0.0	14.652	14.620	14.650	14.600	14.652	14.620	14.553	14.591	14.6	14.621
1.0	14.250	14.237	14.248	14.204	14.250	14.237	14.176	14.188	14.2	14.216
2.0	13.863	13.868	13.861	13.826	13.863	13.868	13.811	13.803	13.8	13.829
3.0	13.491	13.512	13.490	13.465	13.491	13.512	13.456	13.435	13.5	13.460
4.0	13.134	13.169	13.133	13.120	13.134	13.169	13.111	13.084	13.1	13.107
5.0	12.791	12.838	12.790	12.790	12.791	12.838	12.778	12.748	12.8	12.770
6.0	12.462	12.519	12.460	12.475	12.462	12.519	12.456	12.426	12.5	12.447
7.0	12.145	12.213	12.144	12.173	12.145	12.213	12.144	12.118	12.2	12.139
8.0	11.842	11.917	11.841	11.883	11.842	11.971	11.843	11.823	11.9	11.843
9.0	11.551	11.633	11.550	11.606	11.551	11.633	11.553	11.540	11.6	11.559
10.0	11.271	11.360	11.270	11.340	11.271	11.360	11.274	11.268	11.3	11.288
11.0	11.003	11.097	11.002	11.085	11.003	11.097	11.006	11.008	11.1	11.027
12.0	10.746	10.844	10.744	10.840	10.746	10.844	10.748	10.758	10.8	10.777
13.0	10.499	10.601	10.497	10.605	10.499	10.601	10.502	10.517	10.6	10.537
14.0	10.262	10.367	10.260	10.378	10.262	10.367	10.266	10.286	10.4	10.306
15.0	10.034	10.142	10.033	10.161	10.034	10.142	10.041	10.064	10.2	10.084
16.0	9.816	9.926	9.814	9.951	9.816	9.926	9.827	9.850	10.0	9.870
17.0	9.606	9.718	9.604	9.749	9.606	9.718	9.624	9.644	9.7	9.665
18.0	9.404	9.518	9.401	9.555	9.404	9.518	9.432	9.446	9.5	9.467
19.0	9.209	9.325	9.207	9.367	9.209	9.325	9.251	9.254	9.4	9.276
20.0	9.022	9.140	9.019	9.186	9.022	9.140	9.080	9.070	9.2	9.092
21.0	8.841	8.961	8.838	9.011	8.841	8.961	8.920	8.891	9.0	8.915
22.0	8.667	8.789	8.664	8.842	8.667	8.789	8.772	8.720	8.8	8.743
23.0	8.498	8.624	8.495	8.679	8.498	8.624	8.634	8.554	8.7	8.578
24.0	8.334	8.464	8.331	8.521	8.334	8.464	8.506	8.393	8.5	8.418
25.0	8.176	8.309	8.172	8.367	8.176	8.309	8.390	8.238	8.4	8.263
26.0	8.021	8.160	8.017	8.219	8.021	8.160	8.285	8.088	8.2	8.113
27.0	7.871	8.015	7.866	8.075	7.871	8.015	8.190	7.943	8.1	7.968
28.0	7.723	7.875	7.719	7.935	7.723	7.875	8.106	7.802	7.9	7.827

TABLE 39 (continued)
Comparison of Dissolved Oxygen Saturation Values from Ten Equations at 0.0 mg/1 Salinity and 1 atm Pressure

Temperature (°C)	Equation number from Table 38								APHA[291]	APHA[294,de]
	1[a]	2	3[b]	4	5[c]	6	7	8[d]		
29.0	7.579	7.739	7.574	7.800	7.579	7.739	8.033	7.666	7.8	7.691
30.0	7.437	7.606	7.432	7.668	7.437	7.606	7.971	7.533	7.6	7.559
31.0	7.298	7.477	7.292	7.539	7.298	7.477	7.920	7.405	7.5	7.430
32.0	7.159	7.350	7.154	7.414	7.159	7.350	7.880	7.281	7.4	7.305
33.0	7.022	7.227	7.016	7.293	7.022	7.227	7.850	7.161	7.3	7.183
34.0	6.885	7.105	6.880	7.174	6.885	7.105	7.832	7.043	7.2	7.065
35.0	6.749	6.986	6.743	7.058	6.749	6.986	7.824	6.930	7.1	6.950
36.0	6.612	6.868	6.606	6.945	6.612	6.868	7.827	6.819	—	6.837
37.0	6.474	6.751	6.468	6.834	6.474	6.751	7.841	6.711	—	6.727
38.0	6.335	6.635	6.329	6.726	6.335	6.635	7.866	6.606	—	6.620
39.0	6.194	6.520	6.188	6.620	6.194	6.520	7.901	6.505	—	6.315
40.0	6.051	6.404	6.045	6.517	6.051	6.404	7.948	6.405	—	6.412

[a] Equation 194 from the text.
[b] Equation 195 from the text.
[c] Equation 203 from the text — the Weiss equation.
[d] Equation 196 from the text — the Benson-Krause equation.

From Bowie, G. L., Mills, W. B., Porcella, D. B., Campbell, C. L., Pagenkopf, J. R., Rupp, G. L., Johnson, K. M., Chan, P. W. H., and Gherini, S. A., Rates, Constants, and Kinetics Formulations in Surface Water Quality Modeling, 2nd ed., EPA/600/3-85/040, U.S. Environmental Protection Agency, Athens, GA, 1985.

TABLE 40

Comparison of the Dissolved Oxygen Saturation Concentrations from *Standard Methods*[294] with those Computed from Henry's Law Constant and by the Montgomery et al. Equation for Negligible Salinity and Standard Atmospheric Pressure[a]

	Temperature (°C)								
	0	**5**	**10**	**15**	**20**	**25**	**30**	**35**	**50**
Henry's Law constant (atm/mol fraction)	2.55	2.91	3.27	3.64	4.01	4.38	4.75	5.07	5.88
Water vapor pressure (mm Hg)	4.58	6.54	9.21	12.78	17.51	23.69	31.71	42.02	92.30
Computed saturation (mg/l)	14.53	12.67	11.23	10.04	9.07	8.20	7.50	6.92	5.56
Equation 204 (mg/l)	14.63	12.79	11.27	10.07	9.08	8.26	7.57	6.98	—
Equation 196 (mg/l)[294]	14.62	12.77	11.29	10.08	9.09	8.26	7.56	6.95	—

[a]   Adapted from References 73 and 260.

comparison of the concentrations (1) computed from Henry's Law constants; (2) computed from Equation 204; and (3) from the compilation in *Standard Methods*,[294] are given in Table 40.

Over the range of 0 to 30°C, the equation by Montgomery et al. agrees with the standard data adopted by *Standard Methods*[294] to within ±0.02 mg/l. This is surprisingly accurate and makes Equation 204 highly useful because of its very simple form. Given this level of accuracy, the equation can be used for not only making "back of the envelope calculations" because of its simple and easy to remember form, but it should also be equally useful in computer codes where accuracy is desirable and salinity corrections are not necessary.

The Henry's Law computation is reasonably comparable to the Benson-Krause data. At 5°C, the greatest difference occurs as a 0.10 mg/l discrepancy. This is close enough to confirm the general applicability of the Henry's Law constants compiled in other handbooks for estimating oxygen saturation or equilibrium concentrations of other gases. Over a more limited range of 10 to 30°C, there is better agreement (i.e., the difference is 0.02 to 0.06 mg/l). The agreement, however, is not sufficiently close to justify the use of older tabulations of Henry's Law constants in place of the tabulations in *Standard Methods*. Before this is considered, the source of the Henry's Law constants for dissolved oxygen in various tabulations should be investigated. It may well be that the Henry's Law constants may be computed from the data of Whipple and Whipple.

The importance of differences on the order 0.1 mg/l has not been fully confirmed, but there is an indication from at least one waste load allocation study that the magnitude of saturation values can have significant differences on the projected level of treatment required. Whittemore[313] indicates that at least one waste load allocation study for a large paper and pulp mill indicated that the use of two different saturation equations with differences on the order of 0.1 mg/l led to two different waste treatment requirements. The two requirements were reported to differ in cost by an order of $1,000,000. If this is true, it indicates that resolution on the order of 0.05 mg/l or less may be necessary. Therefore, distinctions between the older, out-of-date approaches (Equation 194) and the newer, accurate formulas (Equations 196, 203, and 204) seem to be of importance.

Dissolved solids are less important in streams except at the entrance of rivers into estuaries and where irrigation return flows may be of some importance. Therefore, the effects of salinity on saturation concentrations are less important for stream modeling. Nevertheless, the Benson-Krause and Weiss equations seem adequate when salinity is present. Table 41 shows that these formulas are in agreement to within 0.022 mg/l.

TABLE 41

**Comparison of Dissolved Oxygen Saturation Values
from Selected Equations at a Chloride Concentration of
20,000 mg/l (36.1 ppt Salinity) and 1 atm Pressure**

Temperature (°C)	Equation number			APHA[291]	APHA[294,b]
	7[a]	203	205		
0.0	11.215	11.400	11.492	11.3	11.354
1.0	10.953	11.105	11.203	11.0	11.067
2.0	10.699	10.823	10.924	10.8	10.790
3.0	10.452	10.553	10.653	10.5	10.527
4.0	10.212	10.295	10.391	10.3	10.273
5.0	9.978	10.048	10.139	10.0	10.031
6.0	9.725	9.811	9.895	9.8	9.801
7.0	9.532	9.585	9.661	9.6	9.575
8.0	9.320	9.367	9.435	9.4	9.362
9.0	9.114	9.158	9.218	9.2	9.156
10.0	8.915	8.958	9.011	9.0	8.957
11.0	8.723	8.765	8.812	8.8	8.769
12.0	8.538	8.580	8.623	8.6	8.586
13.0	8.360	8.402	8.442	8.5	8.411
14.0	8.189	8.231	8.270	8.3	8.241
15.0	8.025	8.067	8.108	8.1	8.077
16.0	7.868	7.908	7.954	8.0	7.922
17.0	7.718	7.755	7.809	7.8	7.770
18.0	7.574	7.607	7.674	7.7	7.624
19.0	7.438	7.465	7.547	7.6	7.482
20.0	7.308	7.327	7.429	7.4	7.347
21.0	7.186	7.194	7.321	7.3	7.215
22.0	7.070	7.066	7.221	7.1	7.087
23.0	6.961	6.942	7.130	7.0	6.964
24.0	6.859	6.822	7.049	6.9	6.844
25.0	6.764	6.594	6.976	6.7	6.727
26.0	6.676	6.594	6.912	6.6	6.616
27.0	6.595	6.485	6.857	6.5	6.507
28.0	6.521	6.379	6.812	6.4	6.401
29.0	6.454	6.277	6.775	6.3	6.297
30.0	6.394	6.177	6.747	6.1	6.197
31.0	6.340	6.081	6.729	—	6.100
32.0	6.294	5.987	6.719	—	6.005
33.0	6.254	5.896	6.718	—	5.912
34.0	6.221	5.808	6.726	—	5.822
35.0	6.196	5.722	6.743	—	5.734
36.0	6.177	5.638	6.770	—	5.648
37.0	6.165	5.557	6.805	—	5.564
38.0	6.160	5.477	6.849	—	5.481
39.0	6.162	5.400	6.902	—	5.400
40.0	6.171	5.325	6.965	—	5.322

[a] From Table 38.
[b] Equation 196 from the text.

From Bowie, G. L., Mills, W. B., Porcella, D. B., Campbell, C. L., Pagenkopf, J. R., Rupp, G. L., Johnson, K. M., Chan, P. W. H., and Gherini, S. A., Rates, Constants, and Kinetics Formulations in Surface Water Quality Modeling, 2nd ed., EPA/600/3-85/040, U.S. Environmental Protection Agency, Athens, GA, 1985.

Older data and equations from Table 38 show differences on the order of 0.1 mg/l or more. This includes the equation of Hyter et al.[73] written as

$$C_s = 14.6244 - 0.36134T + 0.004497T^2 - 0.0966S + 0.00205ST + 0.00027S^2 \quad (205)$$

where T is temperature in °C, and salinity, S, is in parts per thousands.

Bowie et al.[73] note that standard methods to measure saturation concentrations do not seem to exist. However, use of the methods employed by Benton and Krause[293] are probably adequate if a need to make such measurements arises.

F. TRACER METHODS: GENERAL DESCRIPTION AND THEORY

One of the most significant recent advances in reaeration and volatilization research involves the development of a tracer technique by Tsivoglou[207] to measure gas transfer at the water surface. Tsivoglou devised procedures and theoretical justification for making gas transfer measurements. He chose tritium as a water tracer that could be used to quantify the effect of dispersion and dilution and measure the time of travel between two or more sampling stations along a stream. Radioactive krypton-85 was chosen to mimic the exchange of oxygen. A dye, usually rhodamine WT, is also used to qualitatively indicate when the tritium and krypton tracers are in the vicinity of a sampling station.

Tritium was chosen because its behavior is expected to be the same as the water being tracked. Other investigators have indicated that it may be conceivable that tritium can be selectively removed from water. In practice, however, tritium seems to be an ideal tracer except for the radiation health effects, which have been minimized in designing field protocols.

Krypton has proven useful because it behaves in a manner very similar to oxygen molecules and is almost completely nonreactive in streams. As a result, the decrease in krypton concentrations, when corrected for the effects of dilution and dispersion using the tritium tracer, can be directly related to the reaeration coefficient.

Since krypton is nonreactive, other processes that cause changes in dissolved oxygen concentrations are not important. This overcomes a serious shortcoming of the dissolved oxygen balance studies that had been attempted in the past to estimate reaeration. To apply the oxygen balance method, simple models were proposed (Streeter-Phelps equation), and all other known parameters were independently estimated, leaving only reaeration to be estimated from the oxygen balance. Unfortunately, this lumped the effect of other processes that were not too well understood (such as sediment oxygen demand and photosynthesis) with the effects of reaeration. As a result, oxygen balance estimates have always been viewed with some deserved suspicion.

In effect, the oxygen balance method seems to involve circular arguments that presuppose the oxygen balance is adequately known so that any measurement of other oxygen balance components contrasted with the measured amounts of dissolved oxygen present could accurately define reaeration. However, like any circular argument, it was never possible to validate the approach for determining K_2. Lacking knowledge of the importance of other unquantified processes, it was never possible to quantify the uncertainty in oxygen balance measurements of reaeration and thus never possible to truly validate oxygen balance models and to exactly define the limitations of these models. Even more unfortunate, now that adequate reaeration measurements are possible, is that there are still processes that are poorly understood and for which measurement techniques are inadequate. Sediment oxygen demand presently seems to occupy the role that reaeration represented prior to the development of tracer measurements. Frequently, sediment oxygen demand is estimated by the oxygen balance method and as a result is quite inaccurate when other measurements are inaccurate or when other processes interfere.

As a result of the circular nature of dissolved oxygen balance studies, applications that were and presently remain a practice must be supplemented and critically supported with well developed engineering judgment. There are a number of types of streams for which it is possible to be relatively certain that the oxygen balance model will adequately represent observed conditions. However, from the nature of of this approach, it is clear that definitive limitations on the use of dissolved oxygen models cannot be devised until all processes are known and quantified. Unfortunately, experience is the only adequate guidance available at this time to indicate when and where the application of the dissolved oxygen model is feasible.

The importance of tracer measurements in this approach is that these independent measurements reduce the degrees of freedom in calibrating a model by one coefficient that was formerly treated as a calibration coefficient. This makes it more likely that dissolved oxygen models can be accurately applied in more streams. In cases where reaeration was the only unknown or unquantified process, it is now possible to perform accurate oxygen balances. In cases where reaeration and one other process such as sediment oxygen demand were unknown, it is now possible to measure reaeration coefficients and assume that the oxygen balance can be used to provide a screening level analysis estimate of sediment oxygen demand or some other poorly understood process. In this regard, tracer measurements are important to reduce model uncertainty and extend the range of the dissolved oxygen model.

Since there are very few cases in which reaeration is unimportant, it is necessary to measure reaeration coefficients in each waste load allocation study focusing on the dissolved oxygen balance. Earlier, it was indicated that predictive equations typically only provide order of magnitude estimates. Therefore, when tracer measurements are not possible, any modeling must tentatively be classified as screening level modeling until it can be demonstrated that all components are known well enough to obtain an accurate oxygen balance.

The only other method that seemed to be available before tracer measurements were devised involved monitoring re-oxygenation of low dissolved oxygen waters downstream of dams[223] and in very small streams where the volume of stream flow was low enough to chemically de-oxygenate the water. In either case, other processes had to be negligible or measurable. These limitations were so strict that only limited data were available and subsequent development of predictive methods was not fully successful (and is still not fully successful for that matter). The lack of adequate field methods is one reason that flume studies were heavily relied upon to estimate reaeration coefficients in the past.

Initially, the tracer procedure was not fully verified. Tsivoglou made a series of measurements in five southern rivers and compared the results to the existing predictive equations to conclude that the tracer method was adequate. This was supplemented with a thorough theoretical development that added credibility but not full verification. Later Tsivoglou and Wallace[207] concluded that existing reaeration coefficient equations were not adequate and used the initial tracer measurements to develop the energy dissipation formulas. Since the original work, a number of laboratory studies (i.e., Wilhelms[314]) with de-oxygenated water and tracers confirmed the validity of the method on a scale larger than Tsivoglou's original beaker tests.

Following the initial work of Tsivoglou, a number of refinements have been attempted to overcome obvious shortcomings of radioactive tracers. The critical disadvantage is that the purposeful introduction of radioactive substances requires a permit in most states of the U.S. and in many parts of the world. In addition, the potential safety hazard to field personnel requires some precaution. In some areas, permits are very difficult to obtain. As a result, at least three other tracers have been developed and tested. These include propane, ethylene, and methyl chloride.

Rathbun et al.[208] developed modified tracer methods that employed propane and ethylene in place of radioactive krypton. These tracers are injected with rhodamine WT dye for measuring the effect of dispersion and dilution. Current meter measurements of discharge are made during the field experiment to measure any loss of dye.

Since the original introduction of ethylene as a tracer, its use has now been discontinued. Subsequent studies indicate that ethylene may be biodegraded in some streams, especially in areas where swamp drainage may be present.

The abandonment of ethylene as a tracer highlights the primary disadvantage of using hydrocarbons as tracers. Biodegradation is always possible and may be probable in streams that receive refinery or natural gas processing wastes. However, Rathbun and his co-workers (personal communication) have investigated biodegradation and have not found any evidence that native flora can acclimate to small hydrocarbon molecules within a few days.[208] At least one long-term experiment did show that acclimation can eventually occur, however.

The other disadvantage of hydrocarbon tracers is that the low solubility requires a continuous injection of bubbling gas to achieve measurable concentrations downstream. This usually limits the application of the hydrocarbon method to streams and small rivers. The measurements by Crawford[315] in the Wabash River (Indiana) at 65.4 to 210 m^3/s (2310 to 7400 ft^3/s) are thought to be near the limit of applicability for the hydrocarbon tracer method. The radioactive tracer is only limited by how long it takes a point injection to mix across the stream if the measurement is to be representative of the stream average K_2.

Compared to the radioactive tracer method, the hydrocarbon method requires a more elaborate set of field equipment to inject the propane (diffusers, tubing, regulators, and gas cylinders — see Figure 103). The radioactive tracers are prepared in the laboratory and instantaneously injected by breaking the glass container under the surface with a metal plunger. In addition, the use of rhodamine WT dye as a quantitative water tracer requires more careful dye sample collection and measurement for the hydrocarbon method. In essence, the hydrocarbon method requires more elaborate methods and equipment, but it does not have public health problems like those associated with the use of radioactive tracers.

In the laboratory, the radioactive tracer must be analyzed with a liquid scintillation counter. The hydrocarbon tracer can be measured at low concentrations using a special trap on a gas chromatograph.

The third tracer that has been used is methyl chloride.[212,213] This method has been applied in New Zealand and relatively little is known about the experience obtained in its use. Like the hydrocarbon tracer method, rhodamine WT dye is used as a quantitative tracer to measure the effect of dispersion and dilution. Methyl chloride is moderately soluble in water and highly soluble in acetone that can be used as a carrier to aid in the injection of reasonable and necessary amounts. Methyl chloride can be detected down to levels of 10^{-12} g with a gas chromatographic technique using electron-capture detection. Background concentrations are expected to be low, especially in New Zealand where the method has been applied. Bioaccumulation and toxicity to aquatic organisms do not seem to be a problem. Hydrolysis is limited (3.8×10^{-9}/s), and sorption is not expected to be significant.[213]

The most significant difference with other tracers is that the relationship between K_2 and the gas transfer coefficient for methyl chloride, K_{CH3Cl}, varies with temperature. Over the temperature range of 5 to 35°C, the relationship has been experimentally defined[213] as

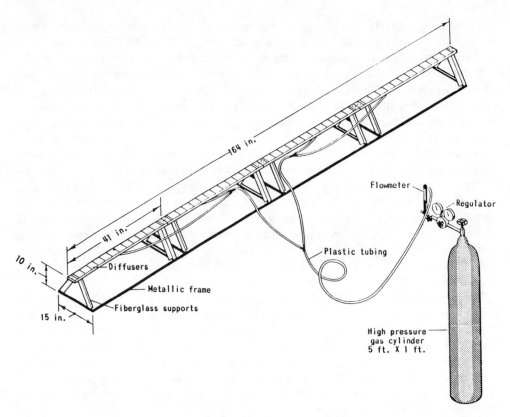

FIGURE 103. Injection equipment used in the hydrocarbon tracer method to measure reaeration coefficients.[315]

$$\frac{K_{CH_3Cl}}{K_2} = 4.323exp\left(\frac{-530.97}{T}\right) \tag{206}$$

Where T is water temperature in °Kelvin. At 20°C, the inverse of Equation 206 is 0.707, as shown in Table 32.

Finally, regarding alternative tracers, Hovis et al.[316] report on the preliminary development of a method to measure nonradioactive krypton for aeration studies in wastewater treatment plants. If it is possible to measure low concentrations of stable krypton, the use of this gas as a tracer will overcome some of the difficulties involved in applying the radioactive and hydrocarbon techniques. The primary difficulty in using nonradioactive krypton may involve access to the proper mass spectrometer.

There have been only a few comparative studies of the radioactive and hydrocarbon tracer techniques. Rathbun and Grant[264] compared the two techniques on a creek and a wastewater effluent channel and found differences of –9 and 4% in one case and 16 and 32% in the other. (A positive percent difference indicates that the results from the hydrocarbon method were larger). The field experiments were not conducted simultaneously, and discharge rates varied significantly (–5 to 3% in one case and 25 to 58% in the other). In addition, some variations of wind conditions were noted but not quantified. Despite the differences in flow rates and wind conditions, the investigators concluded that the methods gave equivalent results. Where discharge was not highly variable, the difference in the results from the two methods were comparable to measurement errors noted in Tsivoglou and Wallace[207] and elsewhere.

NCASI[265] also compared the two tracer techniques on a river with much lower reaeration coefficients (0.2 to 0.4/d vs. 1 to 10/d). These investigators were not

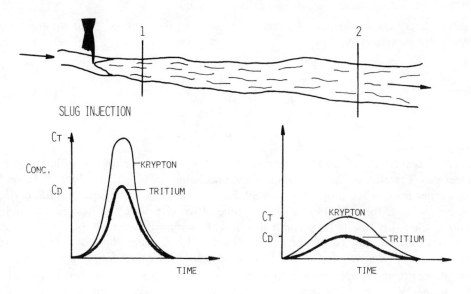

SLUG INJECTION

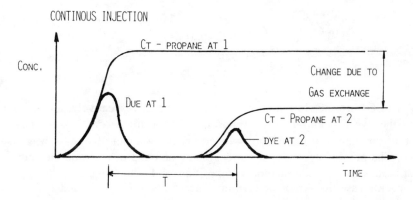

CONTINOUS INJECTION

FIGURE 104. Dye distributions expected downstream of slug and continuous injections. C_T is the peak or plateau gas tracer concentration, and C_D is the peak dye concentration.

associated with the development of either technique as Rathbun and Grant[264] were. They used ethylene as a hydrocarbon tracer and found that the method estimated $K_2 = 0.44/d$ vs. $0.17/d$ for the radioactive tracer. Unfortunately, ethylene was used exclusively as the tracer, and experimental problems occurred that did not seem to be resolved well enough to place much confidence in these results. Therefore, it seems that a careful comparative evaluation is still needed.

Tracers are injected into streams in at least three ways. First, an instantaneous slug injection is employed to introduce the radioactive and methyl chloride tracers. Second, a continuous slug injection is used to inject sufficient quantities of propane. Propane and dye are injected at a constant rate for 30 min to 2 h.[317] Finally, the newest injection method involves a continuous injection of propane until a plateau in concentration is reached at the downstream sampling sites. The general behavior of slug and continuous injections are illustrated in Figure 104.

Reaeration coefficients are computed from field measurements in one of three ways depending on the type of injection and the frequency of sampling to define tracer concentrations. First, the peak method is used when the peak concentrations of the tracers are tracked downstream. In the most general case when the water tracer may not be completely conserved over the reaches of interest, the expression for the reaeration coefficient is written as

$$K_2 = \frac{C}{\Delta t} \ln \left[\frac{\left(\frac{J_u C_{gu}}{C_{wtu}} \right)}{\left(\frac{J_d C_{gd}}{C_{wtd}} \right)} \right] \tag{207}$$

where C is the ratio of the reaeration coefficient to the volatilization coefficient of the tracer gas. These ratios (and estimates of the uncertainty involved) are given in Table 32 for various tracers. For propane, C = 1.39. For krypton, C = 1.20. Equation 206 defines the inverse of C for methyl chloride. The time of travel between an upstream sampling station denoted by the letter u and a downstream station denoted by the letter d is Δt. The peak tracer gas concentration is denoted by C_g and the peak water tracer concentration is denoted as C_{wt}. J_u and J_d are dye loss correction factors at the upstream and downstream sampling sites. These factors are equal to one for tritium. For rhodamine WT dye, these factors represent the loss of dye in transit from the last upstream station and can be expressed as

$$J = \frac{\int_0^\infty Q_2 c_{wt2}(t) d\tau}{\int_0^\infty Q_1 c_{wt1}(t) d\tau} \tag{208}$$

where Q_1 and Q_2 are the flow rates at stations 1 and 2, respectively, and $c_{wt1}(t)$ and $c_{wt2}(t)$ are the dye concentration vs. time curves measured at stations 1 and 2. The integration of the mass flux from 0 to ∞ determines how much dye mass was detected at a station. The ratio of dye mass at the beginning and end of reach defines how much dye is lost because of sorption, photodecomposition, and other effects.

In the specific case of krypton and tritium tracers where C = 1.20 and $J_u = J_d = 1.00$, the expression for the reaeration coefficient is written as

$$K_2 = \frac{1.2}{\Delta t} \ln \left[\frac{\left(\frac{C_{kru}}{C_{tru}} \right)}{\left(\frac{C_{krd}}{C_{trd}} \right)} \right] \tag{209}$$

where C_{kru} and C_{krd} are the peak concentrations of krypton at the upstream and downstream stations, and C_{tru} and C_{trd} are the peak concentrations of tritium at the upstream and downstream stations. Graphically, $K_2 C$ is the slope of the lines between the points $(C_{kru}/C_{tru}, t_1)$ and $(C_{krd}/C_{trd}, t_2)$ on Figure 105, where $\Delta t = t_2 - t_1$.

The area method is a second method of computing the reaeration coefficient. This method is expressed as

$$K_2 = \frac{C}{\Delta t} \ln \left[\frac{J_u Q_u A_u}{J_d Q_d A_d} \right] \tag{210}$$

where Q_u and Q_d are the stream discharges measured at the upstream and downstream of a reach, and A_u and A_d are the areas under the gas tracer concentration vs. time curves at

OBSERVED GAS TRACER VOLATILIZATION FOR THE CHATTAHOOCHEE RIVER AT ATLANTA, GEORGIA

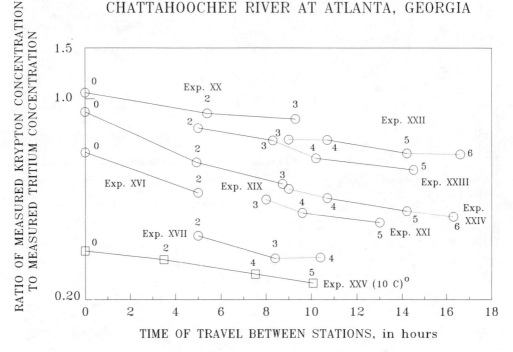

FIGURE 105. The ratio of peak krypton concentrations vs. peak tritium concentrations at various sites on the Chattahoochee River downstream of Atlanta, GA.[207] Note the logarithmic scale for the ratio.

the upstream and downstream stations. For this method, the entire gas and dye concentration curves must be measured as shown in Figure 106.

The third method of computing reaeration coefficients involves measuring the decrease in tracer gas concentration downstream of a continuous injection. In this method, dispersion is not important when flow is steady. The instantaneous slug injection of rhodamine WT dye is only used to determine time of travel from one measurement station to the next. As a result, the tedious dye loss correction is not necessary. Yotsukura et al.[271] report that the reaeration coefficient can be expressed as

$$K_2 = \frac{C}{\Delta t} \ln\left[\frac{Q_u C_u}{Q_d C_d}\right] \tag{211}$$

where C = 1.39 for a continuous injection of propane; Q_u and Q_d are stream discharges measured at the upstream and downstream sites where the injected gas is well mixed across the stream (typically equal, i.e., $Q_u = Q_d$), and C_u and C_d are concentrations of the tracer gas on the plateau shown in Figure 104.

A general procedure for conducting tracer measurements is given in Table 42. For more details on specific field protocols, dose requirements, laboratory analysis techniques, and evaluation of the measurement errors involved, see Tsivoglou and Wallace,[207] Tsivoglou and Neal,[226] Rathbun and Bennett,[233] Rathbun and Grant,[264] Rathbun et al.,[280] Rathbun,[317] NCASI,[265,281] Yotsukura et al.,[271] Bales and Holley,[209] Rainwater and Holley,[211] and Wilcock.[212,213]

The final step in the utilization of the measured reaeration coefficient involves application of the results in a dissolved oxygen model. For the best results, reaeration

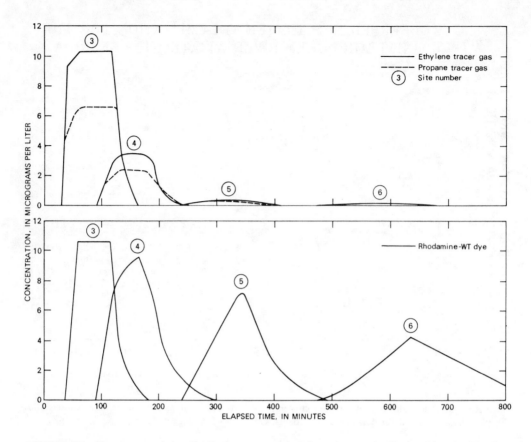

FIGURE 106.   Tracer gas and dye distributions measured at four sites on the Yampa River in CO. The plateau in concentration at site 3 results from a continuous slug injection and does not preclude use of the peak or area methods to compute the reaeration coefficient.[274]

measurements are made at the time data is collected to calibrate the model. In this case, there is no need for extrapolation procedures to adapt measurements to other calibration conditions. However, on many occasions it is not possible to simultaneously collect data for reaeration coefficients and model calibration. When this occurs, two methods have been used to extrapolate the reaeration coefficient estimate to different hydraulic conditions (i.e., changed discharge, velocity, and depth). First, Goddard,[275] Crawford,[315] House and Skavronek,[268] and Terry et al.[278] have used tracer data collected for a specific reach and determined which of the existing K_2 equations best described the data collected. The resulting equation was used to describe the reaeration coefficient when hydraulic conditions were somewhat different. Similarly, McCutcheon[2] used the same general procedure to determine which K_2 equation could best be used to estimate K_2 downstream of the reaches where Tsivoglou and Wallace[207] measured the reaeration coefficient in the Chattahoochee River (downstream of Atlanta, GA). In either case, the limitations on such extrapolations are not known. However, clearly the extrapolation cannot extend to radically different conditions. For example, if increased flow spills out into floodplains, drowns out riffles, or creates whitewater, then measurements made during low flow may not be extended using the equations available (see Table 33). Furthermore, extrapolation from measurements in one reach to estimate K_2 in a reach that was not covered by tracer measurements would be unwise if the geomorphology and conditions of the flow are significantly different.

In the second method, attempts have been made to derive regional equations for K_2. These studies (i.e., the Ohio Environmental Protection Agency[272]) have been attempted

TABLE 42
Procedures for Measuring Reaeration Coefficients

Step	Procedure

1.   Determine injection site and sampling locations from maps and site visits to cover critical reaches where reaeration is most important. The first downstream sampling location is chosen to ensure that vertical and lateral mixing is complete. See previous sections for methods of estimation. Lateral sampling for the water tracer at the first location confirms that the stream is well mixed. Sampling locations should be spaced far enough apart so that a single field crew can collect all samples (i.e., the tracer cloud does not arrive at the next station until it has passed the upstream station). Because this is difficult to arrange in pool-and-riffle streams, arrangements for two sampling crews are typically made.

2.   Determine dosing requirements that will ensure that tracers are present in measureable quantities at the downstream-most sampling location. See Rathbun[317] and Wilcock[213] for guidance.

3.   Inject tracers as required by each procedure. The radioactive tracers, tritium and krypton-85 plus rhodamine dye, are lowered beneath the water surface, and the glass container is broken to achieve an instantaneous slug injection. A solution of methyl chloride in acetone and 20% rhodamine WT dye is also instantaneously injected. The hydrocarbon tracer, propane, is injected continuously and simultaneously with a 20% rhodamine WT dye solution. Both are injected at constant rates unless the plateau method is used for propane, then a slug injection is made of the rhodamine WT dye. Diffusers, tubing, and regulators are required for the propane. A constant discharge pump or constant head vessel is used to inject the dye.

4.   Monitor rhodamine dye with a portable fluorometer to determine when to sample for the other tracers. If the peak method is used, sample only enough to define the peak. If the area method is used, sampling must define the entire tracer cloud as it passes a sampling location. Typically, sampling begins as soon as the dye is detected and continues until the the concentration returns to approximately 1% of the peak concentration.

5.   Samples are collected with a minimum of turbulence to avoid volatilization, and propane samples are fixed with a biocide such as fomulin or refrigerated and returned to the laboratory for analysis. Sample bottles are completely filled to avoid any air contact during transit.

in a number of states, but there is little indication of how successful these attempts have been.

REFERENCES

1. **French, R. H.,** *Open-Channel Hydraulics,* McGraw-Hill, New York, 1985, 34.
2. **McCutcheon, S. C.,** Evaluation of Selected One-Dimensional Stream Water-Quality Models, Tech. Rep. E-83-11, U.S. Army Waterways Experiment Station, Vicksburg, MS, 1983.
3. **Stamer, J. K., Cherry, R. N., Faye, R. E., and Kleckner, R. L.,** Magnitudes, Nature, and Effects of Point and Nonpoint Discharges in the Chattahoochee River Basin, Atlanta to West Point Dam, Georgia, U.S. Geological Survey Water-Supply Paper 2059, U.S. Government Printing Office, Washington, D.C., 1979.
4. **Brown, L. and Barnwell, T. O., Jr.,** The Enhanced Stream Water Quality Models QUAL2E and QUAL2E-UNCAS: Documentation and User Manual, Report EPA/600/3-87/007, U.S. Environmental Protection Agency, Athens, GA, 1987.
5. **Leopold, L. B., Wolman, M. G., and Miller, J. P.,** *Fluvial Processes in Geomorphology,* W. H. Freeman, San Francisco, 1964.
6. **Velz, C. J.,** *Applied Stream Sanitation,* 1st ed., John Wiley and Sons, New York, 1970, 122 (see also 2nd ed., 1984).
7. **McKenzie, S. W., Hines W. G., Rickert, D. A., and Rinella, F. A.,** Steady-State Dissolved Oxygen Model for the Willamette River, Circular 715-J, U.S. Geological Survey, U.S. Government Printing Office, Washington, D.C., 1979.
8. **McKenzie, S. W.,** personal communication, U.S. Geological Survey, Portland, OR, 1987.
9. **McCutcheon, S. C., et al.,** Water Quality and Stream-Flow Data for the West Fork Trinity River in Fort Worth, Texas, U. S. Geological Survey Water Resources Investigation Report 84-4330, NSTL, MS, 1985.

10. **Rantz, S. E., et al.,** Measurement and Computation of Streamflow: Volume 1. Measurement of Stage and Discharge. Volume 2. Computation of Discharge, U.S. Geological Survey Water-Supply Paper 2175, U.S. Government Printing Office, Washington, D.C., 1982, 81.

11. **Edwards, M. D.,** Water data and services available from participants in the national water data exchange, *Water Resourc. Bull.,* 16, 1, 1980.

12. **Hubbard, E. F., Kilpatrick, F. A., Martens, L. A., and Wilson, J. F., Jr.,** Measurement of Time of Travel and Dispersion in Streams by Dye Tracing, Techniques of Water-Resources Investigations, Book 3, U.S. Geological Survey, U.S. Government Printing Office, Washington, D.C., 1982, Chap A9.

13. **Bureau of Reclamation,** *Water Measurement Manual,* 2nd ed., U.S. Department of the Interior, Denver, Colorado, 1967.

14. **U.S. Environmental Protection Agency,** NPDES Compliance Inspection Manual, Office of Water Enforcement and Permits, Washington, D.C., 1984.

15. **American Society for Testing and Materials,** *1974 Annual Book of ASTM Standards,* Part 31, Water, Philadelphia, PA, 1974.

16. **A. Ott Kempten,** C2 Small Current Meter, Manufacturer's Brochure, West Germany, undated.

17. **A. Ott Kempten,** C31 Universal Current Meter, Manufacturer's Brochure, West Germany, undated.

18. **Frazier, A. H.,** Water Current Meters in the Smithsonian Collections of the National Museum of History and Technology, Smithsonian Studies in History and Technology, No. 28, Smithsonian Institution Press, Washington, D.C., 1974.

19. **Smoot, G. F. and Novak, C. E.,** Calibration and Maintenance of Vertical-Axis Type Current Meters, U.S. Geological Survey Techniques of Water Resources Investigations, Book 8, U.S. Government Printing Office, Washington, D.C., 1968, Chap. B2.

20. **Futrell, J. C.,** personal communication, U.S. Geological Survey Hydrological Instrumentation Facility, NSTL, MS, 1987.

21. **Carter, R. W. and Anderson, I. E.,** Accuracy of current meter measurements, *J. Hydraul. Div.,* American Society of Civil Engineers, 89(HY4), 105, 1963.

22. **Corbett, D. M., et al.,** Stream Gaging Procedure: A Manual Describing Methods and Practices of the Geological Survey, U.S. Geological Survey Water-Supply Paper 888, U.S. Government Printing Office, Washington, D.C., 1962, 37.

23. **Marsh-McBirney, Inc.,** Calculating Flow in Open Channels, Gaithersburg, Maryland, 1986.

24. **Garde, R. J. and Ranga Raju, K. G.,** *Mechanics of Sediment Transportation and Alluvial Stream Problems,* New Delhi, John Wiley and Sons, 1977, 122.

25. **McCutcheon, S. C.,** Vertical velocity profiles in stratified flows, *J. Hydraul. Div.,* American Society of Civil Engineers, 107(HY8), 973, 1981.

26. **French, R. H. and McCutcheon, S. C.,** The Stability of a Two Layer Flow Without Shear in the Presence of Boundary Generated Turbulence: Field Verification, Vanderbilt University Dept. Environmental and Water Resources Engr. Technical Report 39, Nashville, TN, 1977, 45.

27. **Majewski, W.,** Modeling of heated discharges into river, in *Int. Symp. Stratified Flows,* Norosybirsk, USSR, 1972, New York, American Society of Civil Engineers, 383.

28. **Parker, F. L. and Krenkel, P. A.,** *Thermal Pollution: Status of the Art,* Rep. 3, Vanderbilt University Dept. Environmental and Water Resources Engineering, Nashville, TN, 1969.

29. **Akiyama, J. and Stefan, H.,** Plunging flow into a reservoir: theory, *J. Hydraul. Eng.,* American Society of Civil Engineers, 110, 484, 1984.

30. **McCutcheon, S. C.,** The Stability of a Two Layer Flow Without Shear in the Presence of Boundary Generated Turbulence: Field Verification, Master's thesis, Vanderbilt University, Nashville, TN, 1977, 53.

31. **McCutcheon, S. C.,** unpublished data collected on the Cumberland River, TN, 1976 to 1977.

32. **Macagno, E. O. and Alonso, C. V.,** Two-layer density stratified flow in an open channel bend, 12th Congr. Internat. Assoc. Hydraulic Research, Aug. 29 to Sept. 3, 1971, Paris, A 25 - 1.

33. **Jarrett, R. D.,** Evaluation of the Slope-Area Method of Computing Peak Discharge, U.S. Geological Survey Water Supply Paper 2310, Denver, CO, 1987.

34. **Phataraphruk, P. and Logan, E., Jr.,** Response of a turbulent pipe flow to a single roughness element, in *Developments in Theoretical and Applied Mechanics,* Proc. 9th SECTAM, Hackett, R. M., Ed., Vanderbilt University, Nashville, TN, May 4-5, 1978.

35. **Wan, Z.,** Bed Material Movement in Hyperconcentrated Flow, Technical University Denmark Institute of Hydrodynamics and Hydrology Engineering Series Paper 31, Lyngby, 1982, 9.

36. **Nowell, A. R. M.,** The benthic boundary layer and sediment transport, *Rev. Geophys. Space Phys.,* American Geophysical Union, 21(5), 1181, 1983.

37. **Hulsing, H., Smith, W. and Cobb, E. D.,** Velocity-Head Coefficients in Open Channels, U.S. Geological Survey Water-Supply Paper 1869-C, U.S. Government Printing Office, Washington, D.C., 1966, C7.

38. **Vanoni, V. A.,** Transportation of suspended sediment by water, *Transactions,* American Society of Civil Engineers, 111, 67, 1946.
39. **Laenen, A. and Smith, W.,** Acoustic Systems for Measurement of Streamflow, U.S. Geological Survey Water-Supply Paper 2213, U.S. Government Printing Office, Washington, D.C., 1983.
40. **Cobb, E.,** Dye-Dilution Gaging Principles, Class Notes, Hydrologic Tracers, U.S. Geological Survey Training Course, Denver, CO, Training Center, 1978.
41. **Cobb, E. D. and Bailey, J. F.,** Measurement of discharge by dye-dilution methods, in *Surface Water Techniques,* Book 1, U.S. Geological Survey, 1965, chap. 14.
42. **Wilson, J. F. Jr.,** Dye-Dilution Discharge Measurements: Wyoming District Procedures, Class Notes, Hydrologic Tracers, U.S. Geological Survey Training Course, Denver, CO, Training Center, 1978.
43. **Yotsukura, N. and Cobb, E. D.,** Transverse Diffusion of Solutes in Natural Streams, U.S. Geological Survey Professional Paper 582-C, U.S. Government Printing Office, 1972.
44. **Fischer, H. B., List, E. J., Koh, R. C. Y., Imberger, J., and Brooks, N. H.,** *Mixing in Inland and Coastal Waters,* Academic Press, New York, 1979.
45. **Morgan, W., Kempf, D., and Phillips, R. E.,** Validation of Use of Dye-Dilution Method for Flow Measurement in Large Open and Closed Channel Flows, paper presented at the NBS Flow Measurement Symposium in Gaithersburg, MD, Central Engineering Laboratories, Santa Clara, CA, 1977.
46. **Leopold and Stevens, Inc.,** *Stevens Water Resources Data Book,* 2nd ed., Beaverton, Oregon, 1974.
47. **American Society of Mechanical Engineers,** *Fluid Meters: Their Theory and Application,* 5th ed., Report of the ASME Research Committee on Fluid Meters, New York, 1959.
48. **Matthai, H. F.,** Measurement of Peak Discharge at Width Contractions by Indirect Methods, Book 3, U.S. Geological Survey Techniques of Water-Resources Investigations, U.S. Government Printing Office, Washington, D.C., 1967, Chap. A4.
49. **Smith, D. J.,** Water Quality For River-Reservoir Systems, U.S. Army Corps of Engineers, Davis, CA, 1978.
50. **Barnwell, T. O., Jr., Brown, L. C., and Marek, W.,** Development of a Prototype Expert Advisor for the Enhanced Stream Water Quality Model QUAL2E, Internal Report, Environmental Research Lab., U.S. Environmental Protection Agency, Athens, GA, 1987.
51. **Ambrose, R. B., Jr. and Vandergrift, S.,** SARAH, A Surface Water Assessment Model for Back Calculating Reductions in Abiotic Hazardous Wastes, Report EPA/600/3-86/058, Assessment Branch, Environmental Research Lab., U.S. Environmental Protection Agency, Athens, GA, 1986 (see also 2nd ed., 1988).
52. **Park, C. C.,** World-wide variations in hydraulic geometry exponents of stream channels: an analysis and some observations, *J. Hydrol.,* 33, 133, 1977.
53. **Richards, K.,** *Rivers: Form and Process in Alluvial Channels,* Methuen, London, 1982.
54. **Chow, V. T.,** *Open-Channel Hydraulics,* McGraw-Hill, New York, 1959, Chap. 5.
55. **Henderson, F. M.,** *Open Channel Flow,* MacMillan, New York, 1966, 98.
56. **Streeter, V. L. and Wylie, E. B.,** *Fluid Mechanics,* 6th ed., McGraw-Hill, New York, 1975, 286.
57. **Gauckler, P.,** Du mouvement de l'eau dans les conduits (The flow of water in conduits), Ann. Ponts Chaussées, 15(4), 229, 1868.
58. **Hagen, G. H. L.,** Untersuchungen uber die gleichformige Bewegung des Wassers (Researches on the uniform flow of water), Berlin, 1876.
59. **Strickler, A.,** Beitrage zur Frage der Geschwindigkeitsformel und der Rauhigkeitszahlen fur Strome, Kanale und geschlossene Leitungen, *Mitteilungen des eidgenossischen Amtes fur Wasserwirtscgaft,* Bern, Switzerland, No. 16, 1923.
60. **Manning, R.,** On the flow of water in open channels, *Trans. Inst. Civil Eng. Ireland,* 20, 161, 1891; Supplement, 24, 179, 1895.
61. **Darcy, H. and Bazin, H.,** "Recherches hydrauliques," 1re partie, Recherches experimentales sur l'ecoulement de l'eau dans les canaux decouverts; 2e partie, Recherches experimentales relatives aux remous et a la propagation des ondes, Academie des Sciences, Paris, 1865.
62. **Cunningham, A. J. C.,** Recent hydraulic experiments, *Proc. Inst. Civil Eng. London,* 71, 1, 1883.
63. **Barnes, H.,** Roughness Characteristics of Natural Channels, U. S. Geological Survey Water-Supply Paper 1849, U.S. Government Printing Office, Washington, D.C., 1967.
64. **Davidian, J.,** Computation of Water-Surface Profiles in Open Channels, Techniques of Water-Resources Investigations of the U.S. Geological Survey, Book 3, U.S. Government Printing Office, Washington, D.C., 1984, Chap. A15.

65. **U. S. Army Corps of Engineers Hydrologic Engineering Center,** HEC-2 Water Surface Profiles: Users Manual, Computer Program 723-X6-L202A, Davis, CA, September, 1982.

66. **Shearman, J. O.,** Computer Applications for Step Backwater and Floodway Analyses, U.S. Geological Survey, Open File Report 76-499, Washington, D.C., 1976.

67. **Dortch, M. and Martin, J. L.,** Water quality modeling of regulated streams, in *Alternatives in Regulated Flow Management,* Gore, J. A., Ed., CRC Press, Boca Raton, FL, 1988.

68. **Schaffranek, R., Baltzer, R. A., and Goldberg, D. E.,** A Model for Simulation of Flow in Singular and Interconnected Channels: Techniques of Water-Resources Investigations of the U.S. Geological Survey, Book 7, U.S. Government Printing Office, Washington, D.C., 1981, Chap. C3.

69. **Wurbs, R. A.,** Military Hydrology, Report 9, State-of-the-Art Review and Annotated Bibliography of Dam-Breach Flood Forecasting, Miscellaneous Paper EL-79-6, U.S. Army Engineer Waterways Experiment Station, Vicksburg, MS, 1985.

70. **DeLong, L. L.,** Extension of the Unsteady One Dimensional Open Channel Flow Equations for Flow Simulations in Meandering Channels with Flood Plains, U. S. Geological Survey Water-Supply Paper 2290, U. S. Government Printing Office, Washington, D.C., 1986.

71. **Fread, D. L.,** DAMBRK: The NWS Dam Break Flood Forecasting Model, Office of Hydrology, National Weather Service, Silver Spring, MD, 1978.

72. **Bedford, K. W., Sykes, R. M., and Libicki, C.,** Dynamic advective water quality model for rivers, *J. Environ. Eng.,* American Society of Civil Engineers, 109(3), 535, 1983.

73. **Bowie, G. L., Mills, W. B., Porcella, D. B., Campbell, C. L., Pagenkopf, J. R., Rupp, G. L., Johnson, K. M., Chan, P. W. H., and Gherini, S. A.,** Rates, Constants, and Kinetics Formulations in Surface Water Quality Modeling, 2nd Edition, EPA/600/3-85/040, U.S. Environmental Protection Agency, Athens, GA, 1985.

74. **Thomann, R. V.,** Effects of longitudinal dispersion on dynamic water quality response of streams and rivers, *Water Resourc. Res.,* 9(2), 355, 1973.

75. **Li, W.,** Effects of dispersion on DO-sag in uniform flow, *J. Sanit. Eng. Div.,* American Society of Civil Engineers, 98(SA1), 169, 1972.

76. **Ruthven, D. M.,** The dispersion of a decaying effluent discharged continuously into a uniformly flowing stream, *Water Res.,* 5, 343, 1971.

77. **Yotsukura, N. and Sayre, W. W.,** Transverse mixing in natural channels, *Water Resourc. Res.,* 12(4), 695, 1976.

78. **Glover, R. E.,** Dispersion of Dissolved or Suspended Materials in Flowing Streams, U.S. Geological Survey Professional Paper 433-K, U.S. Goverrment Printing Office, Washington, D.C., 1964.

79. **Parker, F. L.,** Eddy diffusion in reservoir and pipelines, *J. Hydraul. Div.,* American Society of Civil Engineers, 87(HY3), 151, 1961.

80. **Fischer, H. B.,** The mechanics of dispersion in natural streams, *J. Hydraul. Div.,* American Society of Civil Engineers, 93(HY6), 187, 1967.

81. **Fischer, H. B.,** Dispersion predictions in natural streams, *J. Sanit. Eng. Div.,* American Society of Civil Engineers, 94(SA5), 927, 1967.

82. **Taylor, G. I.,** The dispersion of matter in turbulent flow through a pipe, *Proc. R. Soc. London Ser. A,* 223, 446, 1954.

83. **Elder, J. W.,** The dispersion of marked fluid in turbulent shear flow, *J. Fluid Mech.,* 5(5), 544, 1959.

84. **Krenkel, P.,** Turbulent Diffusion and Kinetics of Oxygen Absorption, Ph. D. thesis, University of California, Berkeley, 1960.

85. **Elhadi, N. and Davar, K.,** Longitudinal dispersion for flow over rough beds, *J. Hydraul. Div.,* American Society of Civil Engineers, 102(HY4), 483, 1976.

86. **Fischer, H. B.,** Methods for Predicting Dispersion Coefficients in Natural Streams, with Application to Lower Reaches of the Green and Dunwamish Rivers, Washington, U.S. Geological Survey Professional Paper 582-A, U.S. Government Printing Office, Washington, D.C., 1968.

87. **Bansal, M. K.,** Dispersion in natural streams, *J. Hydraul. Div.,* American Society of Civil Engineers, 97(HY11), 1867, 1971.

88. **Godfrey, R. G. and Frederick, B. J.,** Streams Dispersion at Selected Sites, U.S. Geological Survey Professional Paper 433-K, U.S. Government Printing Office, Washington, D.C., 1970.

89. **Thackston, E. L.,** Longitudinal Mixing and Reaeration in Natural Streams, Dissertation Presented to Vanderbilt University, Nashville, TN, 1966.

90. **Thackston, E. L. and Krenkel, P.,** Longitudinal mixing in natural streams, *J. Sanit. Eng. Div.,* American Society of Civil Engineers, 93(SA5), 67, 1967.

91. **Miller, A. C. and Richardson, E. V.,** Diffusional dispersion in open channel flow, *J. Hydraul. Div.,* American Society of Civil Engineers, 100(HY1), 159, 1974.

92. **McQuivey, R. S. and Keefer, T. N.,** Simple method for predicting dispersion in streams, *J. Environ. Eng. Div.,* American Society of Civil Engineers, 100(EE4), 997, 1974.

93. **McQuivey, R. S. and Keefer, T. N.,** Dispersion-Mississippi River below Baton Rouge, LA, *J. Hydraul. Div.,* American Society of Civil Engineers, 102(HY10), 1426, 1976.

94. **Liu, H.,** Predicting dispersion coefficient of streams, *J. Environ. Eng. Div.,* American Society of Civil Engineers, 103(EE1), 59, 1977.

95. **Fischer, H. B.,** Discussion of: Simple method for predicting dispersion in streams, *J. Environ. Eng. Div.,* American Society of Civil Engineers, 101(EE3), 453, 1975.

96. **Hays, J. R., Krenkel, P. A., and Schnelle, K. B.,** Mass Transport Mechanisms in Open-Channel Flow, Tech. Report No. 8, Sanitary and Water Resources Engineering, Vanderbilt University, Nashville, TN, 1966.

97. **Thackston, E. L. and Schnelle, K.,** Predicting effects of dead zones on stream mixing, *J. Sanit. Eng. Div.,* American Society of Civil Engineers, 96(SA2), 319, 1970.

98. **Day, T. J.,** Longitudinal dispersion in natural channels, *Water Resour. Res.,* 11(6), 909, 1975.

99. **Day, T. J. and Wood, I. R.,** Similarity of the mean motion of fluid particles disposing in a natural channel, *Water Resour. Res.,* 12(4), 655, 1976.

100. **Liu, H. and Cheng, A. H. D.,** Modified Fickian model for predicting dispersion, *J. Hydraul. Div.,* American Society of Civil Engineers, 106(HY6), 1021, 1980.

101. **Sabol, G. V. and Nordin, C. F.,** Dispersion in rivers as related to storage zones, *J. Hydraul. Div.,* American Society of Civil Engineers, 104(HY5), 695, 1978.

102. **Valentine, E. M. and Wood, I. R.,** Longitudinal dispersion with dead zones, *J. Hydraul. Div.,* American Society of Civil Engineers, 103(HY9), 1977.

103. **Valentine, E. M. and Wood, I. R.,** Dispersion in rough rectangular channels, *J. Hydraul. Div.,* American Society of Civil Engineers, 105(HY12), 1537, 1979.

104. **Rutherford, J. C., Taylor, M. E. U., and Davies, J. D.,** Waikato River pollutant flushing rates, *J. Environ. Eng. Div.,* American Society of Civil Engineers, 106(EE), 1131, 1980.

105. **Beltaos, S.,** Longitudinal dispersion in rivers, *J. Hydraul. Div.,* American Society of Civil Engineers, 106(HY1), 151, 1980.

106. **Beltaos, S,** Dispersion in tumbling flow, *J. Hydraul. Div.,* American Society of Civil Engineers, 108(HY4), 591, 1982.

107. **Bajraktarevic-Dobran, H.,** Dispersion in mountainous natural streams, *J. Environ. Eng. Div.,* American Society of Civil Engineers, 108(EE3), 502, 1982.

108. **Beer, T. and Young, P. C.,** Longitudinal dispersion in natural streams, *J. Environ. Eng.,* American Society of Civil Engineers, 109(5), 1049, 1983.

109. **Jobson, H. E.,** Temperature and Solute-Transport Simulation in Streamflow Using a Lagrangian Reference Frame, U.S. Geological Survey Water Resources Investigations Report 81-2, NSTL Station, MS, 1980.

110. **Jobson, H. E.,** Comment on 'A new collocation method for the solution of the convection-dominated transport equation', *Water Resour. Res.,* 16(6), 1135, 1980.

111. **Jobson, H. E.,** Simulating Unsteady Transport of Nitrogen, Biochemical Oxygen Demand, and Dissolved Oxygen in the Chattahoochee River Downstream of Atlanta, GA, U.S. Geological Survey, Water-Supply Paper 2264, U.S. Government Printing Office, Washington, D.C., 1985.

112. **McBride, G. B. and Rutherford, C.,** Accurate modeling of river pollutant transport, *J. Environ. Eng.,* American Society of Civil Engineers, 110(4), 808, 1984.

113. **Jobson, H. E. and Rathbun, R. E.,** Use of the Routing Procedure to Study Dye and Gas Transport in the West Fork Trinity River, TX, U.S. Geological Survey, Water Supply Paper 2252, U.S. Government Printing Office, Washington, D.C., 1985.

114. **Smart, P. L. and Laidlaw, I. M. S.,** An evaluation of some fluorescent dyes for water tracing, *Water Resour. Res.,* 13(1), 15, 1977.

115. **Wilson, J. F., Cobb, E. D., and Kilpatrick, F. A.,** Fluorometric Procedures for Dye Tracing, Techniques of Water-Resources Investigations of the U.S. Geological Survey, Book 3, U.S. Government Printing Office, Washington, D.C., 1986, Chap. A12.

116. **Grove, P. B., Ed.,** *Webster's Third New International Dictionary of the English Language Unabridged,* Merriam, Springfield, MA, 1981.

117. **Hem, J.,** Study and Interpretation of the Chemical Characteristics of Natural Water, 3rd ed., U.S. Geological Survey Water-Supply Paper 2254, U.S. Government Printing Office, Washington, D.C., 1985.

118. **Pritchard, D. W. and Carpenter, J. H.,** Measurement of turbulent diffusion in estuarine and inshore waters, *Bull. Int. Ass. Sci. Hydrol.,* 20, 37, 1960.

119. **Feuerstein, D. L. and Selleck, R. E.,** Fluorescent tracers for dispersion measurements, *J. Sanit. Eng. Div.,* American Society of Civil Engineers, 89(SA4), 1, 1963.

120. **Yates, W. E. and Akesson, N. B.,** Fluorescent tracers for quantitative microresidue analysis, *Trans. Am. Soc. Agric. Eng.,* 6, 104, 1963.

121. **Watt, J. P. C.,** Development of the dye-dilution method for measuring water yields from mountain watersheds, Master's thesis, Colorado State University, Fort Collins, 1965.

122. **Von Moser, H. and Sagl, H.,** Die Direktmessung hydrologischer Farbtracer im Gelande, *Steirische Beitr. Hydrogeol.,* 18, 179, 1967.

123. **Abood, K. A., Lawler, J. P., and Disco, M. D.,** Utility of radioisotope methodology in estuary pollution control study. I. Evaluation of the use of radioisotopes and fluorescent dyes for determining longitudinal dispersion, Report NYO-3961-1, 197, U.S. Atomic Energy Commission, New York, 1969.

124. **Gowda, T. P.,** Water quality prediction in mixing zones of rivers, *J. Environ. Eng.,* American Society of Civil Engineers, 110(4), 751, 1980.

125. **Abidi, S. L.,** Detection of diethylnitrosamine in nitrite-rich water following treatment with rhodamine flow tracers, *Water Res.,* 16, 199, 1982.

126. **Bandt, H. J.,** Giftig oder ungiftig fur Fische?, *Deut. Fisch. Rundsch.,* 4(6), 170, 1957.

127. **Sowards, C. L.,** Sodium fluorescein and the toxicity of Pronoxfish, *Prog. Fish Cult.,* 20(1), 20, 1958.

128. **Marking, L. L.,** Toxicity of Rhodamine B and fluorescein sodium to fish and their compatibility with antimycin A, *Prog. Fish Cult.,* 31, 139, 1969.

129. **Panciera, M.,** Toxicity of Rhodamine B to eggs and larvae of *Crassostrea virginica, Proc. Natl. Shellfish Assoc.,* 58, 7, 1967.

130. **Woelke, C. E.,** Development of a receiving water quality bioassay criterion based on the 48 hour Pacific oyster *(Crassostrea gigas)* embryo, Technical Report 9, Washington Department Fish, 1972.

131. **Parker, G. G.,** Tests of Rhodamine WT dye for toxicity to oysters and fish, *J. Res.* U.S. Geological Survey, 1(4), 499, 1973.

132. **Worttley, J. S. and Atkinson, T. C.,** personal communication cited by Smart and Laidlaw (1977), 1975.

133. **Akamatsu, K. and Matsuo, M.,** Safety of optical whitening agents (in Japanese), *Senryo To Yakuhin,* 18(2), 2, 1973. (English translation, *Transl. Programme RTS 9415,* British Library, Boston Spa, Yorkshire, England, 1975.)

134. **Keplinger, M. L., Fancher, O. E., Lyman, F. L., and Calandra, J. C.,** Toxicological studies of four fluorescent whitening agents, *Toxicol. Appl. Pharmacol.,* 21, 494, 1974 .

135. **Strum, R. N. and Williams, K. E.,** Fluorescent whitening agents: acute fish toxicity and accumulation studies, *Water Resour. Res.,* 9, 211, 1975.

136. **Chao, A. C., Chang, D. S., Smallwood, C., Jr., and Galler, W. S.,** Effect of temperature on oxygen transfer — laboratory studies, *J. Environ. Eng.,* American Society of Civil Engineers, 113(5), 1089, 1987.

137. **Daniil, E. I., and Gulliver, J. S.,** Temperature dependence of the liquid film coefficient for gas transfer, *J. Environ. Eng.,* American Society of Civil Engineers, 114(5), 1224, 1988.

138. **Rheinheimer, G.,** *Aquatic Microbiology,* John Wiley and Sons, New York, 1980, 191.

139. **Borchardt, J. A.,** Nitrification in the activated sludge process, in *The Activated Sludge Process,* Division of Sanitary and Water Resources Engineering, University of Michigan, Ann Arbor, 1966.

140. **Thornton, K. W. and Lessem, A. S.,** A temperature algorithm for modifying biological rates, *Trans. Am. Fish. Soc.,* 107(2), 284, 1978.

141. **Rathbun, R. E.,** Reaeration coefficients of streams — state-of-the-art, *J. Hydraul. Div.,* American Society of Civil Engineers, 103(HY4), 409, 1977.

142. **Rich, L. G.,** *Environmental Systems Engineering,* McGraw-Hill, New York, 1973, 47.

143. **Dickerson, R. E., Gray, H. B., and Haight, G. P., Jr.,** *Chemical Principles,* Benjamin, Inc., New York, 1970, 787.

144. **Perry, R. H., Sliepcevich, C. M., Finn, D., Green, D. W., Kobayashi, R., Leland, T. W., and Smith, J. M.,** Reaction kinetics, reactor design, and thermodynamics, in *Perry's Chemical Engineers Handbook,* Perry, R. H., Chilton, C. H., and Kirkpatrick, S. D., Eds., New York, McGraw-Hill, 1963, Section 4.

145. **Thomann, R. V., O'Connor, D. J., and DiToro, D. M.,** Mathematical Modeling of Natural Systems, Technical Report, Environmental Engineering and Science Program, Manhattan College, New York, 1976.

146. **Brown, L. C. and Barnwell, T. O., Jr.,** Computer Program Documentation for the Enhanced Stream Water Quality Model QUAL2E, Report EPA 600/3-85/065, U.S. Environmental Protection Agency, Athens, GA, 1985, 95.

147. **Gromiec, M. J., Loucks, D. P., and Orlob, G. T.,** Stream Quality Modeling, in *Mathematical Modeling of Water Quality: Streams, Lakes and Reservoirs,* Orlob, G. T., Ed., John Wiley and Sons, New York, 1983, Chap. 6.

148. **Barnwell, T. O., Jr.,** personal communication, U.S. Environmental Protection Agency Environmental Research Laboratory, Athens, GA, 1986.

149. **Krenkel, P. A. and Novotny, V.,** *Water Quality Management,* Academic Press, New York, 1980.

150. **Mueller, J. A., Paquin, P., and Famularo, J.,** Nitrification in rotating biological contractors, *J. Water Pollut. Control Fed.,* 52(4), 1980.

151. **Zanoni, A. E.,** Secondary effluent deoxygenation at different temperatures, *J. Water Pollut. Control Fed.,* 41, 640, 1969.

152. **Painter, H. A.,** A review of literature on inorganic nitrogen metabolism in microorganisms, *Water Res.,* 4, 393, 1970.

153. **Painter, H. A. and Lovelace, J. E.,** Effect of temperature and pH value on the growth-rate constants of nitrifying bacteria in the activated-sludge process, *Water Res.,* 17(3), 237, 1983.

154. **Buswell, A. M., Shiota, T., Lawrence, N., and Van Meter, I.,** Laboratory studies on the kinetics of the growth of *Nitrosomonas* with relation to the nitrification phase of the B.O.D. test, *Appl. Microbiol.,* 2(1), 1954.

155. **Hansen, J. I., Henriksen, K., and Blackburn, T. H.,** Seasonal distribution of nitrifying bacteria and rates of nitrification in coastal marine sediments, *Microb. Biol.,* 7, 297, 1981.

156. **Downing, A. L.,** Factors to be Considered in the Design of Activated Sludge Plants, in *Advances in Water Quality Improvement,* Gloyna, E. F. and Eckenfelder, W. W., Jr., Eds., University of Texas Press, Austin, 190, 1968.

157. **Stratton, F. E. and McCarty, P. L.,** Prediction of nitrification effects on the dissolved oxygen balance of streams, *Environ. Sci. Technol.,* 1(5), 405, 1967.

158. **Sharma, B. and Ahlert, R. C.,** Nitrification and nitrogen removal, *Water Res.,* 11, 897, 1977.

159. **Department of Scientific and Industrial Research,** Notes on Water Pollution, Number 23, H.M. Stationary Office, Bath, England, 1963.

160. **Boon, B. and Laudelout, H.,** Kinetics of nitrite oxidation by *Nitrobacter Winogradski,* *Biochem. J.,* 85, 440, 1962.

161. **Wong-Chong, G. M. and Loehr, R. C.,** The kinetics of microbial nitrification, *Water Res.,* 9, 1099, 1975.

162. **La Motta, E. J. and Shieh, W. K.,** Diffusion and reaction in biological nitrification, *J. Environ. Eng. Div.,* American Society of Civil Engineers, 105(EE4), 655, 1979.

163. **Laudelout, H. and van Tichelen, L.,** Kinetics of the nitrite oxidation by *Nitrobacter Winogradskyi, J. Bacteriol.,* 79, 39, 1960.

164. **Jones, R. D. and Hood, M. A.,** Effects of temperature, pH, salinity, and inorganic nitrogen on the rate of ammonium oxidation by nitrifiers isolated from wetland environments, *Microb. Ecol.,* 6, 339, 1980.

165. **Downing, A. L., Painter, H. A., and Knowles, G.,** Nitrification in the activated sludge process, *J. Inst. Sewage Purif. (G.B.),* Part 2, 130, 1964.

166. **Stanford, G., Frere, M. H., and Schwaninger, D. H.,** Temperature coefficient of soil nitrogen mineralization, *Soil Sci.,* 115(4), 321, 1973.

167. **Knowles, G., Downing, A. L., and Barrett, M. J.,** Determination of kinetic constants for nitrifying bacteria in mixed culture, with the aid of an electronic computer, *J. Microbiol.,* 38, 263, 1965.

168. **Wild, H. E., Jr., Sawyer, C. N., and McMahon, T. C.,** Factors affecting nitrification kinetics, *J. Water Pollut. Control Fed.,* 43(9), 1845, 1971.

169. **Grenney, W. T. and Kraszewski, A. K.,** Description and Application of the Stream Simulation and Assessment Model: Version IV (SSAM IV), Instream Flow Information Paper, Fish and Wildlife Service, Fort Collins, CO, Cooperative Instream Flow Service Group, 1981.

170. **Wunderlich, W. O.,** Heat and Mass Transfer Between a Water Surface and the Atmosphere, Water Resources Research, Laboratory Report No. 14, Report No. 0-6803, Tennessee Valley Authority, Norris, 1972.

171. **Hamon, R. W., Weiss, L. L., and Wilson, W. T.,** Insulation as an empirical function of daily sunshine duration, *Mon. Weather Rev.,* 82(6), 141, 1954.

172. **Bolsenga, S. T.,** Daily Sums of Global Radiation for Cloudless Skies, U.S. Army Material Command, Cold Regions Research and Engineering Laboratory, Hanover, NH, Technical Report 160, 1964.

173. **Kimball, H. H.,** Measurement of solar radiation intensity and determination of its depletion by the atmosphere, *Mon. Weather Rev.,* 55(1927), 56(1928), 58(1930).

174. **U.S. Geological Survey,** Water Loss Investigations: Lake Hefner Studies, Technical Report, Professional Paper 269, U.S. Government Printing Office, Washington, D.C., 1954.

175. **Ryan, P. J. and Harleman, D. R. F.,** An Analytical and Experimental Study of Transient Cooling Pond Behavior, R.M. Parsons Laboratory, MIT, Technical Report No. 161, 1973.

176. **Hatfield, J. L., Reginato, R. J., and Idso, S. B.,** Comparison of long-wave radiation calculation methods over the United States, *Water Resour. Res.,* 19(1), 285, 1983.

177. **Brunt, D.,** Notes on radiation in the atmosphere, *Q. J. R. Meteorol. Soc.,* 58, 389, 1932.

178. **Morgan, D. L., Pruitt, W. O., and Lourence, F. L.,** Radiation Data and Analyses for the 1966 and 1967 Micrometerological Field Runs at Davis, California, Department of Water Science and Engineering, University of California, Davis, Research and Development Technical Report ECOM 68-G10-2, 1970.

179. **Linsley, R. K., Jr., Kohler, M. A., and Paulhus, J. L. H.,** *Hydrology for Engineers,* 2nd ed., McGraw-Hill, New York, 1975 (see also 3rd ed., 1982).

180. **Edinger, E. and Geyer, J. C.,** Heat Exchange in the Environment, Edison Electric Institute Publication No. 65-902, 1965.

181. **Rodi, W.,** *Turbulence Models and Their Application in Hydraulics,* International Assoc. for Hydraulic Research, Delft, The Netherlands, 1980.

182. **Thackston, E. L.,** Effect of Geographical Variation on Performance of Recirculating Cooling Ponds, U.S. Environmental Protection Agency Report EPA-660/2-74-085, Corvallis, OR, 1974.

183. **Kolher, M. A.,** Lake and Pan Evaporation, in Water Loss Investigations, Lake Hefner Studies, Technical Report U.S. Geological Survey Professional Paper 269, U.S. Government Printing Office, Washington, D.C., 1954.

184. **Zaykov, B. D.,** Evaporation from the water surface of ponds and small reservoirs of the U.S.S.R., *Trans. State Hydrol. Inst. (Tr. Gos. Gidrol. Inst.),* 1949.

185. **Meyer, A. F.,** Evaporation from Lakes and Reservoirs, Minnesota Resources Commission, St. Paul, 1942.

186. **Morton, F. I.,** Climatological estimates of evapotranspiration, *J. Hydraul. Div.,* American Society of Civil Engineers, 102(HY3), 275, 1976.

187. **Rohwer, C.,** Evaporation from Free Water Surfaces, U.S. Department of Agriculture, Washington, D.C., Technical Bulletin Number 271, 1931.

188. **Priestly, C. H. B.,** *Turbulent Transfer in the Lower Atmosphere,* University of Chicago Press, 1959.

189. **Harbeck, G. E., et al.,** Water-Loss Investigations: Lake Mead Studies, U.S. Geological Survey Professional Paper 298, U.S. Government Printing Office, Washington, D.C., 1958.

190. **Turner, J. F., Jr.,** Evaporation Study in a Humid Region, Lake Michie, North Carolina, U.S. Geological Survey Professional Paper 272-G, U.S. Government Printing Office, Washington, D.C., 1966.

191. **Easterbrook, C. C.,** A Study of the Effects of Waves on Evaporation from Free Water Surfaces, U.S. Department of Interior, Bureau of Reclamation, Research Report No. 18, U.S. Government Printing Office, Washington, D.C., 1969.

192. **Jobson, H. E.,** Bed Conductivity Computation for Thermal Models, *J. Hydraul. Div.,* American Society of Civil Engineers, 103(HY10), 1213, 1977.

193. **Jobson, H. E. and Keefer, T. N.,** Modeling Highly Transient Flow, Mass, and Heat Transport in the Chattahoochee River near Atlanta, Georgia, U.S. Geological Survey Professional Paper 1136, U.S. Government Printing Office, Washington, D.C., 1979.

194. **Jobson, H. E.,** Thermal Modeling of Flow in the San Diego Aqueduct, California, and its Relation to Evaporation, U.S. Geological Survey Professional Paper 1122, U.S. Government Printing Office, Washington, D.C., 1980.

195. **Faye, R. E., Jobson, H. E., and Land, L. F.,** Impact of Flow Regulation and Powerplant Effluents on the Flow and Temperature Regimes of the Chattahoochee River — Atlanta to Whitesburg, Georgia, U.S. Geological Survey Professional Paper 1136, U.S. Government Printing Office, Washington, D.C., 1979.

196. **McCutcheon, S. C.,** Unpublished notes on the equilibrium temperature modeling of the West Fork Trinity River, Fort Worth, TX, 1982 (see Reference 9 for the data involved).

197. **Koberg, C. E.,** Methods to Compute Long-Wave Radiation from the Atmosphere and Reflected Radiation from the Water Surface, U.S. Geological Survey Professional Paper 272-F, U.S. Government Printing Office, Washington, D.C., 1964.

198. **Schoellhamer, D. H. and Jobson, H. E.,** Programmers Manual for a One-Dimensional Lagrangian Transport Model, Water Resources Investigation Report 86-4144, U.S. Geological Survey, NSTL Station, MS, 1986; and **Schoellhamer, D. H.,** Lagrangian transport modeling with QUAL II kinetics, *J. Environ. Eng.,* American Society of Civil Engineers, 114 (2), 386, 1988.

199. **Bauer, D. P. and Yotsukura, N.,** Two-Dimensional Excess Temperature Model for a Thermally Loaded Stream, Report WRD-74-044, U.S. Geological Survey, Water Resources Division, Gulf Coast Hydroscience Center, NSTL, MS, 1974.

200. **Novotny, V. and Krenkel, P. A.,** Heat Transfer in Turbulent Streams, Research Report Number 7, Department of Environmental and Water Resources Engineering, Vanderbilt University, Nashville, Tennessee, 1971.

201. **Wentz, D. A. and Steel, T. D.,** Analysis of Stream Quality in the Yampa River Basin, Colorado and Wyoming, U.S. Geological Survey, Water Resources Investigations Report 80-8, Lakewood, CO, 1980.

202. **Mills, W. B., Porcella, D. B., Ungs, M. J., Gherini, S. A., Summers, K. V., Mok, L., Rupp, G. L., and Bowie, G. L.,** Water Quality Assessment: A Screening Procedure for Toxic and Conventional Pollutants in Surface and Ground Water Part 1, (Revised 1985), EPA/600/6-85/002a, U.S. Environmental Protection Agency, Athens, GA, 1985.

203. **Mackay, D. and Shiu, W.Y.,** Physical-chemical phenomena and molecular properties, in *Gas Transfer at Water Surfaces,* Brutsaert, W. and Jirka, G. H., Eds., D. Reidel, Boston, 1984, 3.

204. **Thibodeaux, L. J.,** *Chemodynamics: Environmental Movement of Chemicals in Air, Water, and Soil,* John Wiley and Sons, New York, 1979.

205. **Bird, R. B., Stewart, W. E., and Lightfoot, E. N.,** *Transport Phenomena,* John Wiley and Sons, New York, 1960.

206. **Rathbun, R. E. and Tai, D. Y.,** Volatilization of organic compounds from streams, *J. Environ. Eng. Div.,* American Society of Civil Engineers, 108(EE5), 973, 1982.

207. **Tsivoglou, E. C. and Wallace, J. R.,** Characterization of Stream Reaeration Capacity, EPA-R3-72-012, U.S. Environmental Protection Agency, Washington, D.C., 1972.

208. **Rathbun, R. E., Stephens, D. W., Shultz, D. J., and Tai, D. Y.,** Laboratory studies of gas tracers for reaeration, *J. Environ. Eng. Div.,* American Society of Civil Engineers, 104(EE2), 215, 1978.

209. **Bales, J. D. and Holley, E. R.,** Flume tests on hydrocarbon reaeration tracer gases, *J. Environ. Eng.,* American Society of Civil Engineers, 112(4), 695, 1986.

210. **Rainwater, K. A. and Holley, E. R.,** Laboratory Studies on the Hydrocarbon Gas Tracer Technique for Reaeration Measurement, Technical Report CRWR-189, Center for Research in Water Resources, University of Texas, Austin, 1983.

211. **Rainwater, K. A. and Holley, E. R.,** Laboratory studies on reaeration tracer gases, *J. Environ. Eng. Div.,* American Society of Civil Engineers, 110(EE1), 1984.

212. **Wilcock, R. J.,** Methyl chloride as a gas-tracer for measuring stream reaeration coefficients. I. Laboratory studies. II. Stream studies, *Water Res.,* 18(1), 47, 1984.

213. **Wilcock, R. J.,** Reaeration studies on some New Zealand rivers using methyl chloride as a gas tracer, in *Gas Transfer at Water Surfaces,* Brutsaert, W. and Jirka, G. H., Eds., D. Reidel, Boston, 1984, 413.

214. **Ljubisavljevic, D.,** Carbon dioxide desorption from the activated sludge at the waste water treatment plants, in *Gas Transfer at Water Surfaces,* Brutsaert, W. and Jirka, G. H., Eds., D. Reidel, Boston, 1984, 613.

215. **Hill, J., IV, Kollig, H. P., Paris, D. F., Wolfe, N. L., and Zepp, R. G.,** Dynamic Behavior of Vinyl Chloride in Aquatic Ecosystems, EPA-600/3-76-001, U.S. Environmental Protection Agency, Athens, GA, 1976.

216. **O'Conner, D. J. and Dobbins, W. E.,** Mechanism of reaeration in natural streams, *Trans. Am. Soc. Civ. Eng.,* 2934, 641, 1958.

217. **O'Conner, D. J.,** The Measurement and Calculation of Stream Reaeration Rates, Seminar on Oxygen Relationships in Streams, Robert A. Taft Sanitary Engineering Center Technical Report 58-2, 35, 1958.

218. **Dobbins, W. E.,** BOD and oxygen relationships in streams, *J. Sanit. Eng. Div.,* American Society of Civil Engineers, 90(SA3), 53, 1964.

219. **Krenkel, P. A. and Orlob, G. T.,** Discussion and Closure of 'Turbulent diffusion and the reaeration coefficient', *Trans. Am. Soc. Civ. Eng.,* 128(3), 293, 1963.

220. **Krenkel, P. A. and Orlob, G. T.,** Turbulent diffusion and the reaeration coefficient, *J. Sanit. Eng. Div.,* American Society of Civil Engineers, 88(SA2), 53, 1962.

221. **Thackston, E. L. and Krenkel, P. A.,** Reaeration prediction in natural streams, *J. Sanit. Eng. Div.,* American Society of Civil Engineers, 95(SA1), 65, 1969.

222. **Cadwaller, T. E. and McDonnell, A. J.,** A multivariate analysis of reareation data, *Water Res.,* 2, 731, 1969.

223. **Churchill, M. A., Elmore, H. L., and Buckingham, R. A.,** The prediction of stream reaeration rates, *J. Sanit. Eng. Div.,* American Society of Civil Engineers, 88(SA4), 1, 1962.

224. **Owens, M., Edwards, R. W., and Gibbs, J. W.,** Some reaeration studies in streams, *Int. J. Air Water Pollut.,* 8, 469, 1964.

225. **Parkhurst, J. E. and Pomeroy, R. D.,** Oxygen absorption in streams, *J. Sanit. Eng. Div.,* American Society of Civil Engineers, 98(SA1), 1972.

226. **Tsivoglou, E. C. and Neal, L. A.,** Tracer method of reaeration. III. Predicting the reaeration capacity of inland streams, *J. Water Pollut. Cont. Fed.,* 48(12), 2669, 1976.

227. **Grant, R. S.,** Reaeration-Coefficient Measurements of 10 Small Streams in Wisconsin Using Radioactive Tracers...With a Section on the Energy-Dissipation Model, U.S. Geological Survey, Water Resources Investigations Report 76-96, Madison, WI, 1976.

228. **Grant, R. S.,** Reaeration Capacity of the Rock River Between Lake Koshkonong, Wisconsin and Rockton, Illinois, U.S. Geological Survey, Water Resources Investigations Report 77-128, Madison, WI, 1978.

229. **Shindala, A. and Truax, D. D.,** Reaeration Characteristics of Small Streams, Engineering and Industrial Research Station, Mississippi State University, 1980.

230. **Neal, L. A.,** Reaeration measurement in swamp streams: radiotracer case studies, in *Gas Transfer at Water Surfaces,* Brutsaert, W. and Jirka, G. H., Eds., D. Reidel, Boston, 597, 1984.

231. **McCutcheon, S. C. and Jennings, M. E.,** Discussion of: Stream reaeration by Velz Method, *J. Environ. Eng. Div.,* American Society of Civil Engineers, 108(EE1), 218, 1982.

232. **Hirsh, R. M.,** The Computation of an Equivalent Reaeration Coefficient from the Velz Rational Method and Comparison with Other Methods, U. S. Geological Survey, Systems Analysis Group, Water Resources Division, Reston, VA, 1979.

233. **Bennett, J. and Rathbun, R. E.,** Reaeration in Open-Channel Flow, U.S. Geological Survey Professional Paper 737, U.S. Government Printing Office, Washington, D.C., 1972.

234. **Lau, Y. L.,** A Review of Conceptual Models and Prediction Equations for Reaeration in Open-Channel Flow, Technical Bulletin 61, Inland Waters Branch, Department of the Environment, Ottawa, Canada, 1972. [Also see **Lau, Y. L.,** Prediction equation for reaeration in open-channel flow, *J. Sanit. Eng. Div.,* American Society of Civil Engineers, 98(SA6), 1061, 1972.]

235. **Krenkel, P. A.,** Turbulent Diffusion and the Kinetics of Oxygen Absorption, Ph. D. dissertation, University of California Department of Sanitary Engineering, Berkeley, 1960.

236. **Ice, G. G. and Brown, G. W.,** Reaeration in a Turbulent Stream System, Prepared for Office of Water Research and Technology, Washington, D.C., 1978.

237. **Streeter, H. W. and Phelps, E.,** A Study of the Pollution and Natural Purification of the Ohio River, Public Health Bulletin 146, U.S. Public Health Service, U.S. Government Printing Office, Washington D.C., 1925.

238. **Langbein, W. B. and Durum, W. H.,** The Aeration Capacity of Streams, U.S. Geological Survey Circular 542, U.S. Government Printing Office, Washington D.C., 1967.

239. **Streeter, H. W., Wright, C. T., and Kehr, R. W.,** Measures of natural oxidation in polluted streams. Part III. An experimental study of atmospheric reaeration under stream flow conditions, *Sewage Works J.,* 8(2), 283, 1936.

240. **Gameson, A. L. H., Truesdale, G. A., and Downing, A. L.,** Reaeration studies in Lakeland Beck, *J. Inst. Water Eng.,* 9, 571, 1955.

241. **Isaacs, W. P. and Gaudy, A. F.,** Atmospheric oxidation in a simulated stream, *J. Sanit. Eng. Div.,* American Society of Civil Engineers, 94(SA2), 319, 1968.

242. **Isaacs, W. P. and Maag, J. A.,** Investigation of the effects of channel geometry and surface velocity on the reaeration rate coefficient, *Eng. Bull. Purdue Univ. Eng. Ext. Ser.,* 53, 1969.

243. **Negulescue, M. and Rojanski, V.,** Recent research to determine reaeration coefficient, *Water Res.,* 3(3), 189, 1969.

244. **Padden, T. J. and Gloyna, E. F.,** Simulation of Stream Processes in a Model River, University of Texas, Austin, Report Number EHE-70-23, CRWR-72, 130, 1971.

245. **Bansal, M. K.,** Atmospheric reaeration in natural streams, *Water Res.,* 7, 769, 1973.

246. **Long, E. G.,** Letter to Ray Whittemore, National Council for Air and Stream Improvement, Tufts University, Medford, MA, Texas Department of Water Resources, 1984.

247. **Foree, E. G.,** Reaeration and velocity prediction for small streams, *J. Environ. Eng. Div.,* American Society of Civil Engineers, 102(EE5), 937, 1976.

248. **Foree, E. G.,** Low-flow reaeration and velocity characteristics of small streams (update), *J. Environ. Eng. Div.,* American Society of Civil Engineers, EE5, 1977.

249. **Committee on Sanitary Engineering Research,** Effects of water temperature on stream reaeration, *J. Sanit. Eng. Div.,* American Society of Civil Engineers, 87(SA6), 59, 1961.

250. **Streeter, H. W.,** The rate of atmospheric reaeration of sewage polluted streams, *Trans. Am. Soc. Civ. Eng.,* 89, 1351, 1926.

251. **Phelps, E.,** *Stream Sanitation,* John Wiley and Sons, New York, 1944.

252. **Downing, A. L. and Truesdale, G. A.,** Some factors affecting the rate of solution of oxygen in water, *J. Appl. Chem.,* 5, 570, 1955.

253. **Truesdale, G. A. and Vandyke, K. G.,** The effect of temperature on the aeration of flowing waters, *Water Waste Treat. J.,* 7(9), 1958.

254. **Metzger, I.,** Effects of temperature on stream aeration, *J. Sanit. Eng. Div.,* American Society of Civil Engineers, 94(SA6), 1153, 1968.

255. **Tsivoglou, E. C.,** Tracer Measurement of Stream Reaeration, Federal Water Pollution Control Administration, Washington, D.C., PB-229, 923, 1967.

256. **American Public Health Association, et al.,** *Standard Methods for the Examination of Water and Wastewaters,* Washington, D.C., 1960.

257. **Thackston, E. L. and Krenkel, P. A.,** Discussion of atmospheric oxygenation in a simulated stream, *J. Sanit. Eng. Div.,* American Society of Civil Engineers, 95(SA2), 354, 1969.

258. **Elmore, H. L. and West, W. F.,** Effects of water temperature on stream reaeration, *J. Sanit. Eng. Div.,* American Society of Civil Engineers, 87(SA6), 59, 1960.

259. **Montgomery, H. A. C.,** Discussion of atmospheric oxygenation in a simulated stream, *J. Sanit. Eng. Div.,* American Society of Civil Engineers, 95(SA2), 356, 1969.

260. **Landine, R. C.,** A note on the solubility of oxygen in water, *Water Sewage Works,* 118(8), 242, 1971.

261. **Metzger, I. and Dobbins, W. E.,** The role of fluid properties in gas transfer, *Environm. Sci. Technol.,* 1, 57, 1967.

262. **Jahne, B., Munnich, K. O., Bosinger, R., Dutzi, A., Huber, W., and Libner, P.,** On the parameters influencing air-water gas exchange, *J. Geophys. Res.,* 92(C2), 1937, 1987.

263. **Wilson, G. T. and MacLeod, N.,** A critical appraisal of empirical equations and models for the predictions of the coefficient of reaeration of deoxygenated water, *Water Res.,* 8, 341, 1974.

264. **Rathbun, R. E. and Grant, R. S.,** Comparison of the Radioactive and Modified Techniques for Measurement of Stream Reaeration Coefficients, U.S. Geological Survey Water-Resources Investigations Report 78-68, NSTL Station, MS, 1978.

265. **National Council for Air and Stream Improvement,** An Assessment of the Limitations of the Radiotracer Technique in Measuring Stream Reaeration Rates, Technical Bulletin Number 374, New York, 1982.

266. **Kwasnik, J. M. and Feng, T. J.,** Development of a Modified Tracer Technique for Measuring the Stream Reaeration Rate, Water Resources Research Center, University of Massachusetts at Amherst, Publication Number 102, 1979.

267. **Grant, R. S. and Skavroneck, S.,** Comparison of Tracer Methods and Predictive Equations for Determination Stream-Reaeration Coefficients on Three Small Streams in Wisconsin, U.S. Geological Survey, Water Resources Investigations Report 80-19, Madison, WI, 1980.

268. **House, L. B. and Skavroneck, S.,** Comparison of the Propane-Area Tracer Method and Predictive Equations for Determination of Stream-Reaeration Coefficients on Two Small Streams in Wisconsin, U.S. Geological Survey Water-Resources Investigations Report 80-105, Madison, WI, 1981.

269. **Zison, S. W., Mills, W. B., Deimer, D., and Chen, C. W.,** Rates, Constants, and Kinetics Formulations in Surface Water Quality Modeling, Report EPA/600/3-78-105, U.S. EPA, Athens, GA, 1978.

270. **Covar, A. P.,** Selecting the proper reaeration coefficient for use in water quality models, in *Proc. U.S. EPA Conf. on Environmental Simulation and Modeling,* Cincinnati, OH, 1976.

271. **Yotskura, N., Stedfast, D. A., Draper, R. E., and Brutsaert, W. H.,** An Assessment of Steady-State Propane-Gas Tracer Method for Reaeration in Cowaselon Creek, New York, U.S. Geological Survey Water-Resources Investigations Report, 1983.

272. **Ohio Environmental Protection Agency,** Determining the Reaeration Coefficient for Ohio Streams, draft, 1983.

273. **Fortescue, G. E. and Pearson, J. R. A.,** On gas absorption into a turbulent liquid, *Chem. Eng. Sci.,* 22, 1163, 1967.

274. **Bauer, D. P., Rathbun, R. E., and Lowham, H. W.,** Traveltime, Unit-Concentration, Longitudinal-Dispersion, and Reaeration Characteristics of Upstream Reaches of the Yampa and Little Snake Rivers, Colorado and Wyoming, U.S. Geological Survey Water-Resources Investigations Report 78-122, Lakewood, CO, 1979.

275. **Goddard, K. E.,** Calibration and Potential Uses of a Digital Water-Quality Model for the Arkansas River in Pueblo County, Colorado, U.S. Geological Survey Water-Resources Investigations Report 80-38, Lakewood, CO, 1980.

276. **Gameson, A. L. H., Van Dyke, K. G., and Oger, C. G.,** The effect of temperature of aeration at weirs, *Water Wastes Eng.,* London, 1958.

277. **Tsivoglou, E. C., Cohen, J. B., Shearer, S. D. and Godsil, P. J.,** Tracer measurement of stream reaeration. II. Field studies, *J. Water Pollut. Cont. Fed.,* 40(2), 285, 1968.
278. **Terry, J. E., Morris, E. E., Petersen, J. C., and Darling, M. E.,** Water-Quality Assessment of the Illinois River Basin, Arkansas, U.S. Geological Survey Water-Resources Investigations Report 83-4092, Little Rock, AR, 1984.
279. **Hren, J.,** Measurement of the Reaeration Coefficients of the North Fork Licking River at Utica, Ohio, by Radioactive Tracers, U.S. Geological Survey Water Resources Investigations Report 83-4192, Columbus, OH, 1983.
280. **Rathbun, R. E., Shultz, D. J., and Stephens, D. W.,** Preliminary Experiments with a Modified Tracer Technique for Measuring Stream Reaeration Coefficients, U.S. Geological Survey Open File Report 75-256, 1975.
281. **National Council for Air and Stream Improvement,** A Comparison of Reaeration Estimation Techniques for the Ouachita River Basin, Technical Bulletin Number 375, New York, 1982.
282. **Gameson, A. L. H.,** Weirs and aeration of rivers, *J. Inst. Water Eng.,* 6(11), 477, 1957.
283. **Jarvis, P. J.,** A Study in the Mechanics of Aeration at Weirs, Ph. D. thesis, University of Newcastle, Tyne, England, 1970.
284. **Holler, A. G.,** The Mechanism Describing Oxygen Transfer from the Atmosphere to Discharge Through Hydraulic Structures, Proc. XIV Congr. Int. Assoc. Hydraul. Res., Paper A45, 373, 1971.
285. **Department of the Environment,** Notes on water pollution, in *Aeration at Weirs,* Department of the Environmental Water Research Laboratory, Elder Way, Stevenage, Herts, England, 1973, 61.
286. **Nakasome, H.,** Derivation of aeration equation and its verification study on the aeration at falls and spillways, *Transactions,* J.S.I.D.R.E., 42, 1975.
287. **Avery, S. T. and Novak, P.,** Oxygen transfer at hydraulics structures, *J. Hydraul. Div.,* American Society of Civil Engineers, 104(HY11), 1521, 1976.
288. **Butts, T. A. and Evans, R. L.,** Small stream channel dam aeration characteristics, *J. Environ. Eng.,* American Society of Civil Engineers, 109(3), 555, 1983.
289. **Brutsaert, W. and Jirka, G. H., Eds.,** *Gas Transfer at Water Surfaces,* D. Reidel, Boston, 1984.
290. **Brutsaert, W. and Jirka, G. H.,** Measurement of wind effects on water-side controlled gas exchange in riverine systems, in *Gas Transfer at Water Surfaces,* Brutsaert, W. and Jirka, G. H., Eds., D. Reidel, Boston, 1984.
291. **American Public Health Association, et al.,** *Standard Methods for the Examination of Water and Wastewaters,* 14th ed., Washington, D.C., 1971.
292. **Elmore, H. L. and Hayes, T. W.,** Solubility of atmospheric oxygen in water, 29th report of the committee on Sanitary Engineering Research, *J. Sanit. Eng. Div.,* American Society of Civil Engineers, 86(SA4), 41, 1960.
293. **Benson, B. B. and Krause, D.,** The concentration and isotopic fractionation of gases dissolved in fresh water in equilibrium with the atmosphere. I. Oxygen, *Limnol. Oceanogr.,* 29(3), 620, 1984.
294. **American Public Health Association, et al.,** Standard Methods for the Examination of Water and Wastewaters, 16th ed., Washington, D.C., 1985.
295. **Baca, R. G. and Arnett, R. C.,** A Limnological Model for Eutrophic Lakes and Impoundments, Battelle Inc., Pacific Northwest Laboratories, Richland, Washington, 1976.
296. **Battelle, Inc.,** Explore-I: A River Basin Water Quality Model, Pacific Northwest Laboratories, Richland, Washington, 1973.
297. **Medina, M. A., Jr.,** Level II-Receiving Water Quality Modeling for Urban Stormwater Management, U.S. Environmental Protection Agency Municipal Environmental Research Laboratory, Cincinnati, OH, EPA-600/2-79-100, 1979.
298. **Di Toro, D. M. and Connolly, J. P.,** Mathematical Models of Water Quality in Large Lakes, Part 2: Lake Erie, U.S. Environmental Protection Agency, Duluth, MN, EPA-600/3-3-80-065, 1980.
299. **Johnson, A. E. and Duke, J. H., Jr.,** Computer Program Documentation for the Unsteady Flow and Water Quality Model — WRECEV, U.S. Environmental Protection Agency, Washington, D.C., 1976.
300. **Roesner, L. A., Monser, J. R., and Evenson, D. E.,** Computer Program Documentation for the Stream Quality Model, QUAL-II, U.S. Environmental Protection Agency, Athens, GA, 1977.
301. **U.S. Army Corps of Engineers,** CE-QUAL-R1: A Numerical One-Dimensional Model of Reservoir Water Quality, U.S. Army Corps of Engineers Instruction Report E-821, U.S. Army Corps of Engineer Waterways Experiment Station, Vicksburg, MS, 1982.

302. **Bauer, D. P., Jennings, M. E., and Miller, J. E., Jr.,** One-Dimensional Steady-State Stream Water-Quality Model, U.S. Geological Survey Water Resources Investigations Report 79-45, NSTL, MS, 1979.

303. **Imhoff, J. C., Kittle, J. L., Jr., Donigian, A. S., Jr., and Johanson, R. C.,** Users Manual for Hydrological Simulation Program-FORTRAN (HSPF), Hydrocomp, Inc., Palo Alto, CA, Report EPA-600/9-80-015, U.S. Environmental Protection Agency, Athens, GA, April 1981.

304. **Duke, J. H., Jr. and Masch, F. D.,** Computer Program Documentation for the Stream Quality Model, DOSAG-3, Water Resources Engineers, Austin, TX, for U.S. Environmental Protection Agency, Washington, D.C., Oct. 1973.

305. **Genet, L. A., Smith, D. J., and Sonnen, M. B.,** Computer Program Documentation for the Dynamic Estuary Model, prepared for U.S. Environmental Protection Agency, Systems Development Branch, Washington, D.C., 1974.

306. **Raytheon Oceanographic and Environmental Services,** New England River Basins Modeling Project — Final Report Vol. III — Documentation Report — Part I — RECEIV-II Water Quantity and Quality Model, for U.S. Environmental Protection Agency, Washington, D.C., December 1974.

307. **Weiss, R. F.,** The solubility of nitrogen, oxygen and argon in water and sea water, *Deep-Sea Res.,* 17, 721, 1970.

308. **American Public Health Association, et al.,** Standard Methods for the Examination of Water and Wastewaters, 15th ed., Washington, D.C., 1980.

309. **U.S. Geological Survey,** Quality of Water Branch Technical Memorandum Number 81.11, 1981.

310. **Mortimer, C. H.,** The Oxygen Content of Air-Saturated Fresh Waters Over Ranges of Temperature and Atmospheric Pressure of Limnological Interest, International Association of Theor. and Appl. Limnol. Communication Number 22, Stuttgart, Germany, 1981.

311. **American Public Health Association, et al.,** Standard Methods for the Examination of Water and Wastewaters, 1965.

312. **Aitken, A. C.,** *Mathematical and Physical Tables,* Clarke Irwin and Co. Ltd., Toronto, 1950.

313. **Whittemore, R. C.,** personal communication, Research Engineer, Northeast Regional Center, National Council of the Paper Industry for Air and Stream Improvement, Inc., Tufts University, Medford, MA, 1986.

314. **Wilhelms, S. C.,** Tracer Measurement of Reaeration: Application to Hydraulic Models, U.S. Army Engineer Waterways Experiment Station, Hydraulics Laboratory, Vicksburg, MS, 1980.

315. **Crawford, C. G.,** Determination of Reaeration-Rate Coefficients of the Wabash River, Indiana, by the Modified Tracer Technique, U.S. Geological Survey Water-Resources Investigations Report 85-4290, Indianapolis, IN, 1985.

316. **Hovis, J. S. and McKeown, J. J.,** Gas transfer rate coefficient measurement of wastewater aeration equipment by a stable isotope krypton/lithium technique, in *Gas Transfer at Water Surfaces,* Brutsaert, W. and Jirka, G. H. Eds., D. Reidel, Boston, 403, 1984.

317. **Rathbun, R. E.,** Estimation of the Gas and Dye Quanities for Modified Tracer Technique Measurements of the Stream Reaeration Coefficients, U.S. Geological Survey, NSTL Station, MS, 1979.

APPENDIX I

SYNOPSIS OF SEVERAL STREAM WATER QUALITY MODELS AVAILABLE FOR USE (FROM AMBROSE ET AL. [1982])

The two major components of stream models are the hydraulics and the water quality routines. There are three levels of sophistication for simulating stream hydraulics that describe most approaches. These include: (1) specifying empirical parameters or the hydraulic variables directly; (2) solving a partial set of the hydraulic equations, such as the kinematic wave approximation; and (3) solving the full hydraulic equations, such as the St. Venant equations. There are three basic levels of sophistication in treating water quality: (1) finding the steady-state solution to the equations; (2) solving the time-varying equations in a quasi-dynamic manner (step changes for inputs after long time periods); and (3) solving the time-varying equations in a fully dynamic manner. For each of these levels, there is a relatively continuous spectrum of sophistication in the number of water quality constituents and associated interactions.

To aid in the description of models, Ambrose et al. (1982) (see also Reference 40, Chapter 1) organized stream-reservoir models into four levels based on the following criteria:

1. Level 1 — Steady-state solution; simple kinetics
2. Level 2 — Steady hydrodynamics, specified or handled empirically; steady or time variable water quality; time resolution on the order of weeks to a month
3. Level 3 — Unsteady hydrodynamics, but simplified solution, such as kinematic routine; simplified reservoir solutions; dynamic water quality; time resolution of less than a day
4. Level 4 — Unsteady hydrodynamics with full equation routing; ability to handle backwater and stratified reservoirs; dynamic water quality; time resolution on the order of hours

The following list of models available was compiled in 1982 and has been only marginally updated. Since that time, a number of the models listed are now out-of-date. These are noted below. In addition, the main text covers the new models developed since 1982. (See Table 2, Chapter 1 for primary references to these models.)

AUTO-QUAL/AUTO-QD

The AUTO-QUAL and AUTO-QD models are Level 2 simple computer models for crude planning analysis of streams and estuaries. The steady-state and quasi-dynamic water quality solutions are based on directly specified hydraulics. Constituents can include CBOD, NBOD, DO, total phosphorus, and total nitrogen.

DOSAG-3

The DOSAG-3 model is a Level 2 simple computer model for crude planning analysis of streams and rivers. Its steady state water quality solutions are based on user specified hydraulics. Constituents include CBOD, dissolved oxygen, ammonia, nitrite, nitrate, phosphorus, algae, and fecal coliform bacteria.

EXPLORE-I

The EXPLORE-I model is a Level 4 advanced computer model for detailed management and design analysis of water quality for rivers and reservoirs. The fully dynamic water quality and hydrodynamic models are applicable not only to rivers, but also to thermally stratified reservoirs and well-mixed tidal rivers. The full hydrodynamic equations are solved, including pressure terms applicable to well-mixed estuaries, but the water quality model excludes dispersive transport. The EXPLORE-I model predicts up to 15 variables, including chemical and physical constituents (conservative substances such as dissolved solids, dissolved oxygen, toxic substances), nutrients (sedimentary phosphorus, soluble phosphorus, organic phosphorus, organic nitrogen, ammonia, nitrite, and nitrate), the carbon budget (total organic carbon, refractory organic carbon), biological constituents (phytoplankton, zooplankton), and biochemical constituents (CBOD, benthic BOD). The EXPLORE-I model has been linked with the sediment and contaminant transport model SERATRA but has not been used extensively. As a result its validity is not well established.

HYDROLOGIC SIMULATION PROGRAM — FORTRAN (HSPF)

The HSPF model is a Level 3 or 4 advanced computer model for detailed management analysis of watersheds that incorporates the effect of watershed processes on stream water quality. The model includes stream reaches, nonstratified reservoir reaches, and a choice of two overland runoff components equivalent to the NPS and ARM models. The stream component of the HSPF model is of intermediate complexity. The HSPF model predicts up to 22 variables, including chemical and physical constituents (conservative substances, dissolved oxygen, total dissolved solids, water temperature), nutrients (organic phosphorus, ortho-phosphate, organic nitrogen, ammonia, nitrite, nitrate), the carbon budget (total inorganic carbon, total organic carbon, carbon dioxide, alkalinity, pH), biological constituents (benthic algae, phytoplankton, chlorophyll *a*, zooplankton), biochemical constituents (BOD), and bacteria (total coliform bacteria, fecal coliform bacteria, fecal streptococci). The HSPF model is supported by the U.S. EPA Center for Exposure Assessment Modeling in Athens, Georgia, and the U.S. Geological Survey Office Surface Water Branch in Reston, Virginia.

MIT DYNAMIC NETWORK MODEL (MIT-DNM)

The MIT-DNM model can be classified a Level 4 advanced computer model for detailed management design of conditions affecting river and estuary water quality, or a Level 3 intermediate computer model for fine planning analysis of well-mixed estuaries. It uses a branching network to simulate tidal hydrodynamics that are coupled to the calculated salinity and temperature regimes through the equation of state. Dispersive transport is based on an internally-calculated two-part dispersion coefficient that accounts for both turbulent shear and density-induced vertical circulation. The solution is performed by a finite element, weighted residual technique giving high accuracy. Several versions of the MIT-DNM model are available, but none are well developed. Only the most experienced users have successfully applied the MIT-DNM model. Most users find that the model has not been adequately developed.

Two versions of the model have been applied with limited success. These include:

MIT-DNM, Potomac Version — The Potomac version includes seven forms of nitrogen linked by 12 bacterially mediated biochemical reactions: ammonia-N, nitrite-N, nitrate-N, phytoplankton-N, zooplankton-N, particulate organic-N, and dissolved organic-N. The model, then, is limited to the study of eutrophication in aerobic, nitrogen-limited environments.

MIT-DNM, St. Lawrence Version — The St. Lawrence version incorporates CBOD, DO, inorganic and organic phosphorus, inorganic and organic nitrogen, phytoplankton, and zooplankton.

QUAL-2e

The QUAL-2e model is a Level 2 simple computer model for crude planning analysis of stream water quality. The steady-state or quasi-dynamic water quality solutions are based on user-specified empirical hydraulic parameters. Constituents include CBOD, dissolved oxygen, organic nitrogen, ammonia, nitrite, nitrate, total phosphorus, orthophosphate, chlorophyll *a,* up to three conservative substances, and one first-order decay constituent that can be specified by the user. Flows and loads must be steady, but temperature, wind speed, and light may vary with time. Numerical solution is by an implicit finite-difference technique using a modified Gaussian elimination algorithm. The QUAL-2e model is the most widely used waste load allocation model available. The model is fully supported by the U.S. EPA Center for Exposure Assessment Modeling in Athens, GA. Its use is mandatory for waste load allocation studies in a number of states.

RECEIV-II

RECEIV-II is an out-of-date Level 3 intermediate computer model that was formerly intended for fine planning analysis of water quality in rivers, shallow reservoirs, and tidal rivers. It was operated in steady-state, quasi-dynamic, or dynamic modes for fine screening, crude planning, or detailed planning activities. Based on a branching, link-node network, its hydrodynamic component solved the one-dimensional equations of momentum and continuity with a second-order, predictor-corrector method. The water quality component excluded dispersive transport. The RECEIV-II model predicted CBOD, dissolved oxygen, ammonia, nitrite, nitrate, total nitrogen, total phosphorus, and chlorophyll *a,* as well as salinity and coliform bacteria. An explicit, first order numerical scheme was employed. This model is now primarily only of historical significance where knowledge of its structure is important to interpret past studies.

RIVSCI

RIVSCI is a Level 3 intermediate computer model for fine planning analysis of water quality in rivers, shallow reservoirs, and tidal rivers. A derivative of the RECEIV model, this model can be operated in a dynamic or a steady-state mode. Based on a branching, link-node network, its hydrodynamic component solves the one-dimensional momentum and continuity equations with a second-order, predictor-corrector method. The water quality component excludes dispersive transport. The RIVSCI model can simulate over 16 constituents, including dissolved oxygen, CBOD, ammonia, nitrite, nitrate, phosphate, phytoplankton, and other arbitrary substances that undergo first-order decay. This model is of unknown validity and has not been used extensively.

SIMPLIFIED STREAM MODELING (SSM)

SSM is a Level 2 simple manual methodology for crude screening analysis of streams, rivers, and well-mixed estuaries. It is composed of steady-state analytical solutions based on user-specified hydraulics. Constituents include CBOD, NBOD, dissolved oxygen, total nitrogen and phosphorus, and fecal coliform bacteria. Solution is by hand calculator.

WATER ANALYSIS SIMULATION PACKAGE (WASP)

WASP can be used as a Level 2 simple computer model for crude planning analysis of streams, lakes and reservoirs, and estuaries. It is considered a Level 2 model when the net flow, quasi-dynamic water quality solution is based on user-specified flows. However, more recently the WASP model has been linked to a one-dimensional link-node hydrodynamics model known as DYNHYD. DYNHYD employs an approximate solution of the dynamic flow equations and is fully adapted to representing a tidal boundary condition. When linked to DYNHYD, the WASP model can be considered a Level 3 or 4 model. The WASP model employs the compartment modeling approach whereby segments can be arranged in a one-, two-, or three-dimensional configuration. Water quality constituents are handled in a modular kinetic subroutine that can be written by the user, or chosen from a group from previous applications. Segment volumes are time variable. Other parameters and forcing functions may vary in space and time. A first-order Euler numerical scheme is employed. This model is fully supported by the U.S. EPA Center for Exposure Assessment Modeling and used extensively in large- and small-scale investigations such as that of Chesapeake Bay and dynamic waste load allocation studies. There are several kinetics routines that have been developed that are interchangeable. These include the EUTRO/WASP and TOXI/WASP subroutines that have been devised to include the most up-to-date eutrophication kinetics (linked with dissolved oxygen and BOD) and toxic chemical kinetics (from the EXAMS modeling framework for chemical fate). Rudimentary food chain algorithms are included as well.

WATER QUALITY ASSESSMENT METHODOLOGY (WQAM)

WQAM is a Level 1 simple manual methodology for crude screening analysis of urban and rural loading, streams, lakes and reservoirs, and well-mixed estuaries. The stream section is composed of a steady-state analytical solution based on a user-specified hydraulic condition. Constituents include CBOD, NBOD, dissolved oxygen, total nitrogen and phosphorus (considered conservative), fecal coliform bacteria, and sediment.

WATER QUALITY FOR RIVER-RESERVOIR SYSTEMS (WQRRS)

WQRRS is a Level 4 advanced computer model for detailed management design of rivers and reservoirs. The fully dynamic water quality and hydrodynamic models are based on a solution of the St. Venant Equations for flow. (Other routing approximations, such as the kinematic wave, are available.) This model includes internal linkage with a stratified reservoir model. The WQRRS model predicts up to 18 constituents, including chemical and physical constituents (dissolved oxygen, total dissolved solids), nutrients (phosphate-P, ammonia, nitrite, and nitrate), carbon budget (alkalinity, total carbon), biological constituents (two types of phytoplankton, benthic algae, zooplankton, benthic animals, three types of fish), organic constituents (detritus, organic sediment), and coliform bacteria. The U.S. Army Hydrologic Engineering Center supports this model, but it is extremely difficult to use. As a result, it has not been widely applied and is not recommended for use except in special circumstances.

WRECEV

WRECEV is an out-of-date Level 3 intermediate computer model for fine planning analysis of rivers, shallow reservoirs, and well-mixed estuaries. It was adapted from the RECEIV model by recasting the transport equations in their mass flux form and adding dispersion. Based on a branching, link-node network, its hydrodynamic component solved the one-dimensional momentum and continuity equations with a second-order predictor-corrector method. The water quality component simulated six constituents, including CBOD, NBOD, dissolved oxygen, and other substances undergoing first-order decay. This model has only historical significance in interpreting past studies in which it was used.

APPENDIX II

METHODS OF CONDUCTING AND EVALUATING MEASUREMENTS OF RIVER FLOWS, INFLOWS, AND TRIBUTARY FLOWS

The following methods of measuring flow in open channels, partially full pipes, and pipes flowing full under pressure are briefly reviewed in Table A1.

Methods
  Open Channel Flows:
    Velocity-area
    Moving boat
    Slope-area
    Tracers
    Width contraction
    Flumes:
      Parshall
      Trapezoidal supercritical flow
      Cutthroat
      H, HS, and HL types
      San Dimas
    Weirs:
      Thin plate:
        Rectangular
        V-Notch
        Trapezoidal
        Cipolletti
        Compound weir
        Other types
      Broad Crested:
        Rectangular notch
        Triangular notch
        Truncated triangular notch
        Parabolic notch
        Trapezoidal notch
        Columbus type control
    Submerged orifice
    Acoustic meter
    Electro-magnetic coils
    Rating navigation locks, dam crests, and gates
    Supper-elevation in bends
  Partially Full Pipe, Culvert, and Sewer Line Flow:
    Velocity-area
    Tracers
    Volumetric
    Slope-area
    Culvert equations
    Palmer-Bowlus flume
    USGS sewer flowmeter

Wenzel flume
Trajectory methods:
    Purdue
    California pipes
    Vertical pipes
Parabolic discharge nozzle
Kennison discharge nozzle
Acoustic meter
Electro-magnetic meter
Pressurized pipe flow:
Differential head meters:
    Venturi throat
    Nozzle
    Orifice
    Centrifugal meter
    Pipe friction meter
Velocity-area
Tracers
Gibson method
Mechanical meters:
    Displacement
    Inferential
    Variable area
Acoustic meter
Electro-magnetic meter
Calibrated pumps, turbines, valves & gates

Following Table A1, Tables A2, A3, and A4 review the characteristics of the most common flow measuring devices for the purpose of checking to be sure these devices are in proper working order during stream surveys or investigations of wastewater treatment plants. Table A2 reviews the operating characteristics of thin plate weirs. Table A3 reviews the characteristics of flumes while Table A4 reviews the characteristics of differential head meters.

TABLE A1

Methods of Measuring River Flows, Inflows, and Tributary Flows

Method[Ref.]	Principle of operation used to define flow rate, Q	Measurement uncertainty	Flow range or criteria (m³/s)	Min. depth or diameter (m)	Fig. no.
		Open-Channel Flows			
Velocity area[1,2]	See Table 8: $Q = AU$. Velocity measured with Price or pygmy meters, pitot tubes, acoustic meter, electromagnetic meters, propeller meters, deflection vanes, or floats. Float and vane measurements require correction to obtain the average velocity.	2.2% using Price meters. 10—25% using floats.	0.085 — 2.43 m/s using Price meters.	0.18	42
Moving boat[2]	Adaptation of the velocity-area method. Current meter moved across the stream at a fixed depth, and velocity is recorded for each subsection used to compute mean flow as $$Q = C_{mb}\Sigma(A_i V_v \sin\alpha)$$	Probably on the order of 3—7%.	Larger flows accessible by boat.	1.5—1.8	48,49
Slope-area[3,4]	Flow related to estimates or measurements of slope, S, or head loss for pipes, channel geometry, and some form of the friction coefficient such as the Manning n (friction factor for pipes): $$Q = (A/n)R^{2/3}S^{1/2}$$	10—20% for pipe lengths of 61—305 m. 20—50% for less accurate estimates of R, A, S, and n.	Used for high Q and preliminary estimates.	Depth and width should be as uniform as possible over channel reaches.	107
Tracers[5]	Salt, dyes, and radioisotopes are used to measure flow from the dilution of a tracer: $$Q = q_{it}(C_t - C_D)/(C_D - C_o)$$ where C_t, C_D, and C_o = tracer concentration being injected and measured upstream and downstream of the injection or from the	2—3% in well defined small channels in treatment plants.	Useful for shallow depth where current meters cannot be used. Not useful for large wide flows with limited lateral mixing.	Limited by lateral mixing.	53

TABLE A1 (continued)
Methods of Measuring River Flows, Inflows, and Tributary Flows

Method[Ref.]	Principle of operation used to define flow rate, Q	Measurement uncertainty	Flow range or criteria (m³/s)	Min. depth or diameter (m)	Fig. no.
	tracer travel time between two stations, Δt: $Q = (L/\Delta t)A$ where length between stations, L, and area, A, must be measured.				
Width contraction[6]	Flow related to head loss through a bridge opening or channel constriction: $$Q = C_c A_c [2g(\Delta h + \alpha(U_1^2/2g) + h_f)]^{1/2}$$ where α = velocity head coefficient, $\Delta h = h_1 - h_3$, and h_f = head loss through the constriction.	Order of 10% or more.	High Q or highly constricted with measurable $\Delta h \simeq$ 0.01—0.08.	Small to moderate where constriction is likely.	108
Flumes[2,4,7-9]	Usually forces the flow into critical or supercritical flow in the throat by width contractions, free fall, or steepening of the flume bottom so that discharge can be empirically determined as a function of depth and flume characteristics. The approach channel must be straight and free of waves, eddies, and surging to provide a uniform velocity distribution across the throat. Has less head loss and solids deposition but is more expensive to construct than weirs. Most common device used in sewage plants.	3—6% or better with calibration.	Higher Q than similar size weir.	—	109
Parshall[2,4,5,7,9-11]	Scaled from standard dimensions of 1 in. to 50 ft (0.025—15 m) throat widths. See Rantz et al.[2] for discharge tables for submerged (h_A/h_B = 0.6 to 0.7) critical depth conditions or French[7] and Leopold and Stevens[8] for empirical equations: $$Q = C_p W h_A^N$$	3—6% (7% when submerged).	0.0001—93.9; see range for each size.	0.03—0.09; see each size.	110, 111

Trapezoidal supercritical flow[2,4,12]	Supercritical depth is maintained in three scaled sizes (throat widths = 1, 3, and 8 ft or 0.3, 0.91, and 2.44 m) See rating charts in Rantz et al.[2]: $Q = (A_c)^{3/2}(T_c)^{1/2}g^{1/2}$	~5% (Simlar Venturi flumes are error prone; 1—10% if head loss is small).	0.02—56 (See range for each size. Accurate at lower flows. Up to 736 m³/s for 37 m wide flumes.)	0.03	112
Cutthroat[4]	More sensitive to lower flow; less sensitive to submergence (<80%); less siltation; fits existing channels and ditches. Flat bottom passes solids better than a Parshall flume and is better adapted for existing channels. Accurate at higher degrees of submergence. For free flow: $Q = KW^{1.025}h_A^{n'}$ (English units).	—	—	0.061	113
H, HS, and HL types[4]	Simply constructed and installed agricultural run-off flumes that are reasonably accurate for a wide range of flows. Flumes can be easily attached to the end of a pipe. Approach flow must be subcritical and uniform. The flume slope must be ≤1%. See Holtan et al.[14] for discharge tables.	1% for 30% submergence. 3% for 50% submergence.	Max. flows HS:0.0023—0.023; H:0.010—0.89; and HL:0.586—3.31 m³/s.	0.061	114
San Dimas[4]	Developed for sediment-laden flows, having a 3% bottom slope to create supercritical flow. Not rated for submerged flow. $Q = 6.35W^{1.04}h^{1.5-n'}$ (English units).	Not sensitive or accurate at low Q.	0.0042—5.7	0.061	115
Weirs[2,12]	Control of flow produces relationship between discharge and head. Inexpensive and accurate but requires at least 0.15 m (0.5 ft) head loss and must be maintained to clean weir plate and remove solids.	Accurate under proper conditions.	Most useful for smaller Q where 0.15 m (0.5 ft) head loss can be afforded.	0.061 + P	—

TABLE A1 (continued)
Methods of Measuring River Flows, Inflows, and Tributary Flows

Method[Ref.]	Principle of operation used to define flow rate, Q	Measurement uncertainty	Flow range or criteria (m³/s)	Min. depth or diameter (m)	Fig. no.
Thin plate[2,5,8,12]	Sharp crest exists when $h/L_c \geq 15$ and air is free to circulate under nappe (jet of water over weir). L_c = weir thickness.	2—3% when calibrated and maintained. 5—10% when silted in.	0.007—1.68. Best when 0.061 m ≤ h ≤ h/2b.	h ≥ 0.061 m for nappe to spring free of weir.	116
Rectangular[7-9,14]	$Q = (2/3)C_1(2g)^{1/2}bh^{1.5}$ where b = horizontal length of crest; neglects correction of −0.001—0.004 m.[7]	3—4%	Best when Q = 0.0081 to 23 m³/s.	0.03 m or 15 L_c and P ≥ 0.1 m.	116
V-Notch[4,5,8,11]	More sensitive to low flows: $Q = (8/15)C_2(2g)^{1/2}[\tan(\theta/2)]h_e^{5/2}$ Tailwater level lower than the vertex of the notch. Common angles are 22.5 45, 60, 90, and 120°.	3—6%	Q = 0.0003—0.39 but usually limited to 0.028.	0.049 ≤ h ≤ 0.061.	116
Trapezoidal[7]	$Q = C_3(2/3)(2g)^{1/2}[b + (4/5)h \tan(\theta/2)]h^{3/2}$	—	—	—	116
Cipolletti[2,4,7,8,12]	Trapezoidal weir having same equation for discharge as rectangular weir and is designed with side slopes of 1 horizontal to 4 vertical lengths to avoid the need for contraction adjustments to C_1.	Less accurate than rectangular or V-notch.	Q = 0.0085 to 272	h = 0.061—0.61—0.5b	116
Compound weir[12]	V-notch weir that changes side slope to a wider angle to handle higher flows at higher h. The Q vs. h relationship is ambiguous as the sides change slope. For a 1 ft (0.3 m) deep 90° cut into rectangular notches of widths 2, 4, and 6 ft. (0.61, 1.2, and 1.8 m):	Has not been fully investigated.	—	—	116

Type	Equation	Remarks	Limits		Ref.
Other types[10]	$Q = 3.9h_m^{1.72} - 1.5 + 3.3bh_m^{1.5}$	Sutro or proportional and approximate linear weirs have an approximately linear relationship between Q and h. Other types include approximate exponential and poebing weirs.	Has not been fully investigated.	—	116
Broad crested[2.7.15]	$Q = C_B b(h + h_v)^{3/2}$; $h_t = h + h_v$ (Weir is broad enough to consider to the fluid above at hydrostatic pressure. Notches for low flow modify equation for Q. Typical streamwise cross-sections are rectangular, triangular, trapezoidal, and rounded.) $C_B = C_b C_v K_n g^{1/2}$ where K_n is a constant depending on notch shape.)	More uncertainty for lower h. Difficult to predict without in place calibration.	$0.08 \le h/L \le 0.50$ (No effect of submergence until $h_t/h > 0.65$ to 0.85 depending on weir profile.)	0.061	117
Rectangular notch[2]	Flat crest: $Q = C_b C_v (2/3)(2g/3)^{1/2} bh^{2/3}$ Trenton type with 1:1 slopes upstream and downstream of the flat crest: $Q = 3.5bh^{1.65}$ (English units). The Crump triangular weir with a 50% slope on the upstream face and 20% slope on the downstream face: $Q = 1.96bh_t^{3/2}$	—	$0.08 < h/L \le 0.33$. 0.33 $\le h/L \le 1.5$ to 1.8 is a short-crested weir that has not been fully investigated. $1.5 \le h/L \le 3$ is usually unstable. $h/L \ge 3$ is similar to sharp crested weirs.	0.061	117
Triangular notch[2.7.14]	$Q = C_b C_v (16/25)(2g/5)^{1/2} [\tan(\theta/2)]h^{5/2}$ where $Q = 2.5bh^{1.65}$ for flat vee.	—	Usually ≥ 28 (1000 cfs)	0.061	117

TABLE A1 (continued)
Methods of Measuring River Flows, Inflows, and Tributary Flows

Method[Ref.]	Principle of operation used to define flow rate, Q	Measurement uncertainty	Flow range or criteria (m³/s)	Min. depth or diameter (m)	Fig. no.
Truncated triangular notch[7]	$Q = C_b\, C_v\, T\, (2/3)\, (2g/3)^{1/2}\, (h - h_b)^{3/2}$	—	$h_t \geq 1.25 h_b$	—	117
Parabolic notch[7]	$Q = C_b\, C_v\, (3fg/4)^{1/2} h_t^2$	—	—	—	117
Trapezoidal notch[7]	$Q = C_b\, (Th_c + mh_c^2)\, [2g(h_t - h_c)]^{1/2}$ where T = top width of flow over weir	—	—	—	117
Columbus type control[2]	Most frequently used gauging control in the US. Requires calibration for the full range of flows. Above h = 0.21 m: $Q = 8.5\ (h - 0.2)^{3.3}$ (English units). Consists of an upward convex notch below h = 0.21 m and slopes 1:5 for 0.611 m on either side of the notch and 1:10 to the remaining distance to the banks.	—	—	—	117
Submerged orifice[12]	Flow is related to the head difference across the orifice: $Q = 0.61\ (1 + 0.15 r_o) A_d [2g(\Delta h)]^{1/2}$ for a contracted or suppressed rectangular orifice.	Properly operated meter gates: 2%, but as much as 18% noted.	0.003—1.43	0.38	118
Acoustic meter[12,16-19]	The difference in travel time, of an acoustic pulse across the flow and back (t_{AB} and t_{BA}) over a path length, L_A, diagonal to the flow is related to average velocity along the path: $Q = AC_A L_A\, (1/t_{AB} - 1/t_{BA})/(2\cos\alpha)$	1—7% for parallel flow. ≤14% in poorly developed flow.	Used in large rivers. No practical upper limit for Q.	Depends on W, density gradients, and allowable error.	50

Method	Equation/Description	Accuracy	Comments	Constraint	Ref.
Electromagnetic coils[2]	where α = angle of path with flow. Experimental method involving coils buried in the bed or suspended in the flow. See pipe method. $Q = C_e AE_m/WH$	—	Generally used in shallow flows.	No practical limits	119
Rating navigation locks, dam crests, and gates[2,14]	Q determined by calibration in model studies or by in place measurement. Generally locks behave like submerged orifices: $Q = C\Delta h$ and dams act like weirs: $Q = C_b \, bh_t^{3/2}$	5—30%	—	—	—
Superelevation in bends[2]	The difference in water surface elevation across a bend is related to Q.	—	Requires high vel. for measurable Δh.	No constraint	120

Partially Full Pipe, Culvert, and Sewer Line Flow

Method	Equation/Description	Accuracy	Comments	Constraint	Ref.
Velocity-area[20]	See pipe and open-channel methods: $Q = UA$. U related to point velocities in at least 4 ways: (1) $U = 0.9u_{max}$ (2) $U = u_{0.4}$ (3) $U = (u_{0.2} + u_{0.4} + u_{0.8})/3$ (4) U = average of 3 vertical profiles at the quarter points across the pipe plus 2 measurements at both walls 1/8 the distance across the flow.	All methods are expected to be accurate to at least 10%.	Should be applicable to all flows. Use method (1) or (2) if the flow is rapidly varying.	Use method (1) if depth ≤ 5 cm (2 in.).	121
Tracers[21]	See open-channel method.	5%	Requires good mixing. Useful for smaller flows, but unique conditions may make application to any flow possible	—	53
Volumetric[10]	Flow from a pipe or channel is diverted to a bucket, tank, sump, or pond of known dimensions and the increase in weight or volume is timed to measure the average flow rate. Orifice buckets may also be useful.[10,22]	1% or better depending on how well dimensions are known.	No practical minimum for flows encountered.		122

TABLE A1 (continued)
Methods of Measuring River Flows, Inflows, and Tributary Flows

Method[Ref.]	Principle of operation used to define flow rate, Q	Measurement uncertainty	Flow range or criteria (m³/s)	Min. depth or diameter (m)	Fig. no.
Slope-area[4,23]	The Manning equation is applied to uniform pipe reaches. See channel method and culvert method. The effects of manholes (changing channel shape, slope, or direction) must be avoided or the increased friction losses taken into account. Design slopes should be verifed.	10—20% or 20—50% for less precise estimates of S and h.	Requires a 200—1000 d uniform approach flow.	—	107
Culvert equations[23]	Q computed from continuity and energy equations.	2—8%	Not readily available	0.15	123
Palmer-Bowlus flume[4,13,21,24,25]	Constriction inserted into existing partially full pipes where relationship exists between Q and h: $$Q = h^{3/2} \left[\frac{w(2L + h)}{8(L_w + h)} \right]^3 g^{1/2}$$ Does not have fully standardized design; more accurate but less resolution than Parshall flume. Fits pipe d = 0.10 to 1.1 m and larger. Portable. For full pipe flow: $$Q = 8.335(\Delta h/d)^{0.512} d^{5/2}$$ (English units). See Kilpatrick et al.[21] for exact flume dimensions.	3—5% depending on care in construction. 10% for low Q.	0.0010 to 0.51 for d = 0.15 to 0.76 m. Submergence ≤85%. Requires uniform approach for 25 times diameter with slope ≤2% and 0.4 ≤ depth ≤ 0.9 d.	0.61 h_{min} = 0.061.	124
USGS sewer flowmeter[2]	A U-shaped fiberglass constriction of standard dimensions is inserted in a pipe to form a Venturi flume where Q = f(Δh, h). Full: $$Q = 5.74 d^{5/2} (\Delta h/d)^{0.52}$$	Depends on field calibration.	Not available	0.61 and 1.52	125

Transition between channel and pipe flow:

$$Q = d^{5/2} [2.6 \pm (|0.590 - h_2/d|/0.164)^{1/2}]$$

Channel flow:
(1) Supercritical:

$$Q = 5.58d^{5/2} (h_1/d)^{1.58}$$

(2) Subcritical — culvert slope <0.020:

(a) $h_1/d \geq 0.30$: $Q = 2.85d^{5/2} (h_1/d - 0.191)^{1.76}$

(b) $h_1/d < 0.30$: $Q = 1.15d^{5/2} (h_1/d - 0.177)^{1.38}$

(3) Subcritical — culvert slope ≥ 0.020:

$Q = 1.07ad^{5/2} (h_1/d)^{2.71}$ $a = 2.15 + (9.49)(10)^{11} (S - 0.008)^{6.76}$

Wenzel flume[26]

(all equations in English units)
Symmetrical or asymmetrical constrictions contoured to the side of the pipe result in unique Q vs. h for partially full and full flows. Open bottom does not trap solids.

$$Q = C_w\{2gA_2\Delta h/[1 - (A_2/A_1)^2]\}^{1/2}$$

$$C_w = \{1 + K_e + [A_2^2A_i^2/(A_i^2 - A_2^2)]$$

$$[(f_1L_1/4R_1A_i^2 + f_3L_3/4R_3A_3^2) \, 3/2]\}^{-1/2}$$

Trajectory methods:[10] The extent of a jet leaving the end of a pipe, and pipe flow geometry are related to discharge.

At least 5%. 25% at low flow.	As much as 30:1variation in flows. Not valid for steep S ≥ 0.020	0.20 (8 in.)	126
	Generally not accurate enough for US EPA NPDES inspections.	—	127

TABLE A1 (continued)
Methods of Measuring River Flows, Inflows, and Tributary Flows

Method[Ref.]	Principle of operation used to define flow rate, Q	Measurement uncertainty	Flow range or criteria (m³/s)	Min. depth or diameter (m)	Fig. no.
Purdue[12]	For level pipe flowing full or partially full where the water surface elevation below the top of the pipe at the outlet (if a/d < 0.8), or at 6, 12, or 18 in (0.15, 0.30 or 0.46 m), the distance from the top of the pipe down to the water surface has been experimentally related to flow. See Bureau of Rec.[12] for tabulation of values.	Q will be underestimated if the pipe slopes downward.	Q = 0.00032 to 0.10	d = 0.051 to 0.15	127
California pipe[12,15]	For a level pipe of length 6d or more, flowing only partially full, discharging freely into air, and having a negligible approach, U: $Q = 8.69(1 - a/d)^{1.88} d^{2.48}$ (a and d measured in ft).	—	Confirmed for d = 0.076 to 0.25 m and a/d ≥ 0.5, but probably useful for larger d.	0.076	127
Vertical pipes[12]	In gal/min, with d and height of jet, H', in in. and H' > 1.4d: $Q = 5.01d^{1.99}(H')^{0.53}$ For H' < 0.37d: $Q = 6.17d^{1.25} (H')^{1.35}$ For $0.37 \leq H'/d \leq 1.4$ see tabulation in Bureau Rec.[12]	—	d = 0.051 to 0.30 and H' = 0.013 to 1.52.	0.051	127
Parabolic discharge nozzle[4,10,11,13]	Attached to the end of a pipe with a free outfall. Q is related to h² by laboratory calibration. Nozzle length is ≈ 4d.	5%, 1% at Q_{max}.	0.0000—0.85	0.15, 0.20, 0.25, 0.3, 0.41, 0.51, 0.61, 0.76, 0.91	128

Kennison discharge nozzle[4,10,11,13]	Attached to the end of a pipe with a free outfall. Q is related to h by laboratory calibration. Nozzle length is ≈ 2d.	5%, 1% at Q_{max}.	≤0.85	Same as parabolic nozzle.	128
Acoustic meter	See pipe method.				
Electromagnetic meter[20,27]	See open-channel and pipe methods. Generally used to measure one or more point velocities for the velocity-area method.	2% under optimine conditions.	U = −1.67 to 6.1 m/s (−5.5—20 fps)	Insertable in most pipes.	129

Pressurized Pipe Flow

Differential head meters[2,5,13]	Flow constrictions produce pressure losses related to flow, Q.	Better results when calibrated.	Min. Δh = 25.4 mm (1 in) for water and 51 mm (2in.) for sewage.	—	130
Venturi throat[5,11,13]	$Q = \dfrac{C_d A_d (2g\Delta h)^{1/2}}{(1 - r^4)^{1/2}}$	0.5—3% depending on calibration.	Well tested for diameters up to 0.81 (32 in.)	0.051	130
Nozzle[11,13]	Same as venturi throat	1—1.5%	—	0.051	130
Orifice[10,13]	Same as venturi throat.	0.5—4.4%	5:1 range for Q unless calibrated.	0.038	130
Centrifugal meter[2,13]	Flow rate related to pressure difference between the inside and outside of a pipe bend: $Q = C_d A_d (2g\Delta h)^{1/2}$ For uncalibrated 90° bends with moderate or higher Reynolds no.	<10% if calibrated. Predicted C_d error ≈10%.	May be installed in any pipe for which Δh in measurable (1 in H_2O).	0.038	120
Pipe friction meter[2,13]	$C_d = [r_f/(2d)]^{1/2}$ Flow rate related to friction loss in fully developed pipe flow: $Q = C_d(\Delta h)^{1/2}$ where $C_d = A_d^{1/2}/fL$ and f is the friction factor. Also see type 6 culvert flow in Figure 123.	Unknown	Requires high flow rates to yield meassurable Δh.	Any pipe.	107

TABLE A1 (continued)
Methods of Measuring River Flows, Inflows, and Tributary Flows

Method[Ref.]	Principle of operation used to define flow rate, Q	Measurement uncertainty	Flow range or criteria (m³/s)	Min. depth or diameter (m)	Fig. no.
Velocity area[2,12,13,28]	$Q = A_d U$, A_d determined from inside diameter; U determined from point velocity measurements by pitot tubes, small propeller meters, electromagnetic probes, etc. Velocities best measured at 0.026, 0.082, 0.146, 0.226, 0.342, 0.6658, 0.774, 0.854, 0.918, and 0.974d and then averaged for round pipes. Relationships between U and a single point velocity can be derived. See channel method as well.	0.5% for pitot tube meas. 5% for some propeller meters.	$d \leqslant 1.5$m and $U_{max} = 1.5$ to 6 m/s.	Depends on pitot tube or velocimeter diameter. $\simeq 0.10$ is the limit for a 3.8 cm diameter pitot tube to make precise measurements.	131
Tracers[2,10]	Salt or fluorescent dyes are used to determine flow rate from the dilution of a known amount of tracer mass at a downstream location where the tracer is fully mixed or from the time of travel between two locations at which the tracer is well mixed. The latter method requires that the distance between station and flow cross-sectional area be measured. See channel method.	3%	Requires good mixing.	Should be applicable to any size pipe.	53
Gibson[2]	Pressure rise following valve closure is related to U: $Q = (Ag/L)(\text{area ABCA})$. See Figure 132 for area ABCA under the pressure-response curve.	Believed to be very accurate.	Requires at least 25d approach length.	Unknown	132
Mechanical meters[2] Displacement	Flow displaces piston or disk. Oscillations are counted as a function of time. Calibration required.	1% new, more with age.	0.001—0.28 Depends on manufacturer's specifications.	Varies Generally used for smaller pipes.	— 133

Type	Description	Accuracy	Range	Size	Ref.
Inferential[2,12]	Flow rotates a turbine whose rate of rotation is related to flow rate by calibration. Requires an approach length of at least 20—30d to develop the velocity profile. Vanes control spiraling from bends.	Needs constant checks: 2—5%	0.15—5.2 m/s (0.5 to 17 ft/sec) but inaccurate below 0.3—0.46 m/s (1—1.5 ft/s).	0.61—2.3 (2—7.5 ft)	—
Variable area[2,12]	Rotometer consists of vertical tapered tube with a metal float that rises as flow increases.	1% at Q_{max}	Small flows.	Depends on manufacturer.	134
Acoustic meter[12,16-18,21]	See channel-flow method for time of travel: $Q = AC_A L_A (1/t_{AB} - 1/t_{BA})/2\cos \alpha$ Reflected doppler meters detect a frequency shift from a signal reflected back from a point where sound beams are crossed in the flow. The shift is related to the point velocity which must be related to the average velocity by calibration to calculate the flow rate. Time of flight acoustic meters determine a turbulent signature of the flow at a station and correlates that with a downstream turbulent signature to determine time of travel over a known distance through a known pipe volume: $Q = \pi d^2 L/4\Delta t$	2—5% and higher at bends.	Time of travel and doppler meters require approach lengths of at least 10—20d of straight pipe or must be calibrated to determine C_A. Time of flight meters require a limited approach of a few diameters.	No practical limit.	135
Electromagnetic meter[2,29-32]	All types can be attached to outside of pipes. Induced voltage, E_m, caused by water flow perpendicular to magnetic field, proportional to velocity, U: $Q = AE_m/HdC_e$ Meter must be in contact with fluid. Regular checking of the electrodes is necessary to avoid fouling. Magnetic coils can be embedded in new pipe or in a ring insert, or a probe can be inserted to measure point velocities or profiles to be used in the velocity-area method.	Larger of 1—2% or 0.002 m/s.	Depends on manufacturer. At least up to 16.7 (590 ft³/sec).	At least 0.10—1.52 m diameter (4—60 in).	136

TABLE A1 (continued)
Methods of Measuring River Flows, Inflows, and Tributary Flows

Method[Ref.]	Principle of operation used to define flow rate, Q	Measurement uncertainty	Flow range or criteria (m³/s)	Min. depth or diameter (m)	Fig. no.
Calibrated pumps, turbines, valves, and gates[2]	Empirical relationships between Q and power, or valve or gate opening for appropriate heads are defined with laboratory models or by calibration in place.	Depends on inplace calibration. Laboratory calibration ≈10% or more for accurate in place calibration.	—	—	—

DEFINITION OF SYMBOLS USED IN THE TABLE

A = Cross-sectional area of channels and pipes. Subscript i refers to the area of individual subsections. Subscript c refers to area of a width constriction or area of a throat of a flume.

U = Cross-sectional average velocity.

C_{mb} = Coefficient relating point velocity measurements to the mean velocity in a cross-section. For large rivers $C_{mb} = 0.87$ to 0.92 and 0.9 is typically used. See French[7] for representative channel and conduit values.

n = Manning's roughness coefficient describing friction loss in channels. See French[7] for values.

R = Hydraulic radius (A/wetted perimeter). Subscripts refer to section numbers such as 1 and 3 for the Wenzel flume.

C_c = Discharge coefficient for a width constriction. See Matthai[6] for guidance on selection based on geometric characteristics of the contraction.

g = Gravitational acceleration.

C_p = Discharge coefficient for Parshall flumes. See French[7] and Leopold and Stevens[8] for standard values.

N = Exponent in the Parshall flume equation defined in French[7] and Leopold and Stevens.[8]

T_c = Top width of the water surface in the trapezoidal flume at the point where critical flow occurs.

K = Free flow discharge coefficient for cutthroat flume. See Grant[4] for values for the standard 1.5-, 3-, 4.5-, and 9-ft (0.46-, 0.91-, 1.37-, and 2.74-m)-long flumes varying from 6.1 to 3.5 (Engligh units).

n' = Free flow discharge exponent for a cutthroat flume. See values in Grant[4] for the standard 1.5-, 3-, 4.5-, and 9-ft (0.46-, 0.91-, 1.37-, and 2.74-m)-long flumes varying from 2.15 to 1.56 (English units).

n'' = Discharge exponent for the San Dimas flume = $0.179 W^{0.32}$ (English units). See Grant.[4]

h = Water surface height of the approaching flow above the weir crest or bottom of the notch. For precise estimates French[7] notes that 0.001 m is added to h to account for the effects of surface tension and viscosity. h is measured at least 3 to 4 times h_{max} upstream of the weir face for sharp-crested rectangular and V-notch weirs; 2 to 3 times h for Cipolletti weirs; and 2 to 3 times h_t for broad-crested weirs. Generally the minimum distance upstream should be $4h_{max}$.[8]

P = Height of lip of weir above stream bottom.

C_1 = Discharge coefficient for rectangular thin-plate weir; f(h/P, b/B, E) for free falling, free discharge that springs free of the plate without clinging (thickness of the weir is ≤1/15 of the depth of flow over the weir). French[7] gives equations of C_1 = f(h/P, b/B). Rantz et al.[2] give graphs of C_1 = f(E) and 2/ $3C_1(2g)^{1/2}$ vs. h/P and b/B. See Rantz et al.[2] and Bureau of Reclamation[12] for approximate corrections for submergence.

C_s = $[1 - (h/D)^m]^{0.385}$ where m = 1.44 for a fully contracted weir and 1.50 for a suppressed weir.

C_2 = Discharge coefficient for a V-notch weir ≃0.58 for a fully contracted weir flow where h/P ≤ 0.4, h/B ≤ 0.2, 0.049 < h ≤ 0.381 m, B ≥ 0.91 m. Values cannot be predicted for partially contracted weir flow without calibration.

h_e = h + K_h. K_h varies from 0.0025 to 0.0085 depending on the angle of the notch of the weir. See French[7] for appropriate values when necessary for precise predictions.

C_3 = $0.63C_v$. C_v = velocity head correction factor. See French[7] for values ranging from 1.00 to 1.20.

C_b = Discharge coefficient for broad-crested weir. For flat-crested weirs, C_b depends on the rounding or slope of the upstream face, h/L, and $h_t/(h_t + P)$. For a horizontal upstream face with a sharp corner, the basic discharge coefficient is C_b = 0.848 if 0.08 ≤ h/L ≤ 0.33 and h/(h + P) ≤ 0.35.[7] For h/ L > 0.33 and h/(h + P) > 0.35, French[7] gives correction factors ranging from 1.00 to 1.15 for h/L ≤ 0.85 and h/(h + P) ≤ 0.60. The value of C_b is corrected for h/L ≤ 1.5 if h/(h + P) ≤ 0.35. A final correction is made for the approach velocity head varying from 1.00 to 1.2 as a function of $C_b A_b$/ A_a where A_b is the area of flow over the control structure and A_a is the area of flow in the approach section where h is measured. For rounding of the upstream corner: C_b = $[1 - 2x(L - r_r)/B][1 - x(L - r_r)/h]^{3/2}$ where x is a parameter that accounts for boundary-layer effects and r_r is the radius of rounding of the upstream corner. For field installation of well-finished concrete, x = 0.005. Where clean water flows over precise cut blocks, x = 0.003.[7] See Rantz et al.[2] for flat-crested rectangular weir coefficient charts for different h/L and slopes of the upstream face of the weir. King and Brater[15] give values of $C_b C_v (2.3)(2g/3)^{1/2}$ in English units for rectangular blocks, rounded blocks, weirs with a slight slope on the crest, triangular blocks, trapezoidal blocks, and weir blocks of irregular shape. C_b varies between 0.85 and 1.00 for triangular weirs having h/L = 0.08 to 0.7.[7]

h_v = $\alpha U^2/2g$ for the approach flow. See King and Brater[15] for additional information about values for a variety of different crest profiles.

C_v = Discharge coefficient correcting for the effect of an approach velocity for weirs and orifices.[7]

r_o = Ratio of suppressed portion of the perimeter of a submerged orifice to the entire perimeter.

C_A = Discharge coefficient for acoustic gauging stations relating cross-sectional average velocity to average velocity along the acoustic path. C_A = f (water depth and tidal condition). See Laenen and Smith[19] and Rantz et al.[2] for estimation methods.

C_e = Discharge coefficient for electromagnetic meters determined in the laboratory for portable meters attached to the outside of a pipe. For open-channel coils, C_e = $[1 + (W\sigma_b/2h\sigma_w)]/\beta$ where σ_b and σ_w are conductivities of the bed and river water, respectively, and β is a correction factor for the end effects of the magnetic field and for incomplete coverage of the cross-sectional area by the field in case of limited coils.[2]

H = Magnetic field intensity in Tesla.

E_m = Electromotive force in volts generated by the movement of a conduction fluid such as slightly contaminated water through a magnetic field.

A_d = Area of venturi throat, orifice, or flow nozzle opening or pipe diameter for bend meters.

d = Pipe diameter.

Δh = Pressure drop or hydraulic head difference measured across a venturi throat or flume, orifice, flow nozzle, bend meter, or along a straight length of pipe. Also the difference in water surface elevations across a submerged orifice or width contraction. See Figures 124, 125, and 126 for the locations across which head difference are measured for the Palmer-Bowlus (other criteria are used as well), USGS sewer flowmeter, and Wenzel flumes.

h = Water depth measurement in partially full pipes where subscripts 1 and 2 refer to locations upstream of and locations in the throat of a flume.

S = Slope of a pipe or channel.

C_w = Discharge coefficient for the Wenzel flume that accounts for energy loss and velocity head corrections. For full pipe flow, C_w is given above. K_e in an entrance loss coefficient of approximately 0.2.[26] f is the Darcy-Weisbach friction factor.[33] L_1 is the length defined in Figure 126. Valid when backwater conditions do not exist. For partially full flow, C_w seems to be ≃1.[26]

TABLE A1 (continued)

Methods of Measuring River Flows, Inflows, and Tributary Flows

C_d = Discharge coefficient for venturi throat, orifice, and flow nozzle or bend meter. Function of Reynolds number, Re, for venturi throat[2,13] when pressure taps are properly located in the straight pipe section of the throat and 0.5 to 0.25 pipe diameters upstream of the meter. C_d = 0.96 to 0.99 as a function of Re ≤ 10[6] for flow nozzle when pressure taps are located one pipe diameter upstream of the beginning of the nozzle and before the end of the nozzle. See American Society of Mechanical Engineers[13] for graphs of C_d vs. Re and pipe radius for pressure taps 1 diameter upstream and at the beginning of the nozzle. For orifice meters with various standard pressure tap locations, see American Society of Mechanical Engineers[13] standard tables. For precise measurements, C_d is determined by calibration for bend meters and for straight pipe sections between pressure taps. The best results are obtained in prediction C_d when the flow meter is well maintained and correctly installed. Indirect flow measurements are used to check the meter rating and to determine if debris lodges in the meter. Proper installation places the meter downstream of at least 10 diameters of straightened flow and avoids solids deposition. Vanes are used to straighten flows if necessary.

r = Ratio of throat diameter, d_2, to pipe diameter, d.

r_b = Radius of the center line of a pipe bend.

f = Friction factor for pipes. See Streeter and Wylie[33] for Moody diagram relating f and the Reynolds number for given pipe roughness also given for standard materials.

L = Length of straight pipe between pressure taps for a pipe friction meter.

TABLE A2
Characteristics of a Properly Installed and Maintained Thin Plate Weir[a]

No.	Characteristic

1. Upstream face of the bulkhead across the flow should be smooth and straight and vertically perpendicular to the flow axis. Repair is required if leaks around or under the bulkhead are evident.

2. The weir plate should be flush with the bulkhead, clean, smooth, and straight. The opening should be centered in the flow. Debris and weeds on the crest and damage to the weir edge changes the discharge relationship with head. Algae and aquatic growth on the upstream face also changes the rating.

3. The crest should be level and the upstream edge should form a sharp 90° angle with the upstream face of the weir plate. The sides of the rectangular, Cipolletti, and V-notch weirs should be equally smooth and sharp and should be set on the correct vertical angle (vertical for the rectangular weir, ratio of 4 vertical to 1 horizontal for the Cipolletti weir, and as specified for the V-notch weir).

4. The crest should have a thickness of 1 to 2 mm. Where the weir plate is thicker, the back side should be beveled at 45° or greater. Knife edges are to be avoided because of the likelihood of damage. The crest may be thicker, but the weir will not properly measure lower flow rates when the nappe does not spring free of the downstream face of the weir plate. The crest should be machined or filed if nicks and dents are obvious but should not be marred with scratches or groves. Under no circumstance should the edges be rounded. Only the protrusions above the surface of the crest should be filed, and the opening of the weir should not be enlarged to remove imperfections.

5. The distance from the crest to the bottom of the approach channel and from the sides of the weir opening to the sides of the approach channel should not be less than twice the maximum possible head over the crest or 0.3 m. The cross-sectional area of the approach channel should be at least 8 times greater that the area of the weir opening, and the approach channel should extend uniformly upstream for a distance of about 15 to 20 times the maximum water depth over the crest unless a weir box is provided where the flow wells up into the box and does not have a significant approach velocity. In the event that this is not possible, the nappe will probably only be partially contracted and the approach velocity head ($U^2/2g$) will not be negligible. This will require special care in calibration and relating discharge to head. Any inspection should confirm that sediment and floating objects have not changed the depth or flow area of the approach channel.

6. The nappe or sheet of water flowing over the weir should touch only the upstream face and top edge of the crest. Air must circulate freely under and along the sides of the nappe to maintain the proper head-discharge relationship.

7. The stage measurement to determine the height of water over the crest should be made at a distance of about 4 times but no more than 10 times the maximum depth of water over the crest to avoid the effect of the draw down of the water surface caused by water flowing over the weir. Inspection should confirm that the 0 datum of the gauge corresponds to the level of the crest. Stilling wells and connections to the approach flow should be clear. Levels in the stilling well can be compared to water levels in the approach flow at the location of the connection of the stilling well to the channel with a carpenter's or surveyor's level. A quick check for blockage of the stilling well connection and responsiveness involves filling the well with clean water and noting how quickly the level returns to the original level. Alternatively, bubbler gauges or pressure transducers must also be examined according to manufacturer's instructions.

8. If employed to measure water levels in the approach channel, the float in a stilling well should be just smaller than the well to avoid lateral drift errors. The total volume of the stilling well should be small compared to the volume in the flume if flow changes rapidly. Response of the recording system can be checked when the flow is steady by pushing the float below the surface and then releasing it to determine if it returns to the original level.

9. Calibration curves can be easily checked with the volumetric, velocity-area, and tracer dilution methods. Calibration in place may not be necessary for standard weirs like the fully contacted V-notch weir, rectangular weirs, and Cipolletti weirs. Less frequently employed weirs should be calibrated in place to achieve comparable accuracy in flow measurement.

[a] Adapted from References 5, 10, and 12.

TABLE A3
Characteristics of a Properly Installed and Maintained Flume[a]

No.	Characteristic

1. The flume should be located in a straight section that does not have bends immediately upstream so that the flow is well distributed across the stream and is free of excessive turbulence, surges, and waves. When possible, high approach velocities should be avoided. The flume generally should not involve width contractions in excess of 1/2 to 1/3 of the original channel width.
2. For the most accurate measurements, submerged conditions should be avoided. Each flume type has varied submergence criteria. Some flumes like the Parshall, trapezoidal, and cutthroat flume are well adapted to submerged conditions up to a maximum limit (i.e., 95% for the larger Parshall flumes).
3. The flume should be checked for sedimentation and cleaned if necessary. Floating objects disturbing the flow must also be removed.
4. Stilling wells and connections to the flume must be free of deposits and objects that interfere with the float or other head measuring devices. The connections to the flume must be properly located. See the standard specifications for each type of flume and confirm the correctness of the location of stilling wells. A carpenter's or surveyor's level can be used to confirm that the water level in the stilling well and flume at the measurement point are equal. Filling with clean water and observing the response of the float serves as quick method of checking for blockage of the well connection.
5. The dimensions of the approach channel, throat, and getaway (diverging downstream section that serves as a transition back to normal width of the channel) sections should meet standard specifications to within tolerances of 0.0095 m (1/32 in). The throat dimensions should meet tolerances of 0.0048 m (1/64 in). It is especially important that smaller flumes be precisely constructed. Precise leveling is very important and should be checked during any inspection.
6. If the calibration needs to be checked then the velocity-area, volumetric, or tracer dilution methods are the best methods that can be employed in most cases. A check of the installation records should show that the flume has been calibrated in place to obtain the most accurate performance specified in Table A1.

[a] Adapted from References 4, 5, and 12.

TABLE A4
Characteristics of a Properly Installed and Maintained Differential Head Meter (Venturi Throat, Orifice, Nozzle, Pipe Bend Meter, and Pipe Friction Meter)[a]

No.	Characteristic

1. Check for any manufacturer's specifications and ensure that these have been followed. This should include specification of the location of pressure taps and other important components. See Table A1 under the definition of C_d for pressure tap locations. It should be possible to determine from observation if the device is of a standard type subject to normally expected accuracy by measuring outside dimensions and reviewing installation records. Otherwise, calibration measurements to observe respones will be necessary.
2. The device may be oriented horizontally, vertically, or inclined, but must be flowing full. Installation of a pipe bend meter at a different orientation than its calibrated orientation is cause for recalibration, however.
3. The flow into the measurement device should have the opportunity to be fully developed. This requires straight approaches of length 5 to 20 times the diameter of the throat of a venturi depending on pipe diameter and whether vanes are used to straighten the flow to avoid nonuniform turbulence, helical flow, and areas of high and low pressure. Lengths of 20 to 60 times the pipe diameter are required for orifices and nozzles. Prediction of flow in pipe friction meter from the friction factor will require lengths of about 60 times the diameter. Bend and pipe meters calibrated in place do not require a specific approach length, but recalibration is required if the approach piping is changed.
4. The pressure taps must be checked to insure that they are unplugged. The mechanical or electrical systems that sense pressure changes must be inspected as well.
5. If the device seems to incorrectly register flow, then calibration can be best checked by the volumetric or tracer dilution method. Extremely accurate results generally require calibration at the time of installation. Quicker less adequate methods to check flow rates involve clamp-on acoustic meters, velocity-area methods using an electromagnetic probe, or current meter inserted in a convenient tap, or existing opening in the pipe or the Gibson method of inferring flow rate from pressure responses when the flow is stopped by closing a valve in the pipe line.

[a] Adapted from References 5 and 10.

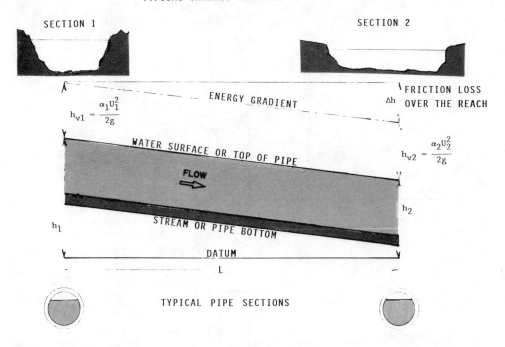

TYPICAL CHANNEL SECTIONS

FIGURE 107.   Illustration of the slope-area method for channels and pipes and the pipe friction meter method for full pipe flow. For best results, the length of the channel pipe or reach, $L \geq 75$ times the depth or pipe diameter and $h_f \geq 0.15m$ or $\geq h_v$ in an approximately uniform reach. Velocity should be well developed in the reach implying that the approach channel should be of the same uniform character for approximately 75 times the depth upstream. $Q = (K_1 K_2 S)^{1/2}$ in irregular channels (U.S. Geological Survey method, see Dalrymple and Benson;[3] other methods such as the U.S. Army Corps of Engineers method involves an arithmetic average); where $K_i = \phi A_i R_{hi}^{2/3}/n$ = conveyance at section 1 or 2; ϕ = constant = 1 for metric units or 1.49 for English units; $R = A/$(wetted perimeter) = hydraulic radius; and n = Mannings roughness coefficient. $S = [(h_1 - h_2) + \Delta h_v - k\Delta h_v]/L$ = slope of the energy grade line. k = expansion or loss coefficient that is difficult to predict for irregular channels. Δh_v = change in velocity head = $\alpha U/2g$ where α = velocity head correction coefficient = 1.0—1.1. See Dalrymple and Benson[3] and French[7] for values and methods of calculation. U is the average velocity. Thus when at all possible this method is used in uniform channels where $k\Delta h_v \cong 0$. For uniform flows, especially sewer lines of constant diameter, $\Delta h_v = 0$ and $S = (h_1 - h_2)/L$.

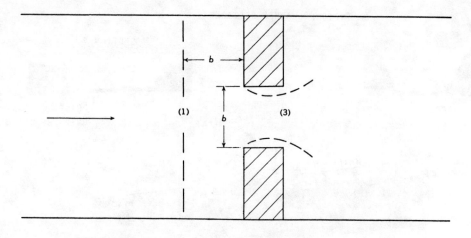

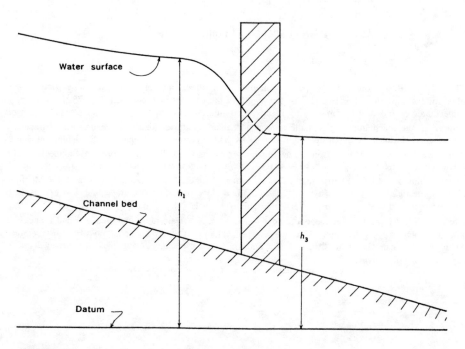

FIGURE 108.   Illustration of the width contraction measurements that are used to estimate discharge.[6]

PLAN VIEW

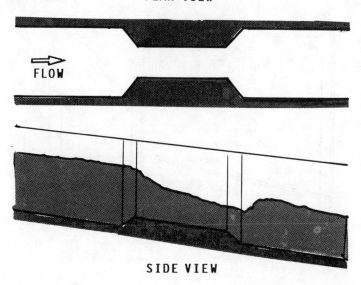

FIGURE 109. General flume configuration where constriction causes a measurable change in the water surface that can be uniquely related to discharge.

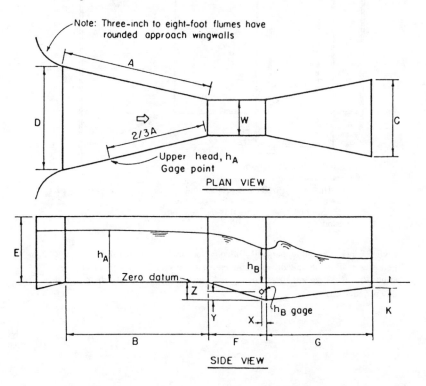

FIGURE 110. Parshall flume.[2] See Rantz et al.,[2] Bureau of Reclamation,[12] French,[7] ASME,[13] Grant,[4] or ASTM[5] for standard lengths of D, A, W, C, E, B, F, G, Z, K, X, and Y for each given size of the flume.

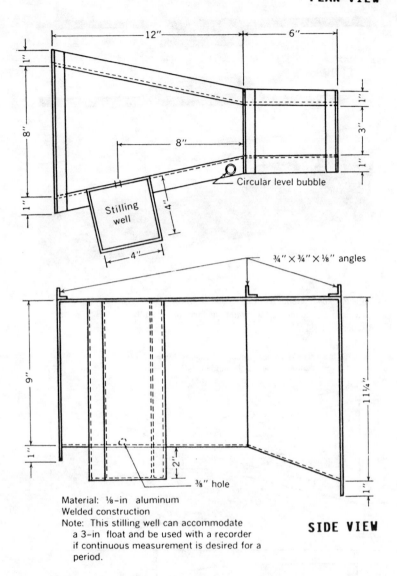

Circular level bubble

Stilling well

¾" × ¾" × ⅛" angles

⅜" hole

Material: ⅛-in aluminum
Welded construction
Note: This stilling well can accommodate
a 3-in float and be used with a recorder
if continuous measurement is desired for a
period.

SIDE VIEW

FIGURE 111.   Portable parshall flume.[2] Note that this flume does not have a getaway section on the downstream end and thus is limited to unsubmerged conditions ($h_B/h_A \leq 0.50$ where h_B + 4 cm (0.125 ft.) = water height at the downstream lip).

Dimensions of trapezoidal supercritical-flow flume									
Flume size, W_T feet	Width of approach reach, W_A feet	Angles		Lengths			Minimum capacity, ft^3/s	Floor slopes	
		Sloping walls θ	Converging walls ϕ	Approach reach, L_A feet	Converging reach, L_C feet	Throat reach, L_t feet		Approach reach, percent	Converging and throat reaches,%
1	5	30°	21.8°	5.0	5.0	5.0	0.7	0	5
3	9	30°	21.8°	variable	7.5	10.0	2.0	0	5
8	14	30°	21.8°	variable	7.5	12.0	6.0	0	5

Note — Height of wall (E) is dependent on magnitude of maximum discharge to be gaged.

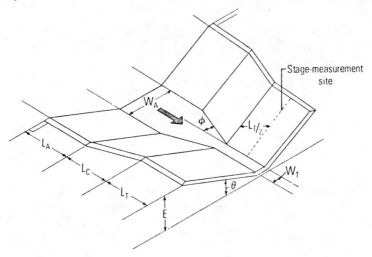

FIGURE 112. Dimensions of the three standard trapezoidal flumes used by the U.S. Geological Survey for stream gaging.[2]

PLAN VIEW

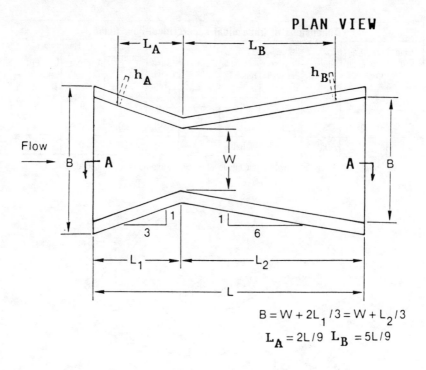

$$B = W + 2L_1/3 = W + L_2/3$$
$$L_A = 2L/9 \quad L_B = 5L/9$$

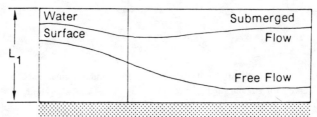

Section A-A
SIDE VIEW

FIGURE 113. Rectangular cutthroat flume. (From Grant, D. M., *ISCO Open Channel Flow Measurement Handbook*, Instrumentation Specialties Co., Environmental Division, Lincoln, NE, 1979. With permission.)

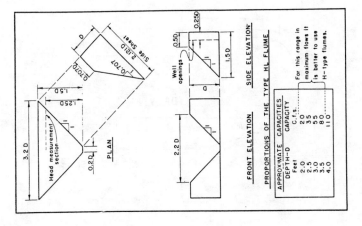

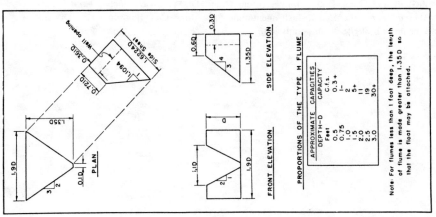

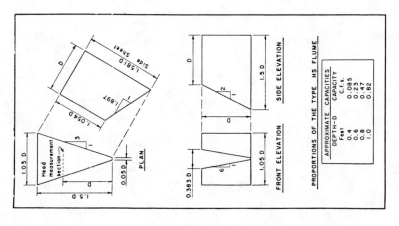

FIGURE 114.   Dimensions and capacities of HS-, H-, and HL-type flumes.[14]

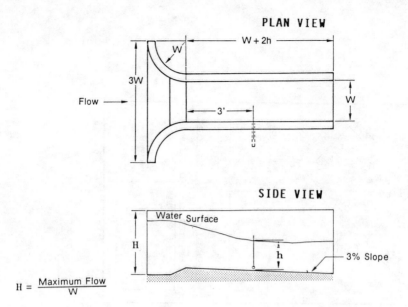

FIGURE 115. San Dimas flume. (From Grant, D. M., *ISCO Open Channel Flow Measurement Handbook*, Instrumentation Specialties Co., Environmental Division, Lincoln, NE, 1979. With permission.)

FIGURE 116. Sharp-crested weirs.

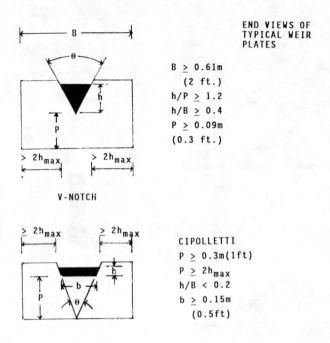

END VIEWS OF
TYPICAL WEIR
PLATES

$B \geq 0.61$m
 (2 ft.)
$h/P \geq 1.2$
$h/B \geq 0.4$
$P \geq 0.09$m
(0.3 ft.)

V-NOTCH

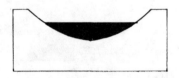

CIPOLLETTI
$P \geq 0.3$m(1ft)
$P \geq 2h_{max}$
$h/B < 0.2$
$b \geq 0.15$m
 (0.5ft)

TRAPEZOIDAL

PARABOLIC

FIGURE 116 (continued)

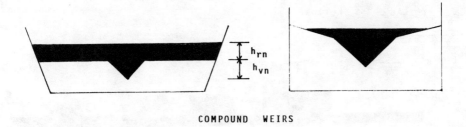

COMPOUND WEIRS

SUTRO

PROPORTIONAL

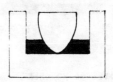

APPROXIMATE LINEAR

APPROXIMATE EXPONENTIAL

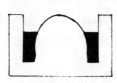

POEBING

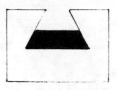

INVERTED TRAPEZOIDAL

FIGURE 116 (continued)

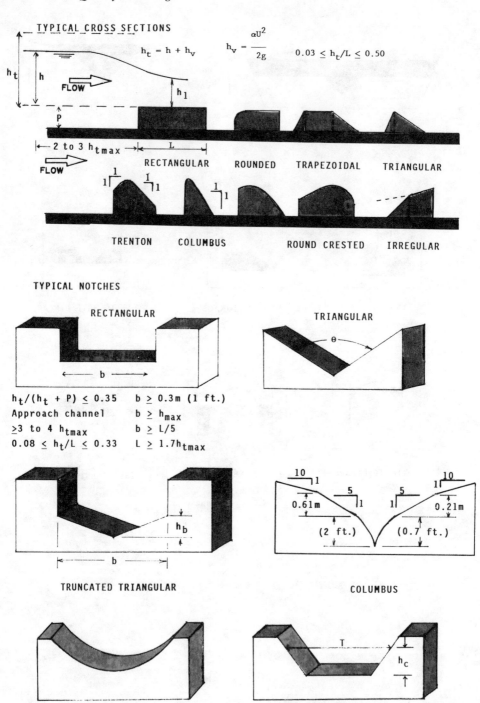

TYPICAL CROSS SECTIONS

$$h_t = h + h_v \qquad h_v = \frac{\alpha U^2}{2g} \qquad 0.03 \le h_t/L \le 0.50$$

FLOW

2 to 3 $h_{t\,max}$

L

FLOW

RECTANGULAR   ROUNDED   TRAPEZOIDAL   TRIANGULAR

TRENTON   COLUMBUS   ROUND CRESTED   IRREGULAR

TYPICAL NOTCHES

RECTANGULAR

b

TRIANGULAR

θ

$h_t/(h_t + P) \le 0.35 \qquad b \ge 0.3m$ (1 ft.)
Approach channel $\qquad b \ge h_{max}$
≥ 3 to 4 $h_{t\,max} \qquad b \ge L/5$
$0.08 \le h_t/L \le 0.33 \qquad L \ge 1.7h_{t\,max}$

h_b

b

TRUNCATED TRIANGULAR

$\frac{10}{1}$ $\frac{5}{1}$ $\frac{5}{1}$ $\frac{10}{1}$
0.61m   0.21m
(2 ft.)   (0.7 ft.)

COLUMBUS

PARABOLIC

T

h_c

TRAPEZOIDAL

FIGURE 117. Broad-crested weirs. The dashed line on the irregular weir profile indicates alternative shapes.

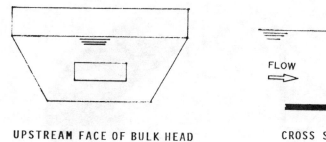

FIGURE 118. Submerged orifice.

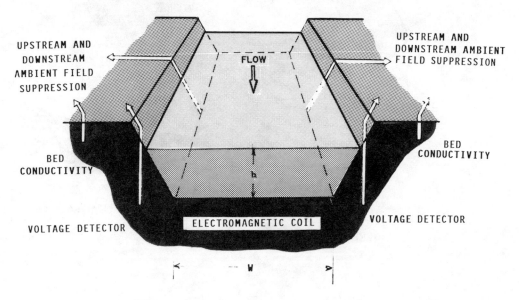

FIGURE 119. Electromagnetic coil for stream gaging.

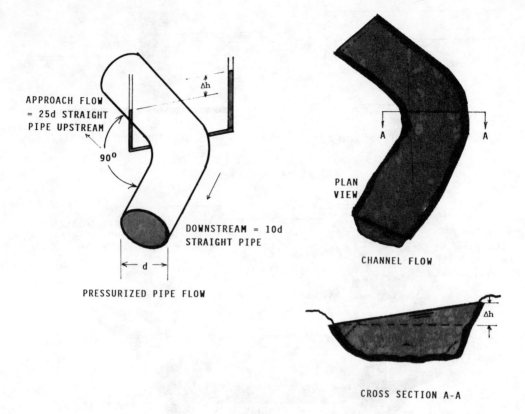

FIGURE 120.   Head differences across bends in flows.

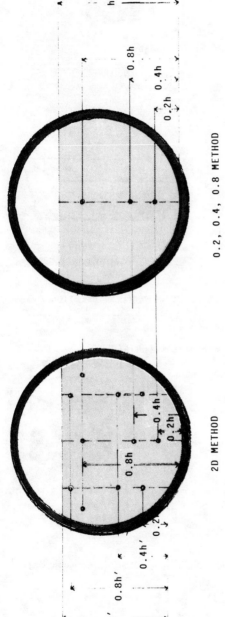

FIGURE 121. Velocity-area method that can be applied to partially full pipe flows.[20]

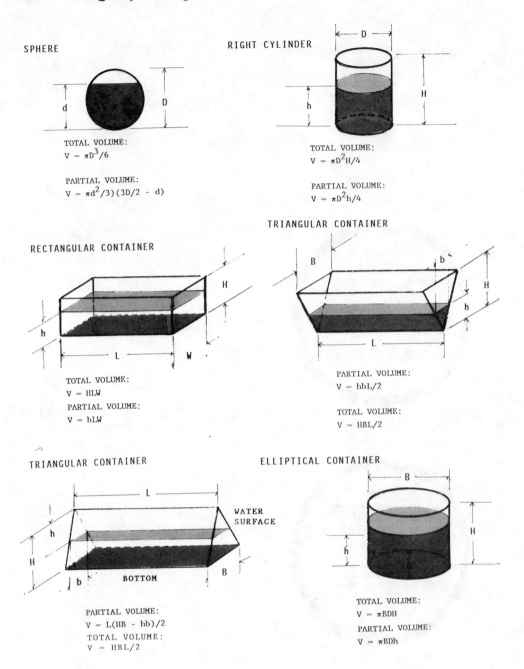

SPHERE

TOTAL VOLUME:
$V = \pi D^3/6$

PARTIAL VOLUME:
$V = \pi d^2/3)(3D/2 - d)$

RIGHT CYLINDER

TOTAL VOLUME:
$V = \pi D^2 H/4$

PARTIAL VOLUME:
$V = \pi D^2 h/4$

RECTANGULAR CONTAINER

TOTAL VOLUME:
$V = HLW$
PARTIAL VOLUME:
$V = hLW$

TRIANGULAR CONTAINER

PARTIAL VOLUME:
$V = hbL/2$

TOTAL VOLUME:
$V = HBL/2$

TRIANGULAR CONTAINER

WATER SURFACE

BOTTOM

PARTIAL VOLUME:
$V = L(HB - hb)/2$
TOTAL VOLUME:
$V = HBL/2$

ELLIPTICAL CONTAINER

TOTAL VOLUME:
$V = \pi BDH$
PARTIAL VOLUME:
$V = \pi BDh$

FIGURE 122. Total and partial volumes of tanks and sumps with regular geometric shapes.[10]

FRUSTUM OF A CONE

 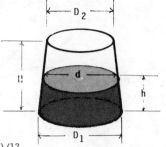

PARTIAL VOLUME:
$$V = \pi h(D_1^2 + D_1 d + d^2)/12$$

TOTAL VOLUME:
$$V = \pi H(D_1^2 + D_1 D_2 + D_2^2)/12$$

CONE

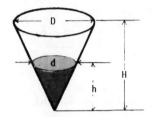

PARTIAL VOLUME:
$$V = \pi d^2 h/12$$

TOTAL VOLUME:
$$V = \pi D^2 H/12$$

PARTIAL VOLUME:
$$V = \pi(D^2 H - d^2 h)/12$$

PARABOLIC CONTAINER

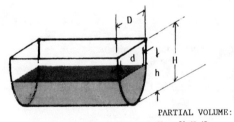

PARTIAL VOLUME:
$$V = 2hdL/3$$

TOTAL VOLUME:
$$V = 2HDL/3$$

PARTIAL VOLUME:
$$V = 2(HD - hd)L/3$$

FIGURE 122 (continued)

CULVERT SIDE VIEWS

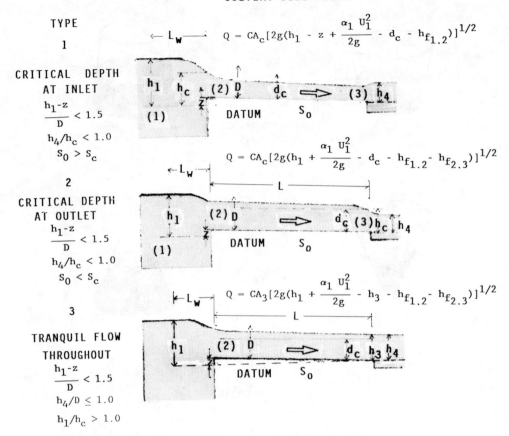

TYPE

1

CRITICAL DEPTH
AT INLET

$$\frac{h_1-z}{D} < 1.5$$

$h_4/h_c < 1.0$

$S_0 > S_c$

$$Q = CA_c[2g(h_1 - z + \frac{\alpha_1 U_1^2}{2g} - d_c - h_{f_{1.2}})]^{1/2}$$

2

CRITICAL DEPTH
AT OUTLET

$$\frac{h_1-z}{D} < 1.5$$

$h_4/h_c < 1.0$

$S_0 < S_c$

$$Q = CA_c[2g(h_1 + \frac{\alpha_1 U_1^2}{2g} - d_c - h_{f_{1.2}} - h_{f_{2.3}})]^{1/2}$$

3

TRANQUIL FLOW
THROUGHOUT

$$\frac{h_1-z}{D} < 1.5$$

$h_4/D \leq 1.0$

$h_1/h_c > 1.0$

$$Q = CA_3[2g(h_1 + \frac{\alpha_1 U_1^2}{2g} - h_3 - h_{f_{1.2}} - h_{f_{2.3}})]^{1/2}$$

FIGURE 123.   Methods of determining flow in culverts.[23]
Type 1: for $(h_1 - z)/D < 1.5$; $h_4/h_c < 1.0$; and $S_0 > S_c$
$$Q = CA_c[2g(h_1 - z + \alpha_1 U_1^2/2g - d_c - h_{f12})]^{1/2}$$
Type 2: for $(h_1 - z)/D < 1.5$; $h_4/h_c < 1.0$; and $S_0 < S_c$
$$Q = CA_c[2g(h_1 + \alpha_1 U_1^2/2g - d_c - h_{f12} - h_{f23})]^{1/2}$$
Type 3: for $(h_1 - z)/D < 1.5$; $h_4/h_c > 1.0$; and $h_4/D \leq 1.0$
$$Q = CA_3[2g(h_1 + \alpha_1 U_1^2/2g - h_3 - h_{f12} - h_{f23})]^{1/2}$$
Type 4: for $(h_1 - z)/D > 1.0$; and $h_4/D > 1.0$
$$Q = CA_0\{2g(h_1 - h_4)[1 + (29C^2n^2L)/R_0^{4/3}]^{-1}\}^{1/2}$$
Type 5: for $(h_1 - z)/D \geq 1.5$; and $h_4/D \leq 1.0$
$$Q = CA_0[2g(h_1 - z)]^{1/2}$$
Type 6: for $(h_1 - z)/D \geq 1.5$; and $h_4/D \leq 1.0$
$$Q = CA_0[2g(h_1 - h_3 - h_{f13})]^{1/2}$$

CULVERT SIDE VIEWS

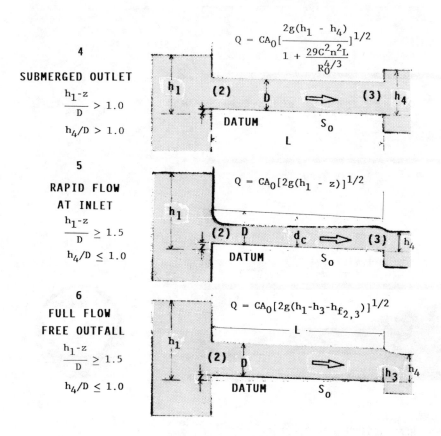

4

SUBMERGED OUTLET

$$\frac{h_1 - z}{D} > 1.0$$

$$h_4/D > 1.0$$

$$Q = CA_0 \left[\frac{2g(h_1 - h_4)}{1 + \frac{29C^2 n^2 L}{R_0^{4/3}}} \right]^{1/2}$$

5

RAPID FLOW AT INLET

$$\frac{h_1 - z}{D} \geq 1.5$$

$$h_4/D \leq 1.0$$

$$Q = CA_0 [2g(h_1 - z)]^{1/2}$$

6

FULL FLOW FREE OUTFALL

$$\frac{h_1 - z}{D} \geq 1.5$$

$$h_4/D \leq 1.0$$

$$Q = CA_0 [2g(h_1 - h_3 - h_{f_{2,3}})]^{1/2}$$

FIGURE 123 (continued)

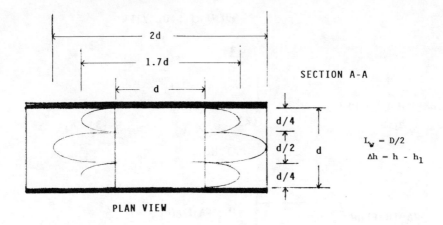

PLAN VIEW

SECTION A-A

$$L_w = D/2$$
$$\Delta h = h - h_1$$

THROAT
CROSS
SECTION

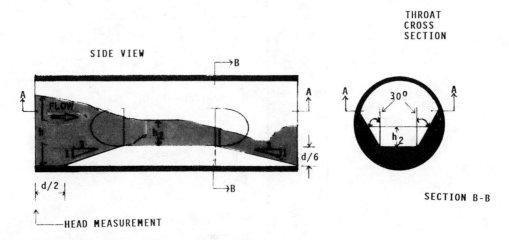

SIDE VIEW

SECTION B-B

HEAD MEASUREMENT

FIGURE 124. Typical Palmer-Bowlus flume for temporary or permanent installation in sewer lines. This design is manufactured by Plasti-Fab.[4]

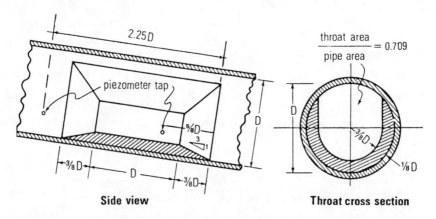

$$\frac{\text{throat area}}{\text{pipe area}} = 0.709$$

piezometer tap

Side view

Throat cross section

FIGURE 125. U.S. Geological Survey sewer flowmeter.[2] Note that D in this figure is equivalent to d in Table A1.

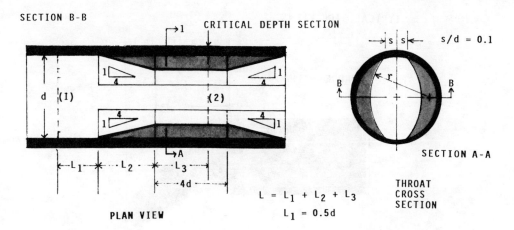

FIGURE 126.   Symmetrical Wenzel flume. The asymmetrical flume in only one of the two inserts.

TYPICAL SIDE VIEWS

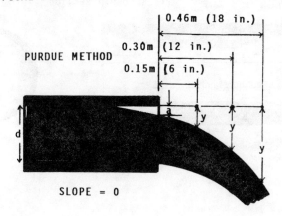

PURDUE METHOD

0.46m (18 in.)
0.30m (12 in.)
0.15m (6 in.)

SLOPE = 0

CALIFORNIA METHOD

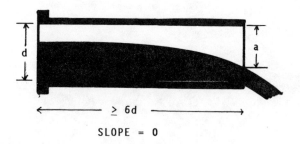

≥ 6d

SLOPE = 0

VERTICAL PIPE METHOD

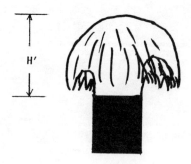

FIGURE 127.   Trajectory methods for pipe flow.

PARABOLIC NOZZLE

END VIEW

OPENING

GAGE

d

SIDE VIEW

≈ 4 d

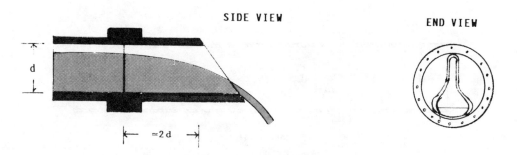

KENNISON NOZZLE

SIDE VIEW

END VIEW

d

≈ 2 d

FIGURE 128.   Nozzles attached to the end of pipes to determine discharge.

TO PROCESSING UNIT

RETAINING BAND

FLOW

SENSOR WITH
ELECTROMAGNETIC
CURRENT METER
AND PRESSURE
TRANSDUCER

INSERT

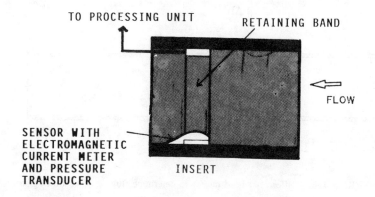

FIGURE 129.   Electromagnetic current meter and pressure transducer
that can be inserted in sewer lines to determine flow rates. The current
meter measures a point velocity that is related to mean velocity, and the
pressure transducer measures depth of flow.[27]

Side Views

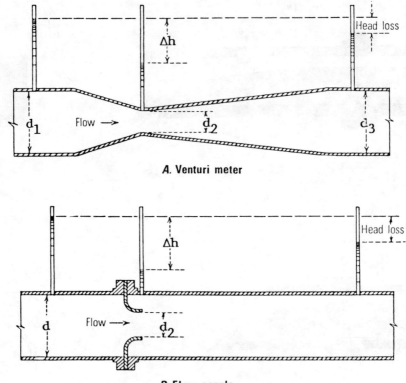

A. Venturi meter

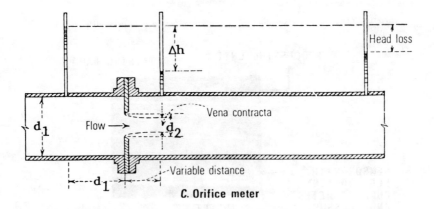

B. Flow nozzle

C. Orifice meter

FIGURE 130. Differential head meters to determine flow in pipes flowing full.[2]

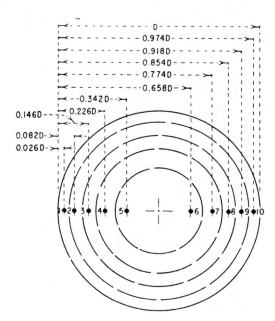

(A). Ten-point system
for circular conduits

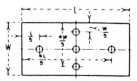

(B). System for rectangular
conduits, where at least
16 divisions must be used

(C). Additional points for data
in areas around periphery
of the rectangular conduit

FIGURE 131. Locations to measure velocity in pipes flowing full to determine discharge by the velocity-area method.[2] Originally, Rantz et al. obtained this figure from British Standard 1042: Section 2.1: published in 1943 and more recently in 1983. Extracts from British Standard 1042 are reproduced here with permission of BSI. Complete copies of the standard can be obtained from national standard bodies. These locations were originally designed for pitot tube measurements but work equally well for any velocity sensors of similar size. The number of positions are chosen to ensure the best precision. Measurements at fewer locations may be equally accurate but less precise.

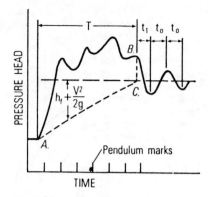

FIGURE 132. Typical pressure-variation graph obtained for a pipe where the flow has been shut off by a valve closure. As part of the Gibson method, the area ABCA is determined using a trial and error refinement in calculating the flow rate at the time of closure of the valve.[2] (From Rouse, H., Ed., *Proc. 4th Hydraulics Conf.*, John Wiley & Sons, New York, 1950. With permission.)

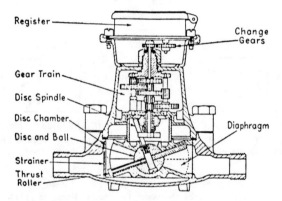

FIGURE 133. Cross section of a typical nutating disk flow meter that must be read at frequent intervals to note average flow rates over the period between readings.[13] Note that the same figure has been redesignated as Figure I-4-7 in the 6th ed. published in 1981.

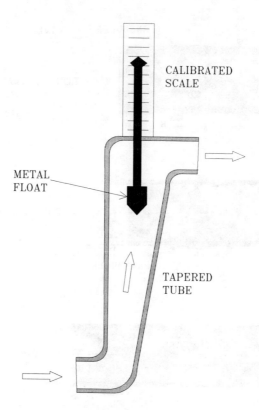

FIGURE 134. Typical rotometer that must be calibrated in the laboratory or in place to achieve accurate measurements.

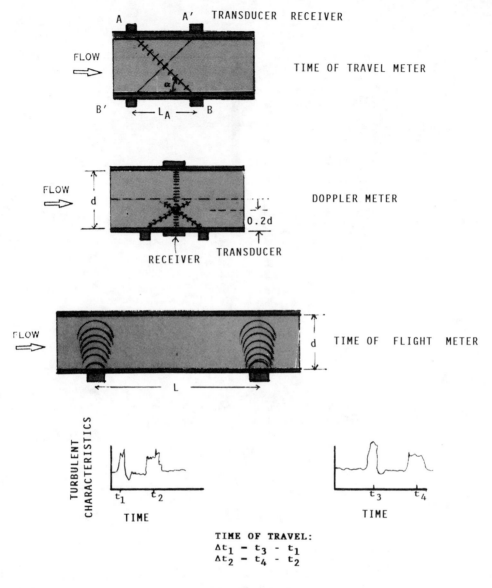

FIGURE 135. Acoustic flow meters that are based on the time of travel, Doppler, and time of flight principles. The time of travel meter requires no calibration in symmetrical flows. $C_A = 1.00$. Development of symmetrical flows requires lengths of the uniform approach flow to be 10 to 20 times the pipe diameter. Pulses are transmitted from A to B and then from B to A. Alternatively, pulses are transmitted simultaneously from A to B and A′ to B′, especially where asymmetrical flows may be present. The Doppler meter measures one or more point velocities that must be related to average velocity by calibration in the laboratory or in place. Usually the flow must be fully developed, and this requires uniform flow lengths of 20 to 60 times the pipe diameter. Calibration in the laboratory requires that the same flow conditions be reproduced in the field. When the flow is fully developed and the sonic beams are crossed at 0.2 d from either wall, then the point velocity should be approximately equal to the average velocity and $C_A = 1.00$. Note that Marsh-McBirney[29] recommends (1/8)d (= 0.125 d) instead. The time of flight principle determines travel time by correlating turbulent signatures at two different stations and determining the time it takes for a water mass with a given signature to arrive at the downstream station. As a result, the meter does not require calibration, and because it is not tied to symmetrical flow or fully-developed flow, approach conditions are not important except that it has been observed that the rapidly varying flow conditions in a pipe bend do lead to errors in excess of 5%.[16] Avoiding the backflow eddies just downstream of a bend seems to be the only criteria to govern placement of the time of flight meter. Thus the time of flight meter may be ideal for clamping on existing pipes where approach flows are usually very limited.

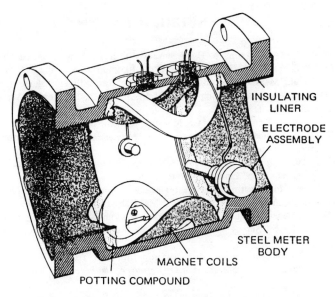

INSULATING
LINER

ELECTRODE
ASSEMBLY

STEEL METER
BODY

MAGNET COILS

POTTING COMPOUND

A

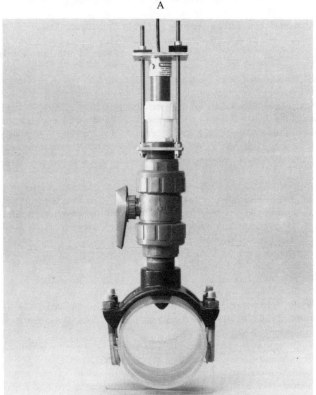

B

FIGURE 136. Two types of electromagnetic flow meters in general use. (A) Spool piece meter that must be built into the pipeline and calibrated in the laboratory or in place.[10] (B) Insertable probe used to measure one or more point velocities as part of the velocity-area method. This particular probe is the Marsh-McBirney model 280 that can be retracted from the flow for protection from debris and for cleaning, thus overcoming two serious practical difficulties in the use of the electromagnetic meter. (Photograph [B] courtesy of Marsh-McBirney, Inc.)

REFERENCES

1. **Carter, R. W. and Anderson, I. E.,** Accuracy of current meter measurements, American Society of Civil Engineers, *J. Hydr. Div.,* 89, HY4, 105, 1963.
2. **Rantz, S. E., et al.,** Measurement and Computation of Streamflow: Volume 1. Measurement of Stage and Discharge. Volume 2. Computation of Discharge, U.S. Geological Survey Water-Supply Pap. 2175, Washington, D.C., 1982, 81.
3. **Dalrymple, T. and Benson, M. A.,** Measurement of peak discharge by the slope-area method, in U.S. Geological Survey Techniques of Water-Resources Investigations, Book 3, U.S. Geological Survey, Washington, D.C., 1967, chap. A2.
4. **Grant, D. M.,** *ISCO Open Channel Flow Measurement Handbook,* Instrumentation Specialties Co., Environmental Division, Lincoln, NE, 1979.
5. **American Society for Testing and Material,** *1974 Annual Book of ASTM Standards,* Part 31 — Water, American Society for Testing and Materials, Philadelphia, 1974.
6. **Matthai, H. F.,** Measurement of peak discharge at width contractions by indirect methods, in Book 3, U.S. Geological Survey Techniques of Water-Resources Investigations, Washington, D.C., 1967, chap. A4.
7. **French, R. H.,** *Open-Channel Hydraulics,* McGraw-Hill, New York, 1985, 34.
8. **Leopold and Stevens, Inc.,** *Stevens Water Resources Data Book,* 2nd ed., Beaverton, OR, 1974.
9. **Mougenot, G.,** Measuring sewage flow using weirs and flumes, in *Water and Sewage Works,* 1974, 78.
10. **U.S. Environmental Protection Agency,** NPDES Compliance Inspection Manual, Office of Water Enforcement and Permits, Washington, D.C., 1984.
11. **American Society of Civil Engineers and Water Pollution Control Federation,** *Wastewater Treatment Plant Design,* New York, 1977, 38.
12. **Bureau of Reclamation,** Water Measurement Manual, 2nd ed., U.S. Department of Interior, Bureau of Reclamation, Washington, D.C., 1967.
13. **American Society of Mechanical Engineers,** *Fluid Meters: Their Theory and Application,* 5th ed., Report of the ASME Research Committee on Fluid Meters, New York, 1959.
14. **Holtan, H. N., Minshall, N. E., and Harrold, L. L.,** Field Manual for Research in Agricultural Hydrology, USDA ARS Agriculture Handbook No. 224, Washington, D.C., 1962.
15. **King, H. W. and Brater, E. F.,** *Handbook of Hydraulics,* 5th ed., McGraw-Hill, New York, 1963.
16. **Cordes, E.,** U.S. Geological Survey, Hydrologic Instrumentation Facility, personal communication, 1987.
17. **Cordes, E.,** Correlation flowmeter: series CFM-P, unpublished manuscript, 1987.
18. **Schuster, J. C.,** Measuring water velocity by ultrasonic flowmeter, *J. Hydraul. Div.,* American Society of Civil Engineers, 101, HY12, 1503, 1975.
19. **Laenen, A. and Smith, W.,** Acoustic systems for measurement of streamflow, U.S. Geological Survey Water-Supply Pap. 2213, Washington, D.C., 1983.
20. **Marsh-McBirney, Inc.,** Calculating flow in open channels, unpublished manual, Gaithersburg, MD, 1986.
21. **Kilpatrick, F. A. and Schneider, V. R.,** Use of flumes in measuring discharge, in U.S. Geological Survey Techniques of Water Resources Investigations, Book 3, U.S. Geological Survey, Washington, D.C., 1983, Chap. A14.
22. **Smoot, C. W.,** Orifice bucket for measurement of small discharges from wells, in *Water Resourc. Div. Bull.,* Illinois Water Survey, Champaign, IL, Nov. 1963.
23. **Bodhaine, G. L.,** Measurements of peak discharge at culverts by indirect methods, in Techniques of Water-Resources Investigations, Book 3, U.S. Geological Survey, Washington, D.C., 1968, Chap. A3.
24. **Palmer, H. K. and Bowlus, F. D.,** Adaptation of venturi flumes to flow measurements in conduits, *Am. Soc. Civil Eng. Trans.,* Paper No. 1948, 1935, 1195.
25. **Wells, E. A. and Gotaas, H. B.,** Design of venturi flumes in circular conduits, in *Am. Soc. Civil Eng. Trans.,* Paper No. 2937, 123, 749, 1958.
26. **Wenzel, H. G.,** Meter for Sewer Flow Measurement, *J. Hydraul. Div.,* American Society of Civil Engineers, 101, HY1, 115, 1975.
27. **Marsh-McBirney, Inc.,** FLO-TOTE^{TM}: Portable open channel flowmeter system, brochure, Gaithersburg, MD, 1987.
28. **Folsom, R. G.,** Review of the Pitot Tube, *Trans. Am. Soc. Mech. Eng.,* 1447, 1956.
29. **Marsh-McBirney, Inc.,** Pulp stock: flowmeter model 280, application note 82-010, Gaithersburg, MD.

30. **Newfeld, M. K.,** Reviewing the advantages of closed-pipe magnetic flowmeters, *Water World News,* American Water Works Assoc., 1, 3, 1985.
31. **Moussiaux, J. J.,** Flow determination and field calibration of pulp stock process flow using a point measurement electromagnetic flow meter, in *Proc. 1982 Joint Symposium Instr. Soc. Am.,* Columbus, OH, 1982, 307.
32. **Metheny, H. M.,** Solving wastewater flow problems, *Measurements & Control,* April 1982.
33. **Streeter, V. L. and Wylie, E. B.,** *Fluid Mechanics,* 6th ed., McGraw-Hill, New York, 1975.

INDEX

A

Acid yellow 7 dye
  background interference and, 151
  chemical quenching of, 159
  excitation and emission spectra of, 148-149
  filters recommended for, 150
  photochemical decay of, 157-158
  sorption and, 153
  temperature correction curves for, 155
Acoustic meter, depiction of, 322
Acoustic method, inflow and outflow hydrographs for
        dynamic flow conditions defined via, 90, 101-
        102, 107, 282, 286, 289
Advective-dispersive equation
  for concentration of constituent, 85
  for dynamic stream modeling, 38, 41
  solution of, 55
Air temperature, equilibrium temperature approxima-
        tion and, 206
Amino G acid
  characteristics of and recommended filters, 146
  as fluorescent water tracing dye, 145
  pH effect on, 156
  photochemical decay of, 158
  sorption and, 152-153
  temperature correction curves for, 155
Analytical models, characterization of, 14
Asellus aquaticus, 163
Atmospheric scattering and absorption term,
        empirical expression for, 189
Atmospheric transmission coefficient, mean, 189
AUTO-D model, applications of, 271
AUTO-QD model, as stream model, 48
AUTO-QUAL model
  applications of, 48, 271
  evaluation of, 50
AUTO-QUAL,QD model, as stream model, 47

B

Background interference, tracers subjected to, 150-
        151
Back radiation, from water body, 191-192
Backwater analysis, stream flow velocity assessed
        via, 127-129
Backwater hydraulic solution, for stream flow
        routing, 31
Bansal equation, reaeration coefficients and, 225, 231
Basin-wide studies, screening level, 17
Bauer-Bennett model, as stream model, 48
Bauer-Yotsukura model, as useful water model, 68-69
Bed conduction, diel temperature of stream predicted
        via, 200
Bennett-Rathbun equation, reaeration coefficients
        and, 223, 225, 230-231

Bensen-Krause equation, gas saturation and, 240,
        244, 246-247, 249
Biochemical oxygen demand (BOD)
  carbonaceous, 179, 182
  cause-and-effect relationship definition and, 3
  change in coefficients of, 22
  models of, 14, 18
Biochemical processes, see specific processes
Blue fluorescent dyes, as water tracing dyes, 144-145,
        148, 151, 153
BOD, see Biochemical oxygen demand
Bookkeeping operations, characterization of, 35
Bottom elevation, sawtooth representation of, 122
Boundary conditions
  characterization of, 15-16, 18-21
  flow routing models and, 130
Boundary layer flows, unidirectional, 38, 40
Boundary layer model, of gas transfer process at
        water surface, 13, 214
Bowen ratio, expression of, 200
Branched networks, of one-dimensional elements, 11
BRANCH model, applications of, 70, 130-131
Broad crested weirs
  depiction of, 306
  flow measurement in, 275, 280
Brunt's equation, expression of, 191

C

Cadwaller-McDonnell equation, reaeration coeffi-
        cients and, 220, 231
Calibration, for water quality models, 1-2, 5, 74-78
Calibration data, characterization of, 15-16
Carassius auratus, 163
Carbon, mass balance volume and organic, 3
Cascade-type models, calibration of, 76
Cause-and-effect relationships, definition of, 1, 3
CE-QUAL-R1 model, applications of, 132, 241
CE-QUAL-RIV1 model, applications of, 68-70
Channel data, as component of modeling system, 7
Channels, flow measurement in, 106, 157, 159, 275
Chapman-Enskog formula, gas transfer and, 215
Chemical quenching, of dye tracers, 157, 159
Chezy equation, force balance and, 120
Churchill et al. equation, reaeration coefficients and,
        223, 230-233, 235
Classifications, of water quality models, 5
Cleon diperum, 163
Complex branch-looped system, characterization of,
        11
Computational element
  in finite difference solution scheme, 10
  in Streeter-Phelps-type model discretization
        scheme, 21
Computational grid network, definition of, 16, 20-26
Computer code or program, definition of term, 5-8

T

U